THE CONSERVATION OF INSECTS AND THEIR HABITATS

THE CONSERVATION OF INSECTS AND THEIR HABITATS

Edited by

N. M. Collins
World Conservation Monitoring Centre, Cambridge, England

J. A. Thomas
Institute of Terrestrial Ecology, Furzebrook Research Station, Wareham, Dorset, England

15th Symposium of the
Royal Entomological Society of London
14–15 September 1989
at the
Department of Physics Lecture Theatre
Imperial College, London

ACADEMIC PRESS

Harcourt Brace Jovanovich, Publishers
London San Diego New York
Boston Sydney Tokyo Toronto

ACADEMIC PRESS LIMITED
24/28 Oval Road, London NW1 7DX

United States Edition published by
ACADEMIC PRESS, INC.
San Diego, CA 92101

British Library Cataloguing in Publication Data
is available

ISBN 0–12–181370–3

Typeset by Setrite Typesetters Ltd
Printed in Great Britain by St Edmundsbury Press,
Bury St Edmunds, Suffolk

Contents

Contributors . . . xi

Foreword . . . xiii
MARTIN W. HOLDGATE

Acknowledgements . . . xv

Opening Remarks . . . xvii
J. P. DEMPSTER

1. Observe Extinction or Conserve Diversity?
N. W. MOORE

I. Introduction: Insects are Difficult to Conserve . . . 1
II. Insects, Nature Reserves and their Matrix . . . 2
III. Special Species . . . 4
IV. Communication with Farmers . . . 5
V. Communication among Entomologists and Conservation Organizations . . . 6
VI. Conclusion . . . 7
References . . . 8

2. North American Problems and Perspectives in Insect Conservation
P. A. OPLER

I. Introduction . . . 9
II. Status of Conservation Programmes . . . 11
III. Status Surveys . . . 14
IV. Progress in Listing Candidates . . . 16
V. Protection and Habitat Acquisition . . . 18
VI. Management and Recovery . . . 20
VII. Prognosis . . . 28
VIII. Conclusion . . . 30
References . . . 30

3. Australia: Conservation of a Continental Insect Fauna
P. GREENSLADE AND T. R. NEW

I. Introduction . . . 33
II. Australian Environments and their History . . . 34
III. The Insect Fauna . . . 39
IV. Current Threats to Insects . . . 49

V. The History and Status of Insect Conservation in Australia 55
VI. Conclusion 63
References 66

4. The Conservation of Island Insects and their Habitats
F. G. HOWARTH AND G. W. RAMSAY

I. Introduction 71
II. Islands 72
III. Island Conservation 77
IV. Recommendations and Solutions 93
References 101

5. The Conservation of Insects and their Habitats in Northern and Eastern Europe
K. MIKKOLA

I. Introduction 109
II. The Red Data Lists of Northern and Eastern Europe 111
III. Protection of Insects and their Habitats in Northern and Eastern Europe 112
IV. Causes of Threat and the Similarity of Species Lists in Finland and Sweden 114
V. Conclusion 117
References 119

6. Mediterranean Insect Conservation
E. BALLETTO AND A. CASALE

I. Introduction 121
II. The Mediterranean Insect Fauna 123
III. Pleistocene Refugia 128
IV. Insect Conservation in the Mediterranean Area 129
V. Threatened Environments of the Mediterranean Area 133
References 140

7. Fragmentation, Isolation and Mobility of Insect Populations
J. P. DEMPSTER

I. Introduction 143
II. Isolation and Population Dynamics 144
III. Examples of Insects responding to Increased Isolation 146
IV. Conclusions 151
References 153

8. Woodlands: Past, Present and Potential for Insects
M. S. WARREN AND R. S. KEY

I. Introduction ... 155
II. Historical Background ... 156
III. Changes in the Status of Woodland Insects ... 167
IV. Ecological Specialization among Insects of Mature Woodland and Open Spaces ... 176
V. The Management of Woods for Insect Conservation ... 190
VI. The Need for Research and Education ... 200
VII. Summary and Conclusions — the Future for Woodland Insects ... 201
References ... 203

9. Lepidoptera as Indicators of Change in the Semi-natural Grasslands of Lowland and Upland Europe
A. ERHARDT AND J. A. THOMAS

I. Introduction ... 213
II. Intensively Cultivated Grassland ... 215
III. Abandoned Grassland ... 217
IV. Changes in Species and Families of Lepidoptera ... 218
V. Reasons for Rapid Change in Insect Populations ... 222
VI. Discussion and Conclusions ... 231
References ... 234

10. Conserving Insects of Aquatic and Wetland Habitats, with Special Reference to Beetles
G. N. FOSTER

I. Introduction: Sites, Habitat, Communities and Species ... 238
II. International Species Protection ... 238
III. The British Red Data Book — Insects ... 240
IV. Species Rarity and Site Evaluation ... 249
V. The Low Profile of Insect Conservation ... 255
VI. Acceptance of the Need for Insect Conservation by Non-entomologists ... 257
References ... 260

11. Creating New and Successional Habitats for Arthropods
M. B. USHER AND R. G. JEFFERSON

I. Introduction ... 263
II. Arthropods in Ecological Successions ... 264
III. Habitat Creation ... 269
IV. Habitat Re-creation ... 272
V. Habitat Mosaics of Different Successional Age ... 275

VI. Discussion 278
References 280
Appendix I 283

12. The Conservation of Insects on Arable Farmland
J. W. DOVER

I. Introduction 294
II. Background to the Development of Conservation Headlands 294
III. Design and Agronomy of Conservation Headlands 296
IV. Effect of Conservation Headlands on Chick-food Insects and Chick Survival 299
V. Butterflies and Conservation Headlands 301
VI. Conservation Headlands as a Tool in Conservation 309
VII. Conclusions 314
References 315

13. The Utilization and Value of Non-domesticated Insects
M. G. MORRIS, N. M. COLLINS, R. I. VANE-WRIGHT AND J. WAAGE

I. Introduction 319
II. Services from Insects 321
III. Goods from Insects 330
IV. Planning for Resource Management 338
V. Conclusions 343
References 344

14. Conservation of Neotropical Environments: Insects as Indicators
K. S. BROWN Jr

I. Introduction 350
II. Adapting Insects to the Tasks of Defining Options, Ranking of Priorities, Making Surveys and Monitoring Systems 357
III. Tools for the Task: Description, Discrimination, Explanation 360
IV. Case Histories and Some Legends 366
V. A Practical Guide to Conservation Action using Insects 387
VI. Epilogue: Important Considerations for the Agenda, 1990–2010 397
References 401

15. Insects and Tropical Forest Conservation
S. L. SUTTON AND N. M. COLLINS

I. Introduction 405
II. The Size of the Insect Resource 406
III. Do Insects Matter? 408

IV. Pattern and Process ... 409
V. Deforestation ... 409
VI. Identification ... 411
VII. Extending the Protected Areas Network ... 414
VIII. Insects as Tools for Conservation ... 421
IX. Conclusions ... 421
References ... 422

16. Poster Presentations 425

Managing Plantation Woodland on Ancient Woodland Sites for Butterflies and Other Insects
J. N. Greatorex-Davies and M. L. Hall ... 425

The Role of Field Boundaries in Hoverfly (Diptera: Syrphidae) Foraging on Arable land
S. Cowgill ... 427

The Bern Convention and Spanish Lepidoptera
J. L. Viejo, J. Martin, E. García-Barros and M. L. Munguira ... 428

Captive Breeding-A Vital Line of Defence
P. Pearce-Kelly ... 429

Insects and Krakatau: A Conserved Habitat of Long-term Research Interest
I. W. B. Thornton ... 430

The Amateur Role in Insect Conservation
D. Lonsdale ... 431

The Nature Conservancy Council Approach to Invertebrate Conservation
R. Key and the NCC Terrestrial Invertebrate Zoology Branch ... 431

Conservation of British Trichoptera
I. D. Wallace ... 432

Index ... 433

Contributors

Numbers in parentheses indicate the page numbers on which the authors' contributions begin.

E. BALLETO (121), Università di Torino, Dipartimento di Biologia Animale, Via Accademia Albertina, 17, 10123 Torino, Italy

K. S. BROWN Jr (349), Departamento de Zoologia, Instituto de Biologia, Universidade Estadual de Campinas, C.P. 6109, Campinas, Sao Paulo, 13.081, Brazil

A. CASALE (121), Istituto di Zoologia, via Muzoni 23, I−7100, Sassari, Italy

N. M. COLLINS (319, 405), World Conservation Monitoring Centre, 219c Huntingdon Road, Cambridge CB3 0DL, UK

S. COWGILL (427), Department of Biology, The University, Southampton S02 3AH, UK

J. P. DEMPSTER (xvii, 143), Department of Zoology, University of Cambridge, Downing St, Cambridge, UK

J. W. DOVER (293), 50 Queens Road, Farnborough, Hants GU14 6DT, UK

A. ERHARDT (213), Botanisches Institut der Universitat Basel, Schonbeinstrasse 6, 4056 Basel, Switzerland

G. N. FOSTER (237), The Scottish Agricultural College, Auchincruive, Ayr KA6 5HW, UK

E. GARCÍA-BARROS (428), Departamento de Biología, C-XV, Universidad Autónoma de Madrid, 28049 Madrid, Spain

J. N. GREATOREX-DAVIES (425), Institute of Terrestrial Ecology, Monks Wood Experimental Station, Abbots Ripton, Huntingdon, PE17 2LS, UK

P. GREENSLADE (33), c/o CSIRO Division of Entomology, P.O. Box 1700, Canberra City, A.C.T. 2601, Australia

M. L. HALL (425), Newlands, Chatteris Road, Somersham, Huntingdon, UK

M. W. HOLDGATE (xiii), IUCN, Avenue du Mont-blanc, CH-1196 Gland, Switzerland

F. G. HOWARTH (71), J. Linsley Gressitt Center for Research in Entomology, Bishop Museum, P.O. Box 19000-A, Honolulu, Hawaii 96817, USA

R. KEY (155, 431), Nature Conservancy Council, Northminster Road, Peterborough PE1 1UA, UK

R. G. JEFFERSON (263), Department of Biology, University of York, York, YO1 5DD, UK

D. LONSDALE (431), Amateur Entomologists' Society, 54 Cherry Way, Alton GU34 2AX, UK

J. MARTÍN (428), Departamento de Biología, C-XV, Universidad, Autónoma de Madrid, 28049, Madrid, Spain

K. MIKKOLA (109), Department of Zoology, Division of Morphology and Ecology, University of Helsinki, P. Rautatiekatu 13, SF-00100 Helsinki, Finland

N. W. MOORE (1), The Farm House, Swavesey, Cambridge CB4 5RA, UK

M. G. MORRIS (319), Institute of Terrestrial Ecology, Furzebrook Research Station, Wareham, Dorset BH20 5AS, UK

M. L. MUNGUIRA (428), Departamento de Biología, C-XV, Universidad Autónoma de Madrid, 28049 Madrid, Spain

T. R. NEW (33), La Trobe University, Bundoora, Victoria, Australia 3083

P. A. OPLER (9), Editorial Office, US Fish and Wildlife Service, Room 259 Aylesworth Hall, Colorado State University, Fort Collins, Colorado 80523, USA

P. PEARCE-KELLY (429), The Zoological Society of London, Regent's Park, London NW1 4RY, UK

G. W. RAMSAY (71), Division of Plant Protection, Department of Scientific and Industrial Research, Private Bag, Auckland, New Zealand

S. L. SUTTON (405), Department of Pure and Applied Biology, Leeds University, Leeds LS2 9JT, UK

J. A. THOMAS (213), Institute of Terrestrial Ecology, Furzebrook Research Station, Wareham, Dorset BH20 5AS, UK

I. W. B. THORNTON (430), Zoology Department, La Trobe University, Bundoora, Victoria, Australia 3083

M. B. USHER (263), Department of Biology, University of York, York YO1 5DD, UK

R. I. VANE-WRIGHT (319), Department of Entomology, The Natural History Museum, Cromwell Road, London SW7 5BO, UK

J. L. VIEJO (428), Departamento de Biología, C-XV, Universidad Autónoma de Madrid, 28049 Madrid, Spain

J. WAAGE (319), CAB International Institute of Biological Control, Imperial College Field Station, Silwood Park, Ascot, Berkshire, UK

I. D. WALLACE (432), Liverpool Museum, William Brown Street, Liverpool, UK

M. S. WARREN (155), 2 Fairside, Higher Ansty, Dorchester, Dorset DT2 7PT, UK

Foreword

Entomologists have long contended that the subject of their concern is the most important animal group in the world, and that it is only the prejudice of the mammals they are addressing that prevents this fact from being generally recognized! Certainly in numerical terms nobody doubts that insects make up the overwhelming bulk of the world's species. It is also widely accepted that the insect fauna of the world exists in greatest profusion and diversity in the tropical forests which cover only some 6% of the earth's land surface and bear the brunt of contemporary devastation. Because of the lack both of exploration and taxonomic follow up, we do not know how many species we are losing, or at what rate, but we have grounds for believing that both figures are large.

Insects are of immense ecological significance. The loss of the species that make up so significant a part of the world's biological diversity is a matter of concern to a much wider audience than entomologists. We must welcome this concern, but accept that it exposes us to a question we at present find difficult to answer — how should we take action to conserve the entomological heritage of the planet? Insect conservation has hitherto been the concern only of an active minority, and wider conservation strategies have generally been based on the assumption that if plant communities and habitats are protected, together with the large mammals and birds that are taken as indicators of the health of the ecosystems as a whole, the insect fauna will be protected likewise. While there is some truth in this assertion, it is not good enough for entomologists. We would all wish to see an approach to conservation that had built into it a more directly informed entomological component.

That is why this Symposium is so important. It brings together a series of papers which reflect the diversity of entomological habitats and ecological roles around the world, emphasize the value of insects to humanity, and set out some practical proposals for conservation, especially in tropical forests and on islands, where their diversity is greatest.

IUCN, as the World Conservation Union, welcomes this Symposium and hopes that the group of specialists it will bring together and the ideas that it will generate will constitute a continuing force for conservation within the wider movement to protect the biological diversity of the earth.

Martin W. Holdgate
Director General, IUCN
September 1989

The Royal Entomological Society of London is a member of the International Union for Conservation of Nature and Natural Resources, often known as the World Conservation Union.

Acknowledgements

Many people contributed to the preparation of the Royal Entomological Society's 15th Symposium, and as a result of their hard work it was attended by more people than ever before. As Convenors of the meeting, and Editors of these proceedings, we would like to extend our thanks to all participants, whether they were in the audience, attending as speakers, or organizers behind the scenes, for the part they played. In particular, we would like to thank the following: Dr J. P. Dempster, President of the Royal Entomological Society, for opening the Symposium; the 25 authors, including 16 speakers, for their presentations and manuscripts; the four session Chairmen, Professor M. J. Samways, Dr R. C. Fisher, Dr I. F. G. McLean and Dr R. I. Vane-Wright; the RESL Symposium Committee, and particularly its Chairman, Dr Trevor Lewis, for their guidance and advice in the preparation of the meeting; the staff of the Royal Entomological Society, notably the Registrar, Mr G. G. Bentley, but also Mr R. Fairburn, Mrs Rita O'Gorman and Ms Donnah de Villiers, who worked well beyond the call of duty to make the meeting a success; the 15 authors of poster displays, which were well attended and much appreciated by participants; Lionel Hithersay, who organized the poster display; our hosts in the Physics Department, Imperial College, and particularly the projectionist; Andrew Richford and Suzanne Evans at Academic Press, for their help, advice and patience in the preparation of the proceedings.

The 15th Symposium was financially supported by three organizations, whose generosity we are pleased to acknowledge: World Wide Fund for Nature, BASF and Imperial Chemical Industries plc.

N. M. Collins and J. A. Thomas
Convenors

Opening Remarks

J. P. DEMPSTER

Twenty-five years ago we might have had difficulty attracting more than a handful of entomologists to a symposium on insect conservation. Today, it has attracted a larger attendance than any previous symposium of the Society, with over 200 participants, including many from overseas. This reflects the enormous growth in interest and concern in the subject that has developed over that time.

There are, I think, two main features of the biology of insects which make them especially important in wildlife conservation. First, insects are one of the most diverse groups of organisms inhabiting this planet. One can only guess how many species there are, but current estimates suggest that there may be as many as 30 million, most of which are still undescribed and so unknown to science. Not unnaturally, current concern over the loss of species, particularly with the large-scale devastation of the tropical rain forests, has focused strongly on insects.

Second, insect numbers are extremely sensitive to environmental change, so that problems in conservation are often first seen in insects. For example, the widely adopted strategy in conservation of protecting small areas of land as nature reserves, on the assumption that the species within them will survive, has proved far less successful for insects than for long-lived plants and vertebrates. Thus, many of the rarer butterflies have disappeared from Britain's national nature reserves over the past 30 years (Collier, 1966; Thomas, 1984: Heath *et al.*, 1984), and there is no reason to believe that other, less well-known, groups of insects have fared any better.

The problems facing conservationists cannot be overstated. The taxonomic research required to meet the challenge posed by the scale of diversity of insect faunas, especially in the tropics, creates huge problems, at a time when taxonomy is so grossly underfunded. Added to this, our knowledge of the basic ecology of most insects is rudimentary. This makes it essential that we share our knowledge and experience, and it is therefore especially pleasing to see contributors to this symposium from Australia, Brazil, Finland, Italy, New Zealand, Switzerland and the USA.

Successful conservation must depend upon international co-operation, and although this is developing, it is still weak for insects. In IUCN's Species Survival Commission, there are specialist groups on ants, Odonata and Lepidoptera, and groups are planned for Orthoptera and the water beetles. Even so, the majority of insect groups are not covered. We have a long way to go and little time to get there.

Finally, it is particularly appropriate that the Royal Entomological Society should host this symposium. Our Society has always been one of the foci for insect conservation in Britain, mainly through the activities of the Joint Committee for the Conservation of British Insects, and ours was the first entomological society to join IUCN. By bringing together experts to highlight the problems and progress in insect conservation it is hoped that this symposium will generate greater efforts in the research that is needed, if we are to save more than a small proportion of the wealth of insects that we have inherited. Let us hope that this meeting can be a launching pad for new ideas and initiatives in insect conservation.

REFERENCES

Collier, R. V. (1966). Status of butterflies on Castor Hanglands NNR 1961–1965 inclusive. *J. Northampton Nat. Hist. Soc.* **36**, 451–456.

Heath, J., Pollard, E. and Thomas, J. A. (1984). "Atlas of Butterflies in Britain and Ireland". Viking, Middlesex.

Thomas, J. A. (1984). The conservation of butterflies in temperate countries: past efforts and lessons for the future. *In* "The Biology of Butterflies" (R. I. Vane-Wright and P. R. Ackery, eds), pp. 333–353. Academic Press, London.

1

Observe Extinction or Conserve Diversity?

N. W. MOORE

I.	Introduction: Insects are Difficult to Conserve	1
II.	Insects, Nature Reserves and their Matrix	2
III.	Special Species	4
IV.	Communication with Farmers	5
V.	Communication among Entomologists and Conservation Organizations	6
VI.	Conclusion	7
	References	8

I. INTRODUCTION: INSECTS ARE DIFFICULT TO CONSERVE

The title is rhetorical: we may learn a great deal by observing the causes of extinction, but all participants in this symposium are determined to do more than that: we all want to conserve diversity. Conservation is essentially a mixture of science and politics with a small p. Perhaps the greatest challenge for conservation biologists today is to convince politicians that we should conserve species, but at this gathering there is no need to answer the question "Why conserve?", and I shall not attempt to do so.

In this chapter I shall use dragonflies as examples because I am familiar with them, not because they provide the best examples. I agree with Henry Disney (1986) that Diptera and Hymenoptera Parasitica are better indicators.

Plants stay put, and their roots, stocks and seeds can survive abnormal conditions more efficiently than most insects. Most birds and mammals are relatively catholic in their habitat requirements and are good at dispersal. The sheer size, broad similarity and contiguity of marine habitats favour the conservation of marine organisms. By contrast most insects are dependent upon particular species of plants or hosts and upon particular microclimates, not only as adults but also as larvae. Even generalized predators like dragonflies require both the shelter of trees and bushes as adults and particular aquatic habitats as larvae. Many insects are flightless or cannot fly far and so are poor at dispersal. However the main difficulty in conserving insects is their abundance and hence our ignorance of them. Many, perhaps most species have not been given names, and we know very little about the requirements of most of the insects which have been described. No wonder that there has been relatively little progress since their conservation was reviewed by Pyle *et al.* (1981). We owe our congratulations to the organizers of this symposium in attacking this enormous problem.

Natural complexity and scientific uncertainty have prevented effective action in the past. If we are to progress I believe that we must meet the challenge head on and respond to it by devising a simple, explicit strategy.

II. INSECTS, NATURE RESERVES AND THEIR MATRIX

What are the essentials of the scientific problem? Let us take the simplest of models. We are about to invade an uninhabited island covered with one type of forest. We intend to build Utopia there by achieving the best mix of development and conservation. Therefore we have to decide on which part of the island development is to occur and on which part conservation is to be the primary land use. Note that land use strategy is not an extra, but a fundamental part of conservation.

We have to decide how large the conservation area has to be and whether it should be a single large area or several small ones. If the latter is chosen some species will not be able to move between the areas, and if their separate populations are not viable they must be provided with corridors or stepping stones so that their separate populations can be linked. Thus conservation is bound to impinge on the developed area to some extent. To answer the simplest of spatial questions autecological studies must be made of every species on the island.

No wonder that conservationists turn to theoretical ecology in the hope that there are short cuts to solving the basic questions of how big and how many reserves. Many hoped that island biogeography (MacArthur and Wilson, 1967) would help conservationists when selecting reserves. So far we have to admit that this interesting approach has given them little practical help.

Recognition of the vital role that disturbance plays in maintaining ecosystems is an important advance in ecological thinking, as is the related idea that rare events can be very important determinants. Yet

Pickett and White (1985) admit that the science of patch dynamics has not yet reached a stage at which it can give much practical help to conservation. Thus practitioners are hardly supported by theoreticians, even in an ideal world.

We do not live in Utopia. We live in a world where many habitats are already fragmented and often modified, and where biological information is extremely uneven. What can we do?

First we must make a distinction between those areas where the habitat is largely unaffected by man and those where man already has a dominant effect. Most of the untouched areas of the world are in the polar regions and support very few insects, but there are still large areas of rain forest in which lightning strikes, tree falls, abnormal flooding, etc., create enough patches to provide habitats for seral species, and can ensure the survival of climax vegetation without human interference. These areas are extremely rich in insect species. They have the great advantage of being large enough to be self-supporting and the great disadvantage of being poorly known. They are under appalling threat and undoubtedly the most urgent task of conservationists alive today is to establish reserves of these habitats while it is still possible to do so. There are difficulties in selecting areas for conservation but the chief problems are economic and political rather than scientific.

The fragmented, modified habitats characteristic of the temperate regions of the earth present a different sort of problem. Much more is known about their organisms, but since more is known about plants than insects, nature reserves are usually selected by using botanical criteria. How effective is this in conserving insects?

If we use dragonflies in Great Britain as an example it appears to work well. There are 38 species of Odonata which breed regularly in this country. All but two have breeding populations in the Nature Conservancy Council's series of National Nature Reserves (Moore, 1976 and in press), which were selected primarily on botanical grounds (Ratcliffe, 1977). Some groups of insects may not fare so well, especially if they include species with very exacting requirements and therefore depend on very specific management practices. For example, it is not enough to provide the larval food plant of the Adonis blue butterfly (*Lysandra bellargus*); its habitat must be closely grazed (Thomas, 1983). Nevertheless, it is safe to draw the general conclusion that if the main types of plant community with their characteristic mosaics are conserved in reserves, a very large proportion of insect species will be conserved also.

Most of the habitat fragments which survive in the temperate region are too small to be maintained in the long run by chance "disasters" providing natural patches. Therefore the provision of patches by management is not just desirable but essential.

One single large reserve is rarely an option in the temperate region: we have to make do with a scatter of small, hence suboptimal reserves. Far too little attention has been given to measures designed to ensure that they provide adequate conditions for their flora and fauna. Studies on both population genetics and dispersal behaviour are required. I tried to address the problem of dispersal when it was my job to rationalize the system by which the Nature Conservancy Council (NCC) selects Sites of Special Scientific Interest (SSSI) (Moore, 1987a). I devised a system by which the best examples of habitat were selected in an "area of search", which was approximately one of $50 \times 50\,km^2$. This system, subsequently adopted by the NCC (1989) ensures that species with a reasonable power of dispersal should be able to move between suitable sites. It also ensures that if there is a climatic change these species can respond by colonizing new sites made habitable by the climatic change. For example, if the climate became warmer the damselfly *Ceriagrion tenellum*, which is now confined to the heaths of the south-west, might be able to colonize heathland SSSI in midland England, which at present are too cold for it.

Measures like these are extremely crude. We do not know enough about the viability of small populations of insects or about their dispersal behaviour to carry out fine tuning. A small isolated population of the dragonfly *Oxygastra curtisi* lasted for at least 140 years in southern England. Very small isolated populations of acid water dragonflies also seem to be able to exist without replenishment for many years (Moore, 1986), but one must assume that they are under threat of genetic deterioration. Stepping stones could be provided if we knew more about dispersal behaviour. At present we know very little. For example, among the Odonata we cannot do more than guess about the relative importance of dispersal at the post-emergent immature stage and dispersal as mature adults.

How much of limited conservation resources should be given to the establishment of nature reserves and how much to general conservation in the wider countryside which supports, or could support, corridors and stepping stones? This is a practical question for almost every conservation body. Conservation in the matrix, in practice, means conservation on farmland. This raises organizational problems which I shall discuss shortly. Before doing so I must mention special species.

III. SPECIAL SPECIES

So far I have assumed that all insect species are equal. Of course they are not. There are insect equivalents of the coelacanth (*Latimeria*) and the tuatara (*Sphenodon*) which clearly deserve special treatment. For example special measures should be taken to conserve the two surviving members of the once thriving suborder Anisozygoptera. *Epiophlebia superstes* in Japan and *E. laidlawi* in the Himalayas are living fossils which provide a link between the true dragonflies and the damselflies. *Hemiphlebia mirabilis* is another species which deserves special protection. It is essentially a Triassic damselfly with unique specialized features. It is good to report that, following its rediscovery by Allen Davies (1985), G. J. Santard T. R. New are undertaking a thorough study of this species (Sant and New, 1988). Fortunately it occurs in an Australian National Park. A whole group can deserve special measures: the damselfly genus *Megalagrion* in Hawaii rivals Darwin's finches in its evolutionary interest, and should be given similar conservation status. When we turn from the global scene to the national, the number of special species increases. This is because the motivation for their conservation changes from world conservation to local conservation, local interest and enjoyment. For example, the swallow-tail butterfly (*Papilio machaon*) is an extremely common and widespread insect on the world scale, but it is the only Papilionid occurring in Britain, where it is much valued for its beauty. As it is very local and threatened in Britain it deserves the national attention it is receiving (see for example Dempster and Hall, 1980).

IV. COMMUNICATION WITH FARMERS

I must now turn from insects to people. Farmers control most of the land which acts as a matrix for nature reserves, and still provides much good insect habitat and actual and potential corridors and stepping stones. Therefore it is crucial to enlist the help of farmers in conserving insects. Experience with the Farming and Wildlife Trust and its constituent Farming and Wildlife Advisory Groups (FWAG) in the United Kingdom is highly relevant (Carter, 1983). The Farming and Wildlife Trust is an umbrella body embracing both the farming community, farming organizations and conservation organizations. It employs Farm Conservation Advisers who, when asked do so, give advice to individual farmers on conservation matters. They survey farms, inform farmers what is of particular interest on their land and how it can be maintained by appropriate management. They also tell them how they can get technical and financial help for conservation work. Since 1983 over 17 000 farms have been visited by advisers, and as a result many trees and hedges have been planted and hundreds of ponds restored or dug. More importantly, many neglected woods and meadows are now managed for their wildlife. The FWAGs are successful because farmers can identify with them. Recently the Nature Conservancy Council has endeavoured to extend the FWAG principle of partnership to other industries (Nature Conservancy Council, 1988). I am convinced that this is an effective approach with immense potential for insect conservation in the farmed and urban landscapes of the temperate regions.

V. COMMUNICATION AMONG ENTOMOLOGISTS AND CONSERVATION ORGANIZATIONS

We will communicate better with farmers if we communicate better with each other. In the United Kingdom our outstanding knowledge of the distribution of invertebrate species within our country is partly due to the poverty of our fauna, but it is also due to effective co-operation between the Institute of Terrestrial Ecology, the Nature Conservancy Council, specialist societies like the British Dragonfly Society and individual entomologists. Even so, communication has not always been adequate. One of the best woods for butterflies in Britain was lost in the 1950s because the lepidopterist who knew it failed to tell any conservation organization about it until the Forestry Commission had begun to clearfell the wood. By then it was too late.

Some nature reserves have been set up specifically for insects. An outstanding example is the Dragonfly Kingdom at Ikedadani near Nakamura in Japan (Moore, 1987b). It supports 64 species of dragonfly within an area of about 50 ha of marsh and ponds created from rice fields. The Dragonfly Kingdom has done much to capture the imagination of the Japanese and has furthered conservation generally in Japan. However, such initiatives are exceptional and insect conservation must generally climb on the back of initiatives to conserve plants, birds and mammals. This can be done by careful examination of proposals for establishing and managing new reserves. Opportunities can easily be lost. For example, the original boundaries of the Cairngorms National Nature Reserve in Scotland, determined on botanical and ornithological grounds, excluded one of the few localities of the damselfly *Coenagrion hastulatum* in Britain. A small change would have included the site within the reserve. Combining insect with bird conservation is well shown at the Royal Society for the Protection of Birds' reserve at Arne in Dorset. This reserve was established to conserve Dartford Warblers (*Sylvia undata*). Today it caters for dragonflies as well: new ponds have been dug especially for them, and illustrated sign boards have been erected beside the ponds so that visiting bird watchers can identify the dragonflies they see.

It is difficult for large conservation bodies to consult with individual specialists. However the process can be greatly helped by specialist and local societies acting as channels for communication. For example, the British Dragonfly Society, which now has over 600 members, has set up a small Dragonfly Conservation Group which liaises directly with the Nature Conservancy Council, and individuals can work through it. The Joint Committee for the Conservation of British Insects has a particularly important role in co-ordinating the efforts of different societies and groups concerned with entomological conservation.

When so many threats to insects are global and resources are so small it is important to get species conservation in perspective. This is one of the functions of the specialist groups of IUCN's Species Survival Commission: they have to assess global priorities (see p. xviii). In Britain odonatists are much concerned about the decline of *Lestes dryas* and *Coenagrion mercuriale*. We need to be reminded that *L. dryas* is found throughout the Holarctic region and is very numerous in parts of it, while *C. mercuriale* is confined to south-west Europe and north Africa and is threatened over much of its range.

Finally, the international conservation bodies themselves should co-operate much more closely. They can do it: the World Conservation Strategy was a co-operative venture. The original idea of partnership between the International Union for Conservation of Nature (IUCN) as scientific advisers and the World Wide Fund for Nature (WWF) as fund raisers was imaginative. It is extremely sad that they have drifted apart: WWF needs scientific advice and IUCN needs financial support. The late Sir Peter Scott was a pioneer with both organizations. I can think of no better memorial to him than to bring these bodies back into effective partnership.

VI. CONCLUSION

I urge you to take time into account (Moore, 1987a). The risk of doing the wrong thing through lack of scientific knowledge is much less than the risk of delay. We must accept that the difficulties of conserving insects due to the abundance of species, lack of knowledge and lack of appeal to the general public will not go away within the timescale necessary for effective action. While admitting our ignorance we must make it clear that we know enough to act with reasonable hope of success, as this book will show.

Therefore, entomologists should not delay in giving total support to measures designed to set up large reserves or national parks of self-perpetuating habitats — principally tropical rain forest whose wealth of insect species is well known.

Elsewhere, where there is more ecological knowledge, we must accept that reserves will continue to be selected primarily on botanical grounds or as habitats for birds and mammals, and that they will usually be too small to perpetuate themselves without active management. Entomologists should concentrate their efforts on improving liaison with those establishing and managing these reserves. As a result boundaries will be drawn with insects in mind as well as other biota, and management will make full use of what we do know about insect requirements. I believe that we have everything to gain by being explicit about a practical concerted policy along these lines.

REFERENCES

Carter, E. (1983). The Farming and Wildlife Advisory Groups. *Biologist* **30**(3), 124–126.

Davies, D. A. L. (1985). *Hemiphlebia mirabilis* Selys: some notes on distribution and conservation status (Zygoptera: Hemiphlebiidae). *Odonatologica* **14**(4), 331–339.

Dempster, J. P. and Hall, M. L. (1980). An attempt at re-establishing the swallowtail butterfly at Wicken Fen. *Ecol. Entomol.* **5**, 327–334.

Disney, R. H. L. (1986). Assessments using invertebrates: posing the problem. *In* "Wildlife Conservation Evaluation" (M. B. Usher, ed.), pp. 271–293. Chapman and Hall, London.

MacArthur, R. H. and Wilson, E. O. (1967). "The Theory of Island Biogeography". Princeton University Press, Princeton.

Moore, N. W. (1976). The conservation of Odonata in Great Britain. *Odonatologica* **5**(1), 37–44.

Moore, N. W. (1986). Acid water dragonflies in eastern England — their decline, isolation and conservation. *Odonatologica* **15**(4), 377–385.

Moore, N. W. (1987a) "The Bird of Time". Cambridge University Press, Cambridge.

Moore, N. W. (1987b). A most exclusive preserve. *New Sci.* 13 August 1987, 64.

Moore, N. W. (1991) Recent developments in the Conservation of Odonata in Great Britain. *Adv. Odonatol.* **5**, in press.

Nature Conservancy Council (1988). Nature Conservancy Council, 14th Report. Peterborough.

Nature Conservancy Council (1989). "Guidelines for Selection of Biological S.S.S.I.s". Nature Conservancy Council, Peterborough.

Pickett, S. T. A. and White, P. S. (1985). Patch dynamics: a synthesis. *In* "The Ecology of Natural Disturbance and Patch Dynamics" (S. T. A. Pickett and P. S. White, eds), Academic Press, London.

Pyle, R., Bentzien, M. and Opler, P. (1981). Insect Conservation. *Annu. Rev. Entomol.* **28**, 233–258.

Ratcliffe, D. A. (ed.) (1977). "A Nature Conservation Review". Cambridge University Press, Cambridge.

Sant, G. J. and New, T. R. (1988). The biology and conservation of *Hemiphlebia mirabilis* Selys (Odonata, Hemiphlebiidae) in southern Victoria. Technical Report Series No. 82. Arthur Rylah Institute for Environmental Research, Conservation, Forests and Lands, National Parks and Wildlife Division, Victoria.

Thomas, J. A. (1983). The ecology and conservation of *Lysandra bellargus* (Lep. Lycaenidae) in Britain. *J. Appl. Ecol.* **20**, 59–83.

2

North American Problems and Perspectives in Insect Conservation

P. A. OPLER

I. Introduction ... 9
II. Status of Conservation Programmes ... 11
A. Canada ... 11
B. United States ... 12
C. Mexico ... 14
III. Status Surveys ... 14
A. Governmental Efforts ... 14
B. Private and Volunteer ... 15
C. Atlases ... 16
IV. Progress in Listing Candidates ... 16
A. Notice of Review ... 16
B. Recent Addition to Protected Lists ... 18
V. Protection and Habitat Acquisition ... 18
A. Available Legal Protection ... 18
B. Land Acquisition ... 19
VI. Management and Recovery ... 20
A. Recovery Plans ... 20
B. Habitat Conservation Plans ... 21
C. Conservation Agreements ... 22
D. Management on Local Preserves ... 23
E. Case Histories ... 24
VII. Prognosis ... 28
VIII. Conclusion ... 30
References ... 30

I. INTRODUCTION

Concern about the conservation of insects in North America was not aroused in a major way until the passage of the United States Endangered

Species Act of 1973 (US Congress, 1974). Prior to passage of the act, concern had been expressed about the plight of several Lepidoptera (Grimshawe, 1940; Young, 1950; Klots, 1951; Tilden 1956; Pyle, 1967; Hessel, 1972; Covell and Rawson, 1973). In the 1970s after passage of the act, concern about insect conservation in North America was pronounced, and several United States species and subspecies of Lepidoptera were listed as endangered or threatened and protected under United States law. The early stages of insect conservation in North America were summarized by Pyle *et al.* (1981).

In the 1980s the banner of insect conservation was taken up by state agencies and conservation groups in the United States, and concerns for the conservation status of insects in Canada began to stir. Although the United States government had added only six insects to its list from 1981 through 1989 (Table I), recovery plans were completed for 12 listed insects, and substantial progress was made in the conservation of three or four listed species.

In 1984, a notice of review was published in the *Federal Register* (US Fish and Wildlife Service, 1984) that listed more than 600 North American (including Hawaii) insect species that might qualify for listing under the US Endangered Species Act. In 1989, a revision of the notice was published

TABLE I. Insects listed under the US Endangered Species Act of 1973. Modified from US Fish and Wildlife Service, 1989a.

Common name	Scientific name	Status
Delta green ground beetle	*Elaphrus viridis*	T
Kretchmarr Cave mold beetle	*Texamaurops reddelli*	E
Tooth Cave ground beetle	*Rhadine persephone*	E
Valley elderhorn longhorn beetle	*Desmocerus californicus*	T
Bay checkerspot butterfly	*Euphydryas editha bayensis*	T
El Segundo blue butterfly	*Euphilotes battoides allyni*	E
Lange's metalmark butterfly	*Apodemia mormo langei*	E
Lotis blue butterfly	*Lycaeides idas lotis*	E
Mission blue butterfly	*Plebejus icarioides missionensis*	E
Oregon silverspot butterfly	*Speyeria zerene hippolyta*	T
Palos Verdes blue butterfly	*Glaucopsyche lygdamus palosverdesensis*	E
San Bruno elfin butterfly	*Incisalia mossii bayensis*	E
Schaus swallowtail butterfly	*Papilio aristodemus ponceanus*	E
Smith's blue butterfly	*Euphilotes enoptes smithi*	E
Pawnee montane skipper	*Hesperia leonardus montana*	T
Kern primrose sphinx moth	*Euproserpinus euterpe*	T
Ash Meadows naucorid	*Ambrysus amargosus*	T

that included 501 insect species found in the conterminous United States (US Fish and Wildlife Service, 1989a). These notices are used by other federal agencies, interested state agencies, private groups and individual conservationists to communicate what taxa may be of important conservation concern.

In Canada, efforts on behalf of rare and endangered insects did not begin until the late 1970s, but now concern is high amongst Canadian conservationists and entomologists. Efforts to begin insect conservation programmes in Canada are now under way at the provincial, organizational and citizen levels, and it is expected that the inclusiveness of Canada's programme may equal or exceed that of the United States by the mid-1990s.

Insects are not treated under endangered species legislation in Mexico or Central America. Nevertheless, the overwintering sites for monarch butterflies (*Danaus plexippus*) are provisionally protected by the Mexican government, and Mexico has recommended that several of its butterflies be protected under provisions of the Convention on International Trade in Endangered Species of Wild Fauna and Flora (CITES).

II. STATUS OF CONSERVATION PROGRAMMES

A. Canada

There is national endangered species legislation in Canada, but the protection of invertebrates is not specified. The Committee on the Status of Endangered Wildlife in Canada (COSEWIC) tracks the status of endangered Canadian animals. Unfortunately, the committee does not monitor Canadian invertebrates.

Two provinces, New Brunswick and Ontario, have endangered species legislation that permits the listing and protection of invertebrates, but only Ontario lists an endangered insect, the West Virginia white butterfly (*Pieris virginiensis*). Although rare in Canada (Brownell, 1981), in the eastern United States this species is widely distributed in native transition zone woodlands. The Ontario Ministry of Natural Resources is considering the downlisting of the West Virginia white from endangered to rare, but is also considering the listing of three additional insects — *Atrytonopsis hianna* (Hesperiidae, Lepidoptera), *Schinia gloriosa* (Noctuidae, Lepidoptera) and *Erastria cruentaria* (Geometridae, Lepidoptera).

Conservation actions on behalf of Canadian insects by private groups include the purchase of a nature preserve in Ontario specifically for the declining Karner blue butterfly (*Lycaeides melissa samuelis*), and the funding by the World Wildlife Fund of a 1988 study of a population of Weidemeyer's admiral (*Limenitis weidemeyerii*).

The Entomological Society of Canada has appointed an *ad hoc* committee of Canadian entomologists to document species of insects whose continued existence in Canada is endangered or threatened. The committee is also charged with determining which Canadian insect localities or habitat types may be threatened with destruction.

B. United States

The United States has the leading insect conservation programme in North America. The founding of the Xerces Society in 1971 by Robert M. Pyle and the passage of the Endangered Species Act of 1973 acted to galvanize interest and concern for insect conservation amongst federal, state and private agencies (Pyle, 1976).

The Endangered Species Act requires the federal government to recognize and list insects endangered or threatened with extinction. Listed insects must be afforded protection from taking which is defined as "...to harass, harm, pursue, shoot, wound, kill, trap, capture, or collect or attempt to engage in any such conduct". The Endangered Species Act is especially beneficial to insects because taking may be included amongst the impacts of habitat destruction. All federal agencies must use their authority in furtherance of the conservation of listed species that might be affected by their actions, fundings or granting of permits. The US Fish and Wildlife Service, which administers the act for insects, is responsible for preparing listing rules, having recovery plans prepared, consulting with other agencies, enforcing regulations and negotiating co-operative agreements with state agencies.

Currently, there are 17 insects listed under the act's provisions — four beetles, one heteropteran and 12 lepidopterans (Table I). Ten of the species are listed as endangered — the category at greatest risk of extinction — and seven are listed as threatened.

At the state level, most states now have their own endangered species programmes, for which they receive matching funds from the federal government. Although states must have the legal authority to protect and conserve invertebrates to be eligible for a co-operative agreement, only a few states actually give legal protection to insects, and most of these only list the same insects that are protected under the federal act. For example, California and Florida add state protection to all federally listed insects that occur within their borders. Michigan and New York however give protection to the Mitchell's Satyr (*Neonympha mitchellii*) and the Karner Blue, respectively — species that are not included on the federal list.

Several states have sponsored endangered species symposia regarding their native biota. These symposia have produced lists and species accounts for many insects that biologists feel are endangered, threatened or of "special concern" in the individual states, even though some may be common in other portions of their range. These insects, however, usually do not receive protection under state law. Examples of states that have had such symposia are Maryland (Opler, 1984a), Minnesota (Cutler *et al.*, 1988), New Jersey (Boyd 1982), Pennsylvania (Opler, 1984b) and Virginia (Opler, 1978).

Most states now have natural heritage programmes or the equivalent. These programmes were initially devised and sponsored by The Nature Conservancy, a nation-wide non-governmental organization, in an effort to document rare or threatened biota, ecosystems or natural features that were underprotected in each state. Most state natural heritage programmes list insects among the elements that they monitor and recommend for protection.

At least four states, California, Florida, Ohio and Washington, have funded status surveys of parts of their insect fauna in order to determine which insects may be in need of protection. The status of a number of butterflies was surveyed by California. The results of Florida's surveys were published as a series of status reports (Franz, 1982). The survey of Ohio's Lepidoptera is still in progress, although preliminary recommendations have already appeared (Shuey *et al.*, 1987). A status and conservation report on Washington's butterflies was prepared for the Department of Wildlife.

Private organizations are playing an increasingly large role in protection of invertebrates in the United States. The World Wildlife Fund US sponsored studies of the rare Wiest's sphinx moth (*Euproserpinus wiesti*) (Bagdonas, 1981) that helped bring the attention of conservationists to this insect's possible plight. The organization also supported a survey of the status of monarch butterfly wintering colonies in California.

The Nature Conservancy now uses information on the presence of rare insects as one factor when deciding whether to acquire a particular piece of land for its preserve system. The Nature Conservancy has an entomologist on its staff to assist in determining whether rare or declining insects occur on its lands, and to make recommendations on the acquisition of lands not under protection. In California, The Nature Conservancy owns Jepson Prairie — home of the threatened Delta green ground beetle (*Elaphrus viridis*) — and Ring Mountain, the only habitat for the Tiburon blind harvestman (*Sitalcina tiburona*). The Nature Conservancy is also building a data base that will contain the national and state status of many groups of North American insects; its staff has already begun entering occurrence information on Cicindelidae, Lepidoptera and Odonata.

C. Mexico

Currently, although Mexico does not include insects in its endangered species programme, there is a national effort to protect the overwintering sites for the monarch butterfly (*Danaus plexippus*). In 1980, a presidential decree was made to protect the monarch in Mexico; however its overwintering habitats there are still in jeopardy. Subsequently, Pro Mariposa Monarca, a civic group in association with the World Wildlife Fund US, was formed to co-ordinate conservation and research efforts. The state government of Michoacan, the Universidad Nacional Autonoma de Mexico, several federal agencies and conservation groups have all been co-operating to conserve the monarch's habitats in Mexico (Ogarrio, 1984; Rodriguez, 1984).

III. STATUS SURVEYS

One of the biggest problems facing North American insect conservationists is an information gap. For the vast majority of insect orders and families, information on which conservation status may be determined either has never been collected or never been synthesized such that objective decisions on vulnerability may be made.

The systematics and the detailed past and present distributions of many insect species are not well known. If declines have occurred in a species' abundance or range, the causes of such changes are usually unknown. Fair status information exists for certain well-collected insect groups such as dragonflies, butterflies, the larger moths, scarab beetles and tiger beetles.

A. Governmental Efforts

The US Fish and Wildlife Service has played a leading role in supporting surveys of North American insect species. Two major contracts were issued to determine the status of selected insect groups in the conterminous United States and Hawaii. The first contract, to the Smithsonian Institution, was to collect data from all available museum specimens of a broad array of insects from the conterminous United States (Perkins, 1980). The second contract, to the Bishop Museum in Honolulu, was designed to provide the Fish and Wildlife Service with status and listing recommendations on several genera of native Hawaiian insects. Although no publication has yet resulted, the Bishop Museum entomologists did provide lists of Hawaiian insects that should receive further field survey and scrutiny. A number of these are listed in the most recent review of animal candidates (US Fish and Wildlife Service, 1989a, summarized in Table II).

The US Fish and Wildlife Service and other United States governmental agencies, notably the Bureau of Land Management and Forest Service, have funded status surveys for many non-listed species, although candidates or species proposed for listing have most often been the subjects of such research. Perhaps the most extensive surveys have taken place in California and Nevada, where most insect candidates are thought to occur. The California Department of Fish and Game and the Fish and Wildlife Service's Sacramento field station have been important sponsors of surveys. Survey activities elsewhere have been less extensive, although the State of Ohio Department of Natural Resources has sponsored a survey of candidate Lepidoptera (Shuey *et al.*, 1987), and the Iowa State Ecologist, in cooperation with the University of Northern Iowa, sponsored a survey of rare skipper butterflies (family Hesperiidae) found on native prairies in western Iowa.

B. Private and Volunteer Efforts

Private organizations have contributed modestly to the need to survey declining populations of rare insects. Notable exceptions are the World Wildlife Fund US, which sponsored a survey of the apparently rare Wiest's sphinx moth, and The Nature Conservancy, which surveys many of its own preserve lands for the presence of insect candidates. In addition, the Xerces Society's small grants programme, now inactive, funded several small local surveys of selected invertebrate taxa.

Private and volunteer surveys have been largely unpaid efforts taken on by many dedicated entomologists and amateur collectors. For example, several collectors have amassed extensive files on the distribution of various Lepidoptera groups, tiger beetles (Cicindelidae), horseflies and deer flies (Tabanidae) and dragonflies (Odonata), to name a few.

Most notable among those efforts by private organizations are surveys of the Lepidoptera of Kentucky and Ohio, sponsored by the Kentucky Lepidopterists and the Ohio Lepidopterists. Such surveys not only enumerate the species in the area searched, but also provide valuable information for conservationists on the present plight of potential candidates.

C. Atlases

There are no national programmes in North America comparable to that of the United Kingdom's Biological Records Centre, although passage of a national biodiversity bill, now being considered by the US Congress, would create a national centre for monitoring biodiversity. If such a centre could co-ordinate the efforts of the many advanced amateur naturalists and biologists, many impressive distributional atlases could be produced in a relatively short time at minimal cost to the government. The breeding bird atlases already produced by volunteer ornithologists in several states are good examples.

Systematic surveys of native insects using the same grid approach as that used in Britain and Europe are uncommon in North America. However, a survey based on a grid system of the butterflies of Massachusetts has been sponsored by the Massachusetts Audubon Society (Cassie *et al.*, 1986), and a survey based on the Universal Transverse Mercator grid system of the insects of Maryland is being sponsored by the Maryland Entomological Society. Both groups plan to produce atlases and maintain computer data bases of their information.

Atlases that use United States or Canadian counties for their survey and mapping units have been attempted in both Canada and the United States. Most of these have been maps of butterfly distributions. The Rocky Mountain atlas was the first such effort (Stanford, 1977), but an atlas of butterflies known to occur in the eastern United States was produced by Opler (1983). In Canada, an atlas of butterflies of Manitoba was published as individual species accounts in a book on that province's fauna (Klassen *et al.*, 1989). The first atlases for North American moths are now being planned (R. Piegler and P. Opler, unpublished manuscript). Hopefully, similar efforts will be planned for other insect groups and geographic areas. Federal or state sponsorship of such basic research would greatly hasten and possibly lend uniformity to the published results and their utility.

IV. PROGRESS IN LISTING CANDIDATES

A. Notice of Review

In 1984, the US Fish and Wildlife Service issued a notice of review in the *Federal Register*, the daily announcement mechanism of the United States' administrative branch. This notice of review detailed all of the invertebrate animals that were being, or had previously been, considered for listing under provisions of the US Endangered Species Act. The notice listed more than 600 species of insects found in the United States. More recently, a second notice was published that included all US candidate animals (US Fish and Wildlife Service, 1989a).

The insects from the conterminous United States included in the second notice are considered candidates for listing (Table II). Each species is placed in one of five categories – 1, for those species for which adequate information is available to support immediate listing; 2, for those species for which sufficient information to support a listing is not available; 3A, for those species thought to be extinct; 3B, for those species whose original status was based on an invalid taxonomic name; 3C, for those species now thought to be widespread or abundant and under no immediate threat.

The insects listed on the notice compose less than 1% of the roughly 88 000 described North American insects (Arnett, 1983), and, based on the proportion of candidates in vertebrate or plant groups, there are probably at least 1500 insect taxa in some degree of jeopardy. The lack of status information to support listing is shown by 427 (95%) of the insects being in category 2 (insufficient data available). Only nine species

TABLE II. North American Insecta included on Animal Notice of Review (US Fish and Wildlife Service, 1989a).

Order	Taxa included	Category				
		1	2	3A	3B	3C
Collembola	2	0	2	0	0	0
Ephemeroptera	15	0	12	2	1	0
Odonata	22	1	20	0	0	1
Plecoptera	13	0	9	1	1	2
Orthoptera	28	0	26	2	0	0
Hemiptera	6	0	5	0	0	1
Homoptera	2	0	2	0	0	0
Neuroptera	2	0	2	0	0	0
Coleoptera	221	4	206	6	1	4
Mecoptera	2	0	2	0	0	0
Diptera	14	0	10	2	0	2
Lepidoptera	101	1	65	11	3	21
Trichoptera	67	2	59	1	4	1
Hymenoptera	9	1	7	0	0	1
Total	504	9	427	25	10	33

were indicated to have sufficient information to support a listing, and two of these have been proposed in the 11 months following publication of the notice.

There is an imbalance in the number of species from each order included in the notice. This probably stems from the relative state of knowledge about the species in each order rather than their relative vulnerability to habitat destruction. Coleoptera (221 species), Lepidoptera (101 species) and Trichoptera (67 species) are best represented on the notice, and Diptera (14 species) and Hymenoptera (9 species) seem to be greatly under-represented.

As mentioned previously, the Entomological Society of Canada is currently preparing a list of Canadian insects that are believed to be declining or possibly in danger of extinction. No effort has begun in Mexico to compile a list of insects that might be in jeopardy of extinction.

B. Recent Additions to Protected Lists

Most insects (12 species) were added to the US list of endangered and threatened wildlife and plants (US Fish and Wildlife Service, 1989b) during the first 8 years after the act was passed. Since 1981, only six insects have been listed, and two others have been proposed. Many conservationists assert that the Fish and Wildlife Service currently has too few staff dedicated to the addition of species to the US endangered species list.

V. PROTECTION AND HABITAT ACQUISITION

A. Available Legal Protection

The US Endangered Species Act provides strong legal protection on behalf of listed species through prohibitions on take, commercial dealings and import–export. Criminal violators may be subject to 1 year in prison and $50 000 in fines. Several individuals have been investigated or apprehended by Fish and Wildlife law enforcement agents for the taking of listed insect species, but no one has been prosecuted. All US federal agencies must co-operate in the conservation of listed and candidate endangered and threatened species, and no agency may execute, authorize or permit any action that might "jeopardize the continued existence" of any listed species. The 1988 amendments to the Endangered Species Act require the Fish and Wildlife Service to monitor the status of candidates, and the Service could be required to list them on an emergency basis if their status declines precipitously.

States that have co-operative agreements for their own endangered species programmes must afford federally listed species in their states at least equally strong protection as that provided by the federal law. As a result, states may have more stringent prohibitions and permit requirements than those specified in federal rules.

To date, Mexico protects the monarch butterfly under its national legislation, but Canada does not protect insects under national law.

B. Land Acquisition

The US Endangered Species Act provides for the acquisition of habitat for the recovery of listed endangered or threatened species. Land acquisition is usually through the Land and Water Conservation Fund Act. Most endangered insects reside in regions with high land value, and hence land acquisition is often prohibitively expensive. In a landmark action, the US Fish and Wildlife Service acquired the entire habitat for the Lange's metalmark butterfly (*Apodemia mormo langei*). Now referred to as Antioch Dunes National Wildlife Refuge, a unit of the San Francisco National Wildlife Refuge complex, this 24-ha riverine sand dune habitat about 80 km east of San Francisco was zoned for commercial development, and at the time of acquisition it had been seriously degraded by sand mining and the invasion of stabilizing weeds.

In another action, the Fish and Wildlife Service acquired private lands slated for development on North Key Largo, Monroe County, Florida. This action was taken on behalf of the listed American crocodile (*Crocodylus acutus*) and the Schaus swallowtail (*Papilio aristodemus ponceanus*). In particular, the population of the Schaus swallowtail on North Key Largo, formerly its major population centre, had almost disappeared due to housing development and the widespread use of insecticidal fogs for the control of adult mosquitoes.

There has been no acquisition of threatened insect habitats by either the Canadian or Mexican governments. Similarly, state or provincial governments have rarely acquired lands wholly or to a significant degree because of the conservation needs of any insect.

In 1987, California passed legislation recognizing the monarch butterfly's significance and directed the state Department of Fish and Game to develop a management plan that identified candidate overwintering sites for acquistion. In 1988, a bond issue was passed that allocates $2 million for acquisition of monarch sites.

The New York state government purchased a large block of oak–pine scrub habitat near Albany, New York, for the state-listed Karner blue butterfly. The City of Albany also purchased several blocks of private land in the same vicinity. Currently, a mosaic of habitats under different ownership and a lack of co-ordinated management are leading to a rapid decline in the Karner blue's habitat quality there.

Conservation organizations, most notably The Nature Conservancy, have acquired lands for their preserve systems that feature habitats of declining insects (see Section II.B).

Ownership of lands may be transferred to conservation or preserve status as a result of legal considerations stemming from the listing and protection of species under federal, state or provincial laws. As discussed below, habitats of several federally or state-listed endangered or threatened species, including the Karner blue, Mission blue (*Plebejus icarioides missionensis*), San Bruno elfin (*Incisalia mossii bayensis*) and Pawnee montane skipper (*Hesperia leonardus montana*), have been acquired or transferred to public ownership due to the species' listed status.

For candidate species, inclusion in the animal notice of review (US Fish and Wildlife Service, 1989a) serves as an indication to other public agencies and conservation groups that the species are of conservation concern, and, as a result, lands may be purchased or protected on their behalf.

VI. MANAGEMENT AND RECOVERY

The mere listing of a species in no way guarantees the species' continued survival. Even the acquisition and protection of its habitat may fail to do so. Almost always, management and research must be undertaken to ensure the species' perpetuity. In the past, several North American insect species (e.g. the Xerces blue (*Glaucopsyche xerces*), for which the Xerces Society was named) became extinct without management (Opler, 1976). Over the past decade, scientists, managers and conservationists in North America, particularly the United States, have devised novel approaches to the problems of management and recovery. The applications of these approaches to insects are presented below.

A. Recovery Plans

The US Endangered Species Act calls for the preparation of recovery plans for all listed species except those that would not benefit from having a plan prepared.

Recovery plans should synthesize all of the life history and ecological information known about the species, and they should include recovery goals. These goals are expressed in terms of population and habitat conditions necessary to "delist" the species as fully recovered, should the goals be attained. Usually these goals are expressed as the number of secure populations having at least a minimum number of individuals. All of the responsible state and federal agencies and required conservation steps are identified in each plan.

Recovery actions may include habitat acquisition or easements, habitat management, propagation and reintroduction, law enforcement and public awareness. For each recommended action in the plan, a projected schedule, responsible agency and cost estimate are described. Unfortunately, few recovery plans are actually implemented because they are not legally binding on the affected agencies, and the necessary personnel and funds are not often available.

Recovery plans have been devised for 12 of the 17 listed United States insects. Meaningful recovery actions are under way for at least six of the listed species — Mission blue, San Bruno elfin, El Segundo blue (*Euphilotes battoides allyni*), Lange's metalmark, Oregon silverspot (*Speyeria zerene hippolyta*) and Schaus swallowtail.

B. Habitat Conservation Plans

A new approach to endangered species conservation was developed through the effort to seek solutions to seemingly unavoidable conflicts between planned private development and listed endangered species. A group of developers and local agencies seeking to develop privately held habitat on the San Francisco peninsula in central California devised what has become known as the San Bruno Mountain Habitat Conservation Plan (Reid and Murphy, 1986); the plan would conserve the callippe silverspot butterfly (*Speyeria callippe callippe*) and three federally listed endangered species, the endangered Mission blue butterfly, the San Bruno elfin butterfly and the San Francisco garter snake (*Thamnophis sirtalis tetrataenio*). Upon approval of such a plan the Fish and Wildlife Service may grant an "incidental take" permit under Section 10(a) of the Endangered Species Act. The permit allows for a limited take of the listed species and other species of concern in the course of approved development. The permit also provides for research, management and recovery of the remaining portions of the habitat, however, so that the probability of the species' long-term survival is actually better than if no development had taken place.

The San Bruno Mountain plan calls for annual monitoring of the population and its subunits. Graded areas and road verges are being planted with propagated native lupines and stonecrop, the caterpillar hosts of the Mission blue and the San Bruno elfin, and appropriate nectar sources. Almost 320 ha of suitable habitat have been transferred from private ownership to public parkland, and habitat improvement is taking place through the removal of invasive woody exotic plants such as gorse.

The US Congress was impressed with the Habitat Conservation Plan concept developed for San Bruno Mountain, and hence it incorporated the procedure into the Endangered Species Act amendments of 1980. Subsequently, despite the fears of some conservationists, the Habitat Conservation Plan concept has been used sparingly and has enhanced the species' long-term prospects.

A habitat conservation plan has also been developed in Oregon for the Clatsop Plains population of the Oregon silverspot butterfly.

A plan similar to a Habitat Conservation Plan, but not legally mandated by the Endangered Species Act, has been proposed for the Karner blue butterfly populations near Albany, New York. Under the plan, which would allow for the expansion of an existing landfill on to known habitat for the butterfly, a "tipping fee" charged to users of the landfill would be used to help finance a conservation plan for the Karner blue in the Albany pine bush. The hope is that the Karner blue would be better off with the plan than without it. There is opposition to the plan, and it is currently in litigation.

C. Conservation Agreements

Conservation agreements are non-binding memoranda of understanding between private landowners and public agencies or conservation organizations that call for the land use or management of the owners' lands to remain compatible with the conservation interests of the species. Conservation agreements for single-site species may in some cases ensure the continued existence of the species without invoking the expensive listing process of the Endangered Species Act. In instances where it is not necessary to use the authority of the act, this method is the preferred approach. However, listing action may be required where the species is experiencing threats over a wide mosaic of land ownership categories or where the legal punch of the act is needed.

Several conservation agreements regarding listed and unlisted North American butterflies have been negotiated. An agreement between the

Fish and Wildlife Service and Waste Management, Inc., has provided protection for a part of the largest remaining population of the Bay checkerspot near Morgan Hill in central California. The Nature Conservancy has an agreement with private landowners in Washington to conserve the strongest populations of three butterflies that are candidates for state listing.

D. Management on Local Preserves

Often the remaining habitat of listed species or other rare insects of conservation concern occurs on small areas. Furthermore, maintenance of the insects on that land calls for intense management to keep the habitat in the correct seral stage so that the proper plants and other habitat features may be maintained. In August 1989, the Ohio chapter of The Nature Conservancy, the Ohio Lepidopterists, and the Ohio Department of Natural Resources sponsored a workshop that dealt with the problem of managing butterfly and moth populations on small preserves.

In North America, there are several habitat types that often have rare or endangered insect species and that require intensive management. These habitats are native prairies, coastal sand dunes and oak–pine sand barrens. Both the prairies and barrens habitats constitute fire succession habitats; without the occurrence of periodic fires the plant communities on these areas would eventually succeed to dominance by woody plants. Coastal sand dunes do not require fire, but man's physical interference with dynamic processes and the introduction of invasive exotic weeds combine to threaten these delicate ecosystems (Powell, 1981).

The need to burn small, tall-grass prairie preserves on almost an annual basis has created a dilemma for those who wish to retain the native insect communities along with the impressive prairie wildflowers. Principles for managing such preserves call for the use of periodic fire, but every effort should be made not to burn entire habitat types at once, and skipped areas should not be reburnt, as these are often the refuge sites for native insects (Opler, 1981; Panzer, 1988). Despite earlier predictions that prairie preserves of 400 ha or more were necessary to maintain populations of native prairie insects (Opler, 1981), Panzer (1988) has indicated that prairie preserves as small as 50 ha typically support sizeable prairie insect communities that often include rare species. I should note, however, that whenever possible the largest possible preserves should be set aside and managed.

E. Case Histories

In order to illustrate the problems facing those wishing to conserve North American insects, I will review the cases of four insects faced with different kinds of threats: the Bay checkerspot, which has lost most of its populations due to residential development; the El Segundo blue, which lost most of its former habitat to urbanization; the Lange's metalmark, whose habitat was reduced by commercial development; and the Pawnee montane skipper, whose habitat is threatened by construction of a large water impoundment.

A common thread that runs through case histories such as these is the sequence of events leading to the species' conservation and recovery. For most species, the first event is the recognition that the species is in trouble, whether the recognition be through legal listing or more informal means. The second event is the co-operation between affected entities and individuals through their joint efforts on a recovery team, or other management group. The team or group then devises a strategy for the species' recovery. The strategy often includes the same elements: security of the habitat, whether through acquisition or easement; periodic monitoring of the population; setting of recovery goals through application of modern conservation principles (e.g. those recommended by Ehrlich and Murphy, 1987); and implementation of the recommended recovery steps. The process often involves a periodic reassessment of procedures and accomplishments.

1. Bay checkerspot (Euphydryas editha bayensis)

In 1987, after a long battle that included the threat of law suits, and controversy and challenge over its status as a valid subspecies, the Bay checkerspot butterfly (Fig. 1) was listed by the Fish and Wildlife Service as a threatened species. This subspecies of *Euphydryas editha* (Murphy and Ehrlich, 1980) is limited to grasslands on serpentine soil in the San Francisco Bay counties of coastal central California. Through urbanization and associated development, 11 of 16 populations have been extirpated, and only one large reservoir population remains.

The Bay checkerspot is highly susceptible to drought, as it must undergo a larval diapause through the long, dry California summer. It feeds on a small annual (*Plantago erecta*) in its early instars, and on *Plantago* and owl's clover (*Orthocarpus densiflora*) after leaving diapause during the next wet season. Only very large populations on sites with a variety of topographic relief and slope directions have enough environmental variety to allow the butterfly to survive over long periods (Weiss *et al.*, 1988).

Fig. 1. The Bay checkerspot (*Euphydryas editha bayensis*), a threatened species, nectaring at tidy-tips (*Layia platyglossa*). Morgan Hill population, Santa Clara County, California (photo by Paul A. Opler).

One of the Bay checkerspot's colonies has been closely monitored by use of mark–release–recapture techniques for more than 30 years (Ehrlich *et al.*, 1975). Principles learned from this group of small butterfly populations provide a fund of knowledge to those interested in the conservation of invertebrate populations (Ehrlich and Murphy, 1987; Murphy *et al.*, 1990).

The large Morgan Hill population occurs on private land owned by two large corporations (Murphy, 1988; Wells, 1987). One of these companies, Waste Management, Inc., has developed a conservation agreement with the Center for Conservation Biology at Stanford University. This agreement allows for protection and enhancement of the Bay checkerspot population on the upper part of the property, while allowing for the development of a landfill on the lower portion, where no butterflies occur.

2. El Segundo blue (Euphilotes battoides allyni)

The El Segundo blue is a small lycaenid limited to remnants of the El Segundo sand dunes along the Pacific coast of California near Los Angeles.

Most of the coastal sand dunes are now occupied by residential and commercial developments. The only remaining sand dune habitats occur on the Los Angeles International Airport (80 ha) and on a Standard Oil refinery (1 ha).

Because of the threat of extension of the airport or the construction of a golf course on much of the remaining sand dunes at the west end of the runways, the butterfly was added to the US Endangered Species list in 1976. Since then, critical habitat for the butterfly was proposed and withdrawn. No major conservation actions have occurred on the airport property, despite the concerns of local biologists, although the City of Los Angeles has funded population surveys there. In contrast, the Standard Oil Company, after initial encouragement from the Xerces Society (Oppewall, 1976), has contracted with local conservationists to establish a butterfly preserve on its remaining 1 ha. The butterfly's caterpillar host plant and primary adult nectar source is coastal buckwheat (*Eriogonum parvifolium*). These plants have been propagated and reintroduced to the Standard Oil sand dune remnant, and the butterfly's population numbers have increased substantially there.

3. Lange's metalmark (Apodemia mormo langei)

This localized subspecies of the more widespread Mormon metalmark (*Apodemia mormo*) is now limited to about 20 ha of riverine sand dune along the Sacramento–San Joaquin River in central California. The butterfly (Fig. 2) was listed as endangered in 1976 because the remnants of its habitat were threatened by sandmining and industrial development (Opler and Robinson, 1986). In the face of continuing sandmining, most of the remaining habitat was acquired in 1980 and incorporated into the Fish and Wildlife Service's San Francisco Bay National Wildlife Refuge complex. Since acquisition, various recovery steps have been attempted, but the butterfly's population numbers continued to decline until disking of a large sand hardpan area resulted in luxuriant growth of the butterfly's caterpillar host plant and important nectar source, auriculate buckwheat (*Eriogonum nudum*). Over the past three years the butterfly's peak population has roughly tripled, from a low estimate of about 154 in 1986 to 668 in late August 1989. Monitoring of the population was initially by the mark–recapture method, but a modified transect observation method (modified from Pollard, 1977) has been adequate in recent years.

Although the butterfly's numbers have rebounded dramatically, much work remains on the refuge. Recontouring and revegetation of an abandoned grape vineyard with native plants will help by increasing critical habitat for the Lange's metalmark, several other high priority insects and two listed plants.

Fig. 2. The Lange's metalmark (*Apodemia mormo langei*), an endangered species, nectaring at wild buckwheat (*Eriogonum nudum*) — its sole larval host. San Francisco Bay National Wildlife Refuge, Contra Costa County, California (photo by Paul A. Opler).

4. Pawnee montane skipper (Hesperia leonardus montana)

The Pawnee montane skipper (Fig. 3) was listed as a threatened species in late 1987. It is an isolated subspecies of a more widespread prairie butterfly and is threatened by potential water impoundments and recreational development in the upper South Platte River canyon above Denver, Colorado (Two Forks Project).

In order to assess the skipper populations and potential effects of planned water development, an extensive survey and population monitoring project was developed for the Denver Water Department in co-operation with federal and state agencies (ERT, 1986, 1988). Since skippers are not normally suitable subjects for mark–release–recapture experiments, 46 randomly selected, standardized walking transects were established throughout the butterfly's 200-km^2 range. This survey requires the services of at least three entomologists over the skipper's flight period. The amount of nectar source and density of host plant were also monitored. Three years of sampling for the transects above the impoundment level of the proposed reservoir have shown the population size of this skipper to be remarkably stable.

Fig. 3. The Pawnee Montane Skipper (*Hesperia leonardus montana*), a threatened species, nectaring at blazing star (*Liatris punctata*) — its primary nectar source. Platte Canyon, Jefferson County, Colorado (photo by Paul A. Opler).

The Denver Water Department, the Colorado Division of Wildlife and four federal agencies have developed a conservation agreement that would take effect if the water development project were authorized. The agreement includes management actions that would guarantee the Pawnee montane skipper's long-term survival. Currently, the Two Forks project is on hold because the US Environmental Protection Agency has not approved the project over concern for negative environmental effects.

VII. PROGNOSIS

A growing appreciation of the conservation needs of insects and other invertebrates, coupled with a growing body of scientific knowledge of declining and endangered insects, has spawned a new field of invertebrate conservation science. This science incorporates ideas from other areas of biology, including the principles of island biogeography and population genetics. Based on long-term studies of the Bay checkerspot butterfly, Ehrlich and Murphy (1987) discussed the following nascent principles of insect conservation biology:

1. "The movement of individuals between habitat patches or reserves does not necessarily mean that gene flow is occurring. It also does not mean, in case of extinction, recolonization of one patch by individuals from the other will occur readily."
2. "Understanding the scale and structure of populations is crucial to designing and managing preserves."
3. "Understanding the pattern of mating within demographic units is also required if conservation efforts are to succeed, because those patterns may change with density."
4. "Topographic heterogeneity or other aspects of habitat diversity may be key determinants of the habitat suitability of an area for the maintenance of populations of many insect herbivores that otherwise require only small areas of habitat."
5. "If conservation options are limited, top priority should be given to identifying and preserving reservoir populations, if any."
6. "Natural history investigations should be undertaken before a reserve is designed in an attempt to identify subtle habitat requirements that could potentially play key roles in the long-term survival of the population to be protected."
7. "Under severe environmental conditions, the degree of oligophagy may be a crucial determinant of extinction proneness."
8. "Recolonization of empty patches may not occur in most years; many populations may be re-established in rare years of explosive dispersal."
9. "Environmental stochasticity, especially variation in host quantity, quality, and phenology, is the major cause of extinction, particularly in small populations."
10. "Habitat patches and metapopulations (clusters of geographically adjacent but geographically isolated populations between which exchange of individuals may be of varying difficulty) may be difficult to protect."
11. "The ease of maintaining a metapopulation or re-establishing extinct metapopulations by introducing insects to 'empty' habitat is difficult to predict."
12. "Introduced mammalian herbivores may have complex impacts on endangered herbivorous insects."
13. "Endangered populations of insects may be 'studied to death'."

Based on their studies and experience, Ehrlich and Murphy stated that we "should make a best guess about the amount of habitat required to preserve a species, and then at least double that area to reach a recommended minimum reserve size."

The continued study of insect populations in need of conservation should refine and add to the above principles.

VIII. CONCLUSION

Since the early 1970s insect conservation in North America has undergone a rapid evolution. In the United States, conservation programmes are well under way, while those in Canada are just emerging. Mexico, with its remarkable insect diversity, has no programme for endangered insect species, although it is seeking protection for the significant overwintering colonies of the monarch butterfly.

Conservationists have worked with federal, state or provincial, local and private entities to achieve conservation goals. Concerned citizens and biologists must continue to co-operate if long-term goals are to be achieved. More extensive status surveys of insect taxa are necessary, as are more aggressive federal, state and provincial listing and protection efforts.

REFERENCES

Arnett, R. H., Jr (1983). Status of the taxonomy of the insects of America north of Mexico, a preliminary report prepared for the subcommittee for the Insect Fauna of North America Project. Florida State Collection of Arthropods, Gainesville, Florida.

Bagdonas, K. (1981). The life history and ecology of *Euproserpinus wiesti* Sperry. Final report for IUCN/World Wildlife Fund Project 1793. World Wildlife Fund, Washington, District of Columbia, 70 pp.

Boyd, H. (1982). Invertebrates. *In* "New Jersey's Endangered and Threatened Plants and Animals". Center for Environmental Research, Stockton State College, Pomona, New Jersey.

Brownell, V. R. (1981). The West Virginia white butterfly (*Artogeia virginiensis* Edwards) in Canada: a status report. Nongame Program, Wildlife Branch, Ontario Ministry of Natural Resources, Toronto, 63 pp.

Cassie, B., Leahy, C. and Walton, D. (1986). Massachusetts Butterfly Atlas Project. *Atlas Newsletter* **3**, 2 pp.

Covell, C. V., Jr and Rawson, G. W. (1973). Project Ponceanus: a report on first efforts to survey and preserve the Schaus swallowtail in southern Florida. *J. Lepid. Soc.* **27**, 206–210.

Cutler, B., Dana, R., Huber, R., Bright, R. and Pfannmuller, L. (1988). Invertebrates. *In* "Minnesota's Endangered Flora and Fauna" (B. Coffin and L. Pfannmuller, eds), pp. 375–431. University of Minnesota Press, Minneapolis.

Ehrlich, P. R. and Murphy, D. D. (1987). Conservation lessons from long-term studies of checkerspot butterflies. *Conserv. Biol.* **1**, 122–131.

Ehrlich, P. R., White, R. R., Singer, M. C., McKechnie, S. W. and Gilbert, L. E. (1975). Checkerspot butterflies: a historical perspective. *Science* **188**, 221–228.

ERT (1986). 1986 Pawnee montane skipper field studies. Report to Denver Water Department. ERT, 1716 Heath Parkway, Fort Collins, Colorado 80524, 83 pp.

ERT (1988). Pawnee montane skipper 1987 field studies. Report to Denver Water Department. ERT, 1716 Heath Parkway, Fort Collins, Colorado 80524, 28 pp.

Franz, R. (ed.) (1982). "Rare and Endangered Biota of Florida. Vol. 6, Invertebrates". Florida Game and Fresh Water Fish Commission, University Presses of Florida, Gainesville.

Grimshawe, F. M. (1940). Place of sorrow: the world's rarest butterfly and Matecumbe Key. *Nat. Mag.* **33**, 565–567, 611.

Hessel, S. A. (1972). A preliminary scan of rare and endangered Nearctic moths. *Atala* **4**, 19–21.

Klassen, P., Westwood, A. R., Preston, W. B. and McKillop, W. B. (1989). "The Butterflies of Manitoba". Manitoba Museum of Man and Nature, Winnipeg, 290 pp.

Klots, A. B. (1951). "A Field Guide to the Butterflies". Houghton Mifflin Co., Boston, Massachusetts.

Murphy, D. D. (1988). The Kirby Canyon Conservation Agreement: a model for the resolution of land-use conflicts involving threatened invertebrates. *Environ. Conserv.* **15**, 45–48, 57.

Murphy, D. D. and Ehrlich, P. R. (1980). Two California checkerspot butterfly subspecies: one new, one on the verge of extinction. *J. Lepid. Soc.* **34**, 316–320.

Murphy, D. D., Freas, K. E. and Weiss, S. B. (1990). An "environment-metapopulation" approach to population viability analysis for a threatened invertebrate. *Conserv. Biol.* **4**, 41–51.

Ogarrio, R. (1984). Development of the civic group, Pro Monarca, A. C., for the protection of the Monarch butterfly wintering grounds in the Republic of Mexico. *Atala* **9**, 11–12, 33–34.

Opler, P. A. (1976). The parade of passing species: a survey of extinctions in the U.S. *The Science Teacher* **43**, 30–34.

Opler, P. A. (1978). Freshwater and terrestrial arthropods. *In* "Endangered and Threatened Species of Virginia" (D. Linzey, ed.). University of Virginia Press, Charlottesville.

Opler, P. A. (1981). Management of prairie habitats for insect conservation. *J. Nat. Areas Assoc.* **1**(4), 3–6.

Opler, P. A. (1983). "County Atlas of Eastern United States Butterflies". US Fish and Wildlife Service, Washington, District of Columbia.

Opler, P. A. (1984a). Endangered and threatened insects of Maryland. *In* "Threatened and Endangered Plants and Animals of Maryland" (A. W. Norden, D. C. Forester and G. H. Fenwick, eds), pp. 267–272. Maryland Natural Heritage Program Special Publication 84-I, Annapolis, Maryland.

Opler, P. A. (1984b). Invertebrates. *In* "Species of Special Concern in Pennsylvania" (H. Genoways and F. Brenner, eds), Ch. 2, pp. 79–165. Carnegie Museum of Natural History Special Publication 11.

Opler, P. A. and Robinson, L. (1986). Lange's metalmark butterfly. *In* "Audubon Wildlife Report 1986" (R. Di Silvestro, ed.), pp. 910–916. The National Audubon Society, New York.

Oppewall, J. C. (1976). The saving of the El Segundo blue. *Atala* **3**, 25–28.

Panzer, R. (1988). Managing prairie remnants for insect conservation. *Nat. Areas J.* **8**, 83–90.

Perkins, P. D. (1980). North American insect status review. Final report of a research study conducted by Department of Entomology of Smithsonian Institution, contract 14–16–0009–79–052. US Fish and Wildlife Service, Washington, District of Columbia.

Pollard, E. (1977). A method for assessing changes in the abundance of butterflies. *Biol. Conserv.* **12**, 115–134.

Powell, J. A. (1981). Endangered habitat for insects: California coastal sand dunes. *Atala* **6**, 41–55.

Pyle, R. M. (1967). Conservation and the lepidopterist. *Bull. Assoc. Minn. Entomol.* **2**, 1–5.

Pyle, R. M. (1976). Conservation of Lepidoptera in the United States. *Biol. Conserv.* **9**, 55–75.

Pyle, R. M., Bentzien, M. and Opler, P. A. (1981). Insect conservation. *Annu. Rev. Entomol.* **26**, 233–258.

Reid, T. S. and Murphy, D. D. (1986). The endangered Mission blue butterfly: (*Plebejus icarioides missionensis*). *In* "The Management of Viable Populations: Theory, Applications, and Case Studies" (B. A. Wilcox, P. F. Brussard and B. G. Marcot, eds), pp. 1–21. Center for Conservation Biology, Stanford University, Stanford, California.

Rodriguez, J. J. R. (1984). Mariposa Monarca. *Atala* **9**, 9–10, 31–32.

Shuey, J. A., Metzler, E. H., Iftner, D. C., Peacock, J. W., Watkins, R. A., Hooper, J. D. and Babcock, W. F. (1987). Status and habitats of potentially endangered Lepidoptera in Ohio. *J. Lepid. Soc.* **41**, 1–12.

Stanford, R. E. (1977). "Rocky Mountain Butterfly Distribution Maps", 1st edn. R. E. Stanford, 720 Fairfax Street, Denver, Colorado, 80220, viii + 43 pp.

Tilden, J. W. (1956). San Francisco's vanishing butterflies *Lepid. News* **10**, 113–115.

US Congress (1974). "The Endangered Species Act of 1973, P.L. 93–205 (87 Stat 884)". United States Government Printing Office, Washington, District of Columbia.

US Fish and Wildlife Service (1984). Endangered and threatened wildlife and plants; invertebrate notice of review. *Fed. Reg.* **49**, 21664–21675.

US Fish and Wildlife Service (1989a). Endangered and threatened wildlife and plants; animal notice of review. *Fed. Reg.* **54**, 554–579.

US Fish and Wildlife Service (1989b). Endangered and threatened wildlife and plants. US Fish and Wildlife Service, Washington, District of Columbia, 34 pp.

Weiss, S. B., Murphy, D. D. and White, R. R. (1988). Sun, slope, and butterflies: topographic determinants of habitat quality for *Euphydryas editha*. *Ecology* **69**, 1486–1496.

Wells, K. (1987). One company treats a butterfly gruffly, another is friendlier. *Wall Street J.* 23 April **116**(79), 1, 15.

Young, F. N. (1950). Notes on the disappearance of the Florida blue butterfly (*Eumaeus atala florida* Rober). *Lepid. Medlemsblad Lepid.* Forening 1950–1951, pp. 1–2.

3

Australia: Conservation of a Continental Insect Fauna

PENELOPE GREENSLADE AND T. R. NEW

I.	Introduction	33
II.	Australian Environments and their History	34
	A. Climate and Vegetation	34
	B. Cainozoic History	38
III.	The Insect Fauna	39
	A. Distribution, Size and State of Knowledge	39
	B. Nature and Origins	42
	C. Endemism and Diversity	43
IV.	Current Threats to Insects	49
	A. Vegetation Clearance and Degradation of Habitat	49
	B. Other Threats	53
	C. Greenhouse Effects	54
V.	The History and Status of Insect Conservation in Australia	55
	A. History	55
	B. Legislation	57
	C. Federal and State Initiatives	58
	D. The Future	62
VI.	Conclusion	63
	References	66

I. INTRODUCTION

Australia is of particular importance in the conservation of insects for a number of reasons. Nowhere else in the world does a continent under a single national government span such a range of biomes from lowland tropical rain forest through arid formations and sclerophyll forests and woodlands to cold montane environments, even extending to polar ecosystems in Australian claims in Antarctica. It covers 50° of latitude from 10° to 60° south when the outlying subantarctic territories of Macquarie and Heard Islands are included. Although topographical complexity is largely confined to the Great Dividing Range on the eastern seaboard, the predominantly low relief of the interior consists of a mosaic of ancient and recent soil landscapes. Europeans arrived only 200 years ago and the human population has remained low with most urban and arable development taking place on the margins of the continent. Consequently a large proportion of the land area still carries something akin to its original cover of native vegetation (although often much disturbed through grazing by introduced stock).

Furthermore, the continent is believed to have been separated from other land masses through the mid-Tertiary, so that much of the fauna that we see now was able to adapt and radiate in isolation. It is highly endemic at species, generic and higher levels, and contains many phylogenetically primitive forms and higher taxa which are absent or rare in other parts of the world. We see a diverse as well as endemic fauna with much narrow specialization so that species are restricted both geographically and in their habitats and resource needs.

Striking features to an entomologist from the northern hemisphere are surprisingly high insect diversities in sclerophyllous vegetation in seasonally arid climates, and the discovery of how little is known of this fauna, a reflection of a small scientific population and a continental biota. Finally, an important factor in conservation is a two-tiered, state and federal system of government with most control of land management vested in the states.

II. AUSTRALIAN ENVIRONMENTS AND THEIR HISTORY

A. Climate and Vegetation

An arid core (Fig. 1) is surrounded by concentric bands of increasingly moister climates. Monsoonal summer rainfall in the tropics grades into winter rainfall in temperate, mediterranean climates in the south. Subhumid and wetter climates are restricted to near-coastal areas.

The potential native vegetation is shown in Fig. 2 simplified from that of Bridgewater (1987), who identified 34 main associations (but many of them small in extent). Main types to be noted are as follows:

1. In and surrounding the arid core, *Acacia* shrublands (especially mulga, *A. aneura*, and brigalow, *A. harpophylla*) and hummock grasslands (*Triodia* and *Plectrachne* spp.) on soils from sand plains

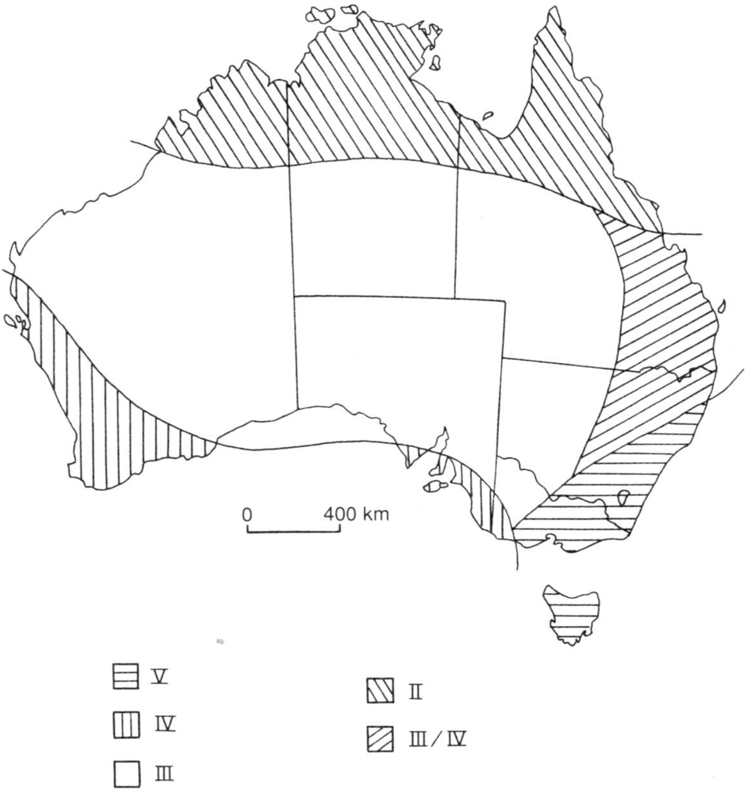

Fig. 1. Australia's climatic zones (after Bridgewater, 1987): II, tropical zone; III, subtropical dry zone; IV, transitional zone with winter rain; V, warm temperate zone.

to skeletal hill soils, with tussock grassland of *Astrebla* species on stony table lands and cracking clays.

2. In the wet–dry tropics to the north, extensive savannah woodlands and, to the south, chenopod shrub steppe giving way to mallee (shrub woodland of eucalypts with multiple stems arising from a lignotuber) centred on semi-arid climates.
3. Eucalypt-dominated forest and woodlands on south-western, south-eastern and eastern seaboards in sub to perhumid climates with extensive heaths on non-calcareous sands.
4. Rain forest, virtually all in the east, of which more than half of that not cleared is in tropical Queensland with another third represented by Tasmania's cool temperate rain forest (Fig. 3).

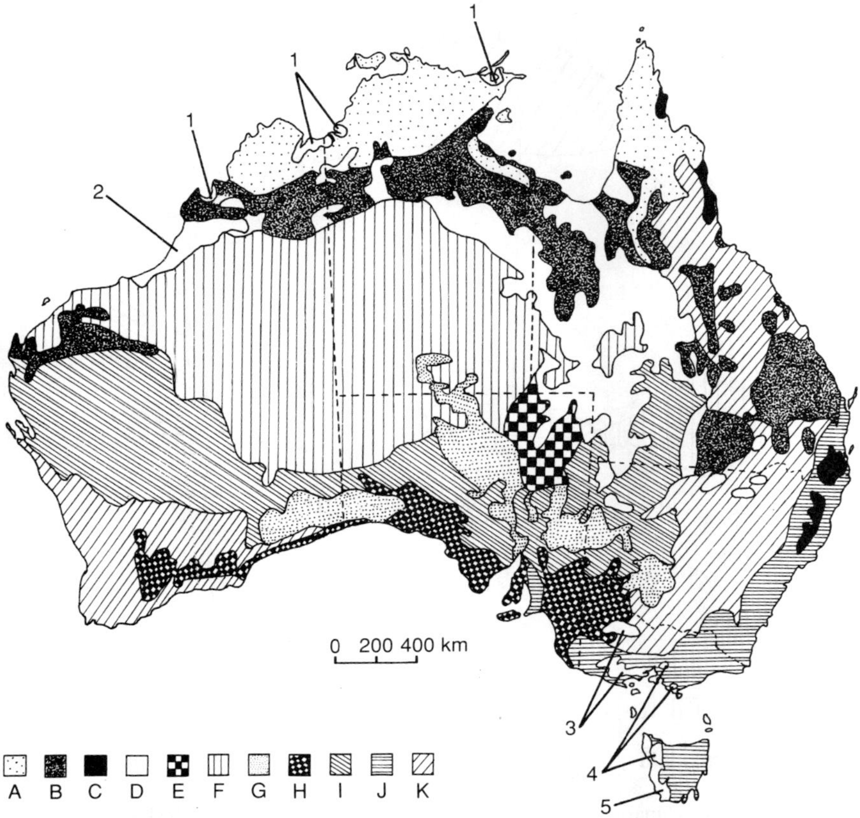

Fig. 2. Potential natural vegetation of Australia, simplified from Bridgewater (1987), where 34 vegetation types are distinguished. Many of the areas shown here are themselves diverse. Tropical/subtropical wooded vegetation: A, *Eudesmia–Corymbia–Blakella–Symphyomyrtus* forest or woodland; B, other woodlands or forests, except rain forest; C, tropical/subtropical rain forest; 1, mangrove forest. Semiarid grassland and shrublands: D, *Dicanthium* or *Astrebla* grassland; E, *Zygochloa* grassland; F, *Triodia* grassland; G, *Atriplex–Maireana* (Saltbush) shrubland; L, H, sclerophyllous shrubland (mallee); I, *Acacia* shrubland (mulga); 2, *Acacia* thicket, with *Triodia* understorey. Temperate forests, grasslands and shrubland; J. *Symphyomyrtus–Monocalyptus* forest/woodland, and localized alpine sclerophyllous shrubland, various forms; K, *Symphyomyrtus* woodland or forest, and some other woodland/shrubland associations; 3, *Stipa* grassland; 4, *Nothofagus* rain forest; 5, *Gymnoschoenus* sedgeland.

5. Alpine vegetation, restricted to Tasmania and the southern Great Dividing Range (with structurally allied formations on the subantarctic islands).

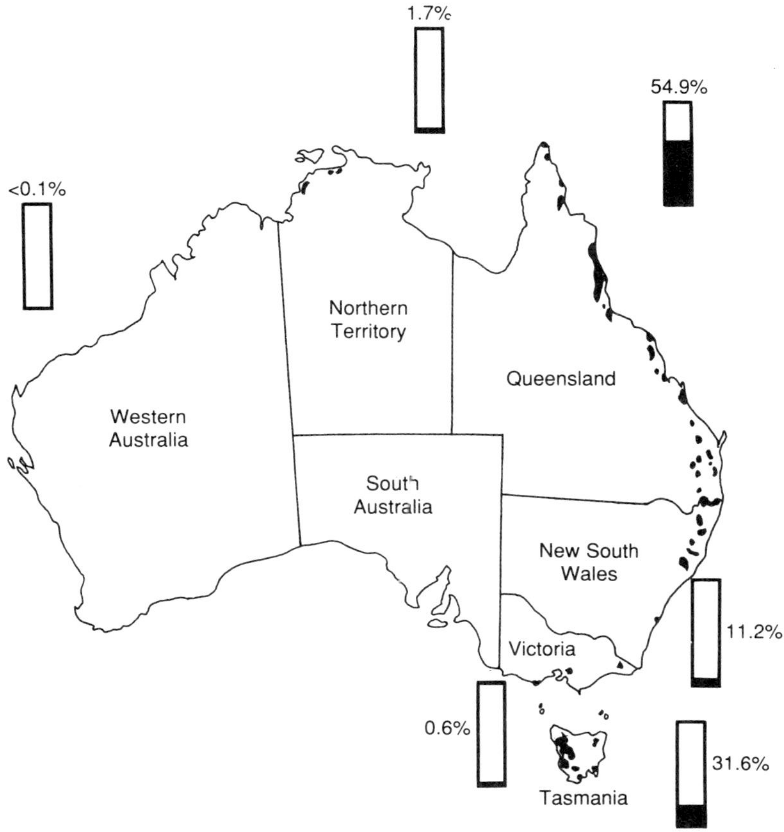

Fig. 3. Distribution of rain forests in Australia, with percentages of total in each state. Major political divisions of Australia are shown (after Working Group on Rainforest Conservation, 1985).

These vegetation types are by no means uniform as much natural fragmentation has occurred in the past during climatic oscillations. This has resulted in interdigitation of two or more vegetation types and development of many different floristic communities within each vegetation. As an example, rain forest in Tasmania has been classified into four main structural types and subdivided into 39 different vegetation associations (Jarman *et al.* 1984, 1987).

Features of terrestrial ecosystems over most of Australia are high evaporation rates, limitation of biological activity by rainfall and the prevalence of sclerophyll vegetation on nutrient-poor soils, although often with remarkably high floristic diversity. The importance of fire and ants and the abundance of termite mounds are features of most landscapes north of the Tropic of Capricorn.

Australian freshwater environments, described by De Deckker and Williams (1986), are concentrated near the eastern seaboard and in Tasmania. Over the greater part of arid and semi-arid Australia, nearly all natural water bodies are temporary and often saline, depending on unpredictable rainfall events. Exceptions are the permanent mound springs occurring in the driest part of the continent fringing the Great Artesian Basin in Queensland and South Australia. In northern Australia, with monsoonal rainfall, aquatic habitat is highly seasonal while large freshwater lakes are cold and restricted to Tasmania and the southern Great Dividing Range.

B. Cainozoic History

Generalizations here are derived from authors included in Dyne and Walton (1987) and Barker and Greenslade (1982). In the late Mesozoic Australia was a part of the southern continent, Gondwanaland, and was continuous with Antarctica up to the end of the Cretaceous. The climate was mild and moist and vegetation seems to have been closed canopy forest, conditions which persisted into the Cainozoic. The Palaeocene–Eocene saw the severance of the Australian tectonic plate and its slow drift northwards to lower latitudes, accompanied by the gradual recession of Cretaceous inland seas. For perhaps 30 million years, through the middle of the Tertiary period, Australia seems to have been separated from all other land masses so that its biota evolved in isolation. During the Tertiary the effects on temperature regimes of movement towards and into the tropics were offset by lowering of the temperature of the southern oceans. As a result, over this time Australia seems to have provided a variety of fairly stable thermal environments ranging from equivalents of today's subtropical to cool temperate climates (Nix, 1982).

The continent-wide aridity that we see now, had its origins towards the end of the Miocene and was accompanied by increasing climatic seasonality with a switch from predominant summer rainfall in the Pliocene to winter rainfall in the Pleistocene. There is palaeobotanical evidence that the closed forest cover of the beginning of the Tertiary became, to some extent, fragmented and interspersed with grasslands and open habitats at an early stage and it has been argued by Smith-White (1982) that at least local, if only edaphic, arid environments were present through most of the Tertiary.

There are uncertainties and factors whose biogeographical implications have yet to be assessed. According to Frakes *et al.* (1987) there may have been three Tertiary phases of cooling and aridity. Audley-Charles (1987) suggests intermittent mid-Tertiary connections with south-east Asia while the earlier drifting of Gondwanan terrains, independently of India and Australia, to form parts of south-east Asia complicates the possible origin and history of Gondwanan insect stocks in Australia.

During the Pleistocene, oscillations between humid and cool, arid phases caused successive expansions and contractions of Australia's concentric arid, semiarid and humid climates. There were cycles of glacial activity in the south-east and in Tasmania. Changes in sea levels led to the formation and breaking of land connections between mainland Australia and New Guinea to the north and Tasmania to the south which were last severed only some 6000–8000 (New Guinea) and 12 000 (Tasmania) years ago. The increase in the incidence of fire with the advent of aboriginal peoples from the north in the late Pleistocene is believed to have shifted the dominant vegetation of some areas from *Casuarina* to more fire-tolerant *Eucalyptus* forests.

III. THE INSECT FAUNA

A. Distribution, Size and State of Knowledge

Australia can be divided into three major biogeographical regions (Fig. 1), with interzones which are objectively defined bioclimatically (Nix, 1982): Bassian (mesotherm–microtherm, seasonal), south-western and eastern Australia and Tasmania; Eyrean (megatherm–mesotherm, arid), the arid centre; Torresian (megatherm, seasonal), the northern wet–dry tropics. Nix (1982) also distinguishes a minor Irian area (megatherm, non-seasonal) covering the lowland tropical rain forests of north Queensland (and also those of New Guinea). The majority of our insects can be regarded as predominantly characteristic of one or other of these regions or interzones. All orders except Zoraptera, Raphidioptera and Grylloblattodea are present.

At present we can make no more than an informed estimate of how many insect species there are in Australia. The figure of 125 000 in Table I is probably very conservative. At least half, perhaps as many as 75%, of species do not yet bear names and almost every revisionary study of the fauna reveals an unexpectedly high number of undescribed species. P. J. M. Greenslade (1985) rated knowledge of the Australian insect fauna at 3 on a global scale extending from 1, insect faunas very imperfectly known

TABLE I. Status of taxonomic study of the Australian fauna (from B. R. Richardson, personal communication).

Taxon	No. of taxonomists	No. of Species	Species/ taxonomist	% species described	Undescribed species/ taxonomist
Tetrapoda	190	3600	17	86	3
Pisces	33	6300	190	82	34
Echinodermata	21	3400	163	61	63
Mollusca	28	15700	560	65	196
Crustacea	45	17800	396	4	222
Insecta	141	125000	884	56	390
Other Arthropoda	37	24700	667	35	434
Annelida	12	15500	1295	35	840
Helminthes	32	48000	1496	16	1260
Coelenterata	13	4800	372	81	130
Porifera	7	1500	214	?	?
Protozoa	29	?	?	?	?

(most of the humid tropics), to 6, probably more that 90% of species described, at least as adults, in the majority of groups (north-western Europe). Furthermore, Australian species which are barely recognizable on morphological grounds alone have scarcely been explored (New, 1983). For example, in the dolichoderine ant genus *Iridomyrmex*, studies of the ecologically important and Australian endemic *purpureus* species group, which included gross morphology of workers, male genitalia, genetics and distributions, revealed 10 species where previously three "varieties" of one species had been recognized (Greenslade and Halliday, 1983).

Diverse suites of phytophagous insects are associated particularly with the genera *Eucalyptus* (>500 Australian species) and *Acacia* (900 species) which together make up about 13% of the Australian angiosperm flora and constitute an unusual floral imbalance on the continent (Beadle, 1981; Bridgewater, 1987). These suites of species are composed of many unrelated taxa, a number of which have radiated to produce numerous species, often closely related and taxonomically intractable instances of "diffuse coevolution" (Fox, 1981; New, 1983). Examples are Eurymelidae and spondyliaspine Psyllidae in the Hemiptera, Pergidae (Hymenoptera) and some groups of Oecophoridae (Lepidoptera). The chrysomelid beetle genus *Paropsis* (*sensu lato*) has hundreds of species on *Eucalyptus* and *Acacia* (Table II). In turn, parasitic Hymenoptera and predators depend on these *Eucalyptus* and *Acacia* insects so that documenting these plant-based communities is a major theme in insect ecology in Australia (New, 1988).

Such radiations are not restricted to plant-dependent taxa. For example,

TABLE II. Examples of phytophagous insect groups which have undergone radiation in association with *Eucalyptus* and/or *Acacia* in Australia (partly after New 1983). × = 50–100 species, ×× = >100 species.

Taxa	*Eucalyptus*	*Acacia*
Hemiptera		
Psyllidae	××	×
Eurymelidae	××	
Thysanoptera		
Phlaeothripidae (gall formers)		×
Coleoptera		
Chrysomelidae	××	××
Curculionidae (seed eaters)		××
Lepidoptera		
Oecophoridae	××	
Tortricidae	××	
Hymenoptera		
Pergidae	×	
Colletidae (oligolectic pollinators)	×	
Halictidae (oligolectic pollinators)	××	

in the Collembola, Australia is the centre of diversity for the sminthurid subfamily Katianninae. There are a number of endemic genera, some undescribed, found mainly in *Eucalyptus* forest and woodland. At least one of these contains around 50 species, only nine of which are described. A number of ant genera (notably *Camponotus, Iridomyrmex, Melophorus* and *Meranoplus*) contain several hundred species, many of them closely related and forming taxonomically difficult complexes. Furthermore, these and other ant genera combine in complex, highly integrated communities (Greenslade, 1979). As a corollary, the Myrmeleontidae (Neuroptera), whose larvae are "ant lions", have perhaps 300 species in arid and semiarid areas (New, 1985).

Inevitably, for the great majority of Australian insect species, detailed distributional and general biological information is virtually non-existent. With a small (Table I) and diminishing population of Australian taxonomists, the intrinsic difficulty of appraising this continental fauna becomes a major impediment to insect conservation. The situation is even more serious since the figure of 141 in Table I for insect taxonomists includes all biologists expressing an interest in systematics and probably only a third are working full time in the discipline. Any approach towards a complete inventory of Australian insects is clearly not practicable with present manpower resources.

A further difficulty is the lack, for whatever reasons, of the amateur entomologist on whom depends, for example, the detailed knowledge of insect distributions that is emerging in western Europe and North America (Heath, 1973; Opler, Chapter 2). The absence of a basic foundation of published natural history information in turn creates problems for beginners, amateur or student. This then means that, without an assured market, avenues for publication of guides to faunas and introductions to groups are limited, even if the information to produce them is available.

B. Nature and Origins

It is relevant to our understanding of this continental fauna and its conservation to consider its biogeographical history. It is derived from four main sources. First there are older stocks inherited from Gondwanaland and so represented on the Australian land mass from the early Tertiary. The oldest Pangaean or Archaean element consists of taxa going back to the Mesozoic, before the break up of Pangaea. Generally primitive, they are known from Mesozoic or earlier fossils and/or have extant representatives in the northern hemisphere. Examples are the rare but widely distributed beetle family Cupedidae with two Australian genera and Permian fossils, the mecopteran family Meropidae (Table III) in North America and Western Australia, again going back to the Palaeozoic, and the hairy cicadas *Tettigarcta tomentosa* and *T. crinita* from montane areas in Tasmania and New South Wales, respectively, with northern hemisphere Mesozoic and Tertiary fossils.

Next are Gondwanan groups some of which are clearly identifiable as such, occuring in southern Australia, New Zealand and South America. Today they range from relict species with Bassian distributions such as the Osmylidae in the Neuroptera, often in "unfavourable" habitats on mountains, in cold lakes or torrential streams (and, like much of the Pangaean fauna of phylogenetic and necessarily conservation significance), to genera that have radiated to become major components of the contemporary fauna.

The Gondwanan ant tribe Melophorini is an example that contains a group of five described Australian genera with an essentially Bassian range (although extending to a varying degree up the east coast even into New Guinea). One of these, *Prolasius*, found also in New Zealand with a closely allied genus in southern South America, has a south-east Australian distribution and is the most frequent genus in the very limited ant fauna of Tasmanian rain forest. The sixth and final genus, *Melophorus*, is an Australian endemic that has achieved the remarkable physiological ability to be active on the soil surface at very high temperatures and has generated hundreds of species throughout arid and semiarid areas (and we have fewer than 30 specific and infraspecific names in the literature!).

A third faunal element is northern or Malesian (i.e. from the Oriental region and New Guinea). It is postulated that in the late Tertiary there were two concurrent events, albeit continuing over millions of years. First, the Australian plate had moved sufficiently far to end its mid-Tertiary isolation and proximity to south-east Asia permitted immigration from the north. At about the same time, Miocene–Pliocene climatic change was accompanied by the expansion of *Eucalyptus* and *Acacia* and by a postulated radiation of the older insect fauna. Consequently, new northern elements seem to have encountered an evolving and coevolving biota already adapting to a changing climate. Different groups invading over a considerable period either contributed substantially to the developing modern fauna by subsequent radiation (Heatwole, 1987) or failed to penetrate existing communities. In the Odonata for example, O'Farrell (1970) distinguished older, morphologically isolated southern forms whose larval habitat is typically cold, permanent water and widespread groups with evident Oriental affinities. Another conspicuous group which succeeded in colonizing Australia from the Orient are the papilionid butterflies. The uncertainties referred to earlier concerning Tertiary land connections and climate suggest that the biogeographic patterns observed may have more complex origins that those suggested here.

Fourthly there is an element of species introduced relatively recently, during the last 200 years of colonization by Europeans. It includes deliberately and accidentally introduced taxa. Well-studied examples include pest species such as the recently arrived spotted alfalfa aphid, *Therioaphis trifolii* F *maculata*, the blow fly, *Lucilia cuprina*, and other cosmopolitan elements such as species of stored products (e.g. Tineidae, Lepidoptera) and *Pieris rapae*, the cabbage white butterfly, which has recognizably different regional populations that have developed in parts of eastern Australia within the last 50 years (Gilbert, 1988). A growing number of biological control agents fall into this category, such as the dung-feeding Scarabaeidae, pollinators such as the leaf cutter bees, and species for which weeds are specific hosts such as *Dialectica scalariella* (Lepidoptera) on Salvation Jane, *Echium plantagineum* (Delfosse and Cullen, 1985).

C. Endemism and Diversity

In general, efforts toward insect conservation have predominantly concentrated on detecting undisturbed sites of high diversity that are representative of a particular community type. In Australia another aspect of conservation

TABLE III. Status of some Australian insects of conservation interest.

Taxon	Common name	Distribution
Mecoptera		
Austromerope poultoni Killington	Western Australian scorpion fly	Widespread in Darling Botanical District
Ephemeroptera		
Tasmanophlebia lacus-coerulei[a] Tillyard	Large Blue Lake butterfly	5 small lakes
Odonata		
Hemiphlebia mirabilis[a] Selys	Small damselfly	<5 ha
Orthoptera		
Cooloola dingo Rentz	Dingo monster	?
Cooloola propator Rentz	Cooloola monster	<10 ha
Cooloola ziljan Rentz	Sugar cane monster	>50 km^2
Dryococoelus australis[a] (Montrouzier)	Lord Howe Is. stick insect	<1 ha?
Plecoptera		
Eusthenia nothofagi[a] Zwick	Otway stonefly	?
Leptoperla cacuminis[a] Hynes	Mt Kosciusko wingless stonefly	1 small stream
Riekoperla darlingtoni[a] (Illies)	Mt Donna Buang wingless stonefly	*ca.* 2 ha
Coleoptera		
Phalacrognathus muelleri (Macleay)	King stag beetle	>100 km^2
Mecynognathus damelii Macleay	Giant Cape York carabid	?
Diptera		
Edwardsina gigantea[a] Zwick	Giant torrent midge	2 streams
Edwardsina tasmaniensis[a] Tonnoir	Tasmanian torrent midge	1 site?
Lepidoptera		
Acrodipsas illidgei (Waterhouse and Lyell)	Moreton Bay butterfly	>100 ha
Paralucia pyrodiscus lucida Crosby	Eltham copper	<3 ha
Paralucia spinifera Edwards and Common	Bathurst copper	<10 ha, 4 small sites
Ornithoptera spp.[b]	Birdwing butterflies	Widespread in Queensland
Papilio ulysses joesa[b] Butler	Ulysses butterfly	Widespread in Queensland
Formicidae		
Nothomyrmecia macrops[a,c] Clarke	Nothomyrmecia ant	<10ha

TABLE III. Cont'd

Taxon	Land Tenure	Special Protection	Conservation status
Mecoptera			
Austromerope poultoni Killington	Varied	None	Apparently safe
Ephemeroptera			
Tasmanophlebia lacus-coerulei[a] Tillyard	National Park	None	Pollution
Odonata			
Hemiphlebia mirabilis[a] Selys	National Park	None	Threatened by prescribed burning
Orthoptera			
Cooloola dingo Rentz	State Forest	None	Unknown
Cooloola propator Rentz	National Parks (2)	None	Safe
Cooloola ziljan Rentz	Private farm land	None	Vulnerable to ploughing for sugar cane
Dryococoelus australis[a] (Montrouzier)	Reserve	None	Probably extinct, eliminated by introduced rats
Plecoptera			
Eusthenia nothofagi[a] Zwick	National Park (part)	None	Forest clearing, stream pollution
Leptoperla cacuminis[a] Hynes	National Park	None	Uncertain
Riekoperla darlingtoni[a] (Illies)	Scenic Reserve	None	Tourism, run-off from car parks
Coleoptera			
Phalacrognathus muelleri (Macleay)	Varied	None	Not endangered but heavily collected
Mecynognathus damelii Macleay	Aboriginal Reserve	None	
Diptera			
Edwardsina gigantea[a] Zwick	National Park	None	Pollution and siltation
Edwardsina tasmaniensis[a] Tonnoir	?Crown land	None	One population wiped out by hydroelectric scheme
Lepidoptera			
Acrodipsas illidgei (Waterhouse and Lyell)	Private land	160 ha reserved	Safe?
Paralucia pyrodiscus lucida Crosby	Council? reserve	2 ha reserved	Needs management
Paralucia spinifera Edwards and Common	Private land	None	Habitat threatened by weeds, goats and clearing
Ornithoptera spp.[b]	Species protected	Not endangered	
Papilio ulysses joesa[b] Butler	Species protected	Not endangered	
Formicidae			
Nothomyrmecia macrops[a,c] Clarke	Private farmland	0.5 ha fenced	Largest colony destroyed by government bulldozer

[a] Species listed by Wells *et al.* (1983). [b] Species protected in Queensland. [c] Species protected in Western Australia.

must be considered separately, that of conserving endemism. The highest proportion of endemic insect taxa at the level of genus or above is found among the Pangaean and Gondwanan elements in relicts of previously more widespread vegetation types such as the rain forests of Tasmania and Queensland, where the isolated pockets are clearly remnant oases for much native biota. This is illustrated by the generic composition of the collembolan fauna of six vegetation types of Tasmania (Fig. 4) where genera endemic to Tasmania were found only in the rain forest (Greenslade, 1987). An example is *Megalanura*, a monobasic collembolan genus whose closest relatives are from the Australian mainland and New Zealand. The IUCN Invertebrate Red Data Book (Wells *et al.*, 1983) lists nine insects all of which could be classified as relicts of Gondwanan or Pangaean

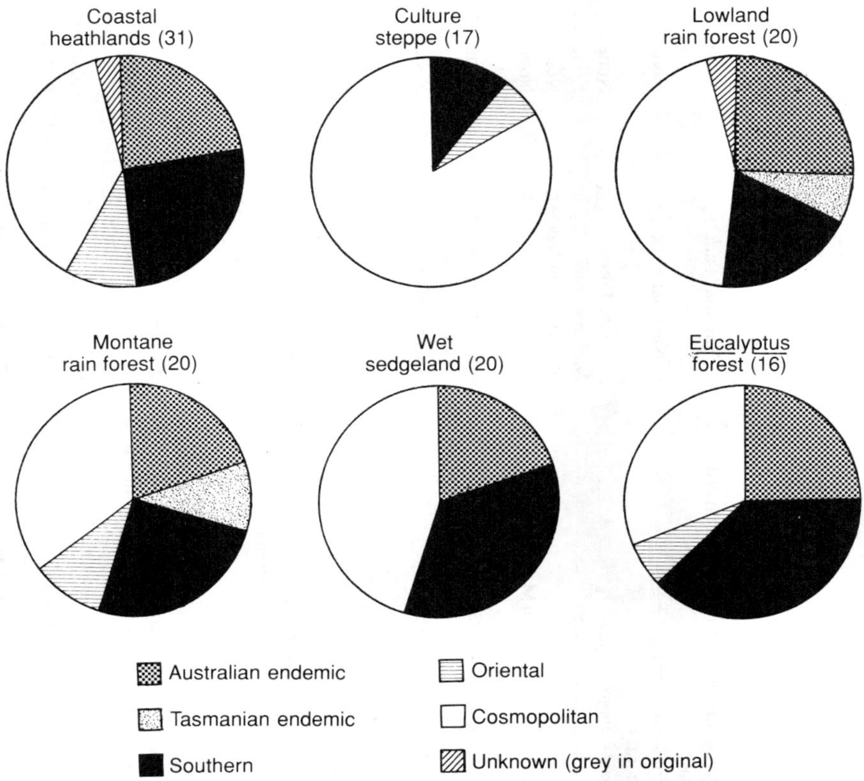

Fig. 4. Number of genera (in brackets) and distribution of Collembola from six vegetation types in Tasmania (after Greenslade, 1978).

faunas and which, for the main part, occur either in aquatic habitats, which are limited in Australia where 80% of the continent is arid, or in similar relict patches of vegetation typically with distributions of 10 ha or less (Table III). The table lists them with a number of other species of conservation interest. It is noticeable that nearly all are relict species isolated taxonomically, with very restricted distributions (except for the Lepidoptera and Lucanidae). The primitive damselfly *Hemiphlebia mirabilis* in Victoria is one example (Sant and New, 1988) for which only a few hectares may be sufficient to conserve its major population. Most of the species listed (Table III) are either unprotected and/or threatened.

The typically Australian radiations which also need to be considered for conservation are found in the *Eucalyptus* forest and woodlands but the requirements for preservation of endemism and diversity differ as far as their habitats and the area required for conservation are concerned. This is illustrated in Fig. 4. Although highest levels of Tasmanian endemism were found in the rain forest, the highest number of all genera occurred in the *Acacia* coastal heath and the highest proportion of Australian endemic genera occurred in *Eucalyptus* forest.

The optimal size of reserve required for invertebrate species in eucalypt forest is being examined in a vegetation patch experiment at Wog Wog in south-eastern New South Wales (Margules, 1987). Here eucalypt forest has been cleared for plantations of exotic pines while blocks of the original vegetation of three sizes have been retained. Invertebrates are being sampled to assess the influence of patch size on the population persisting. The importance of small patches of native vegetation for the conservation of insects was first pointed out by Key (1978). He showed that native grasses surviving in old cemeteries can support narrowly endemic morabine grasshoppers not present in the nearby modified vegetation.

Other research on invertebrates relevant to the problem of reserve size has been carried out in remnants of shrub heath within eucalypt woodland in Western Australia's wheat belt by Main (1987a,b). In a study of 15 species of mygalomorphs (trapdoor spiders) over a 104-ha patch of bushland, she found that likely persistence of species differs according to their foraging, reproductive and dispersal behaviour. Most mygalomorph spiders are relatively immobile and linked, for most of their life history, to burrows in the soil. In the study patch, Main found that the hydrological character of some microhabitats had changed following isolation and this had affected the burrow conditions for one species in particular, *Aganippe castellum*. Adults are long-lived and are persisting on the site, but juveniles cannot disperse into appropriate microhabitats and so the population apparently consists entirely of older individuals. This species may eventually become extinct in the patch because of lack of available microhabitats and reduced recruitment. However, those species which have a clustering dispersal strategy are likely to persist indefinitely. From this study it seems clear that even very small patches are capable of sustaining at least some original representative species.

Higher taxa of insects reach their greatest diversity in various vegetation types, for example, tropical rain forest is the centre of diversity for taxa of Malesian origin such as some Papilionoidea (Kitching, 1981). But in the mallee, semiarid southern Australia, up to 150 ant species can be found within an area of less than 1 ha, a density matched only by the ant faunas of the lowland equatorial tropics. To illustrate this relative species richness in different vegetation types, the number of species of ants per collection site of a few hectares has been compared in a transect from

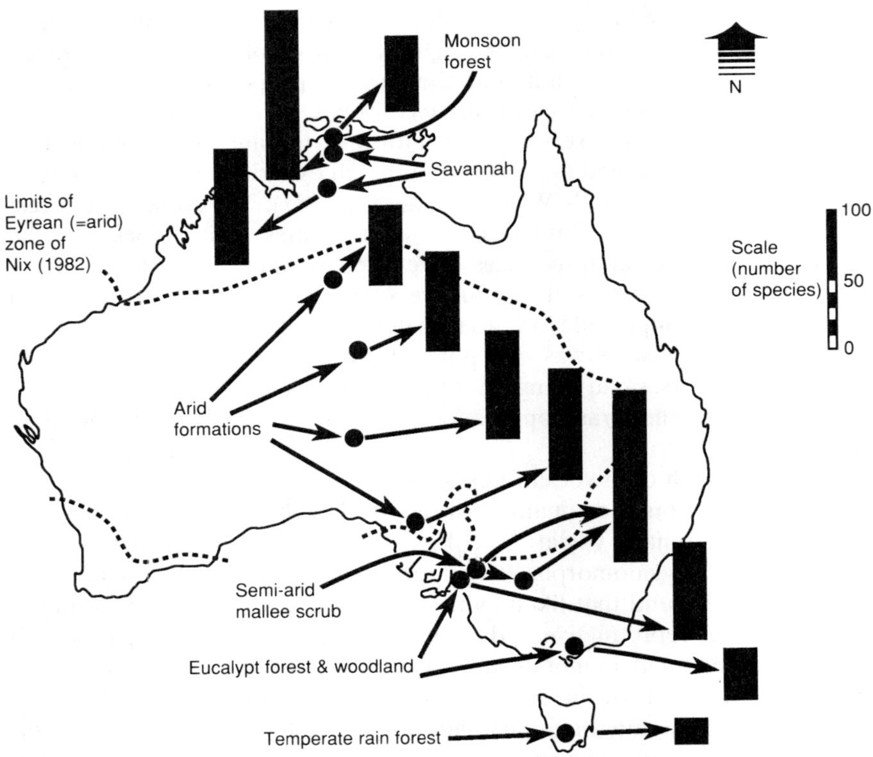

Fig. 5. Schematic map of numbers of species of ants in different vegetation types in a transect across Australia.

north to south (Fig. 5). Maximum diversity is in the northern and southern semiarid woodlands, savannah eucalypt woodland in the tropics of the north, or mallee eucalypt woodland in the south (Greenslade, 1979; Andersen, 1983; Andersen and Yen, 1985; Greenslade and Greenslade, 1989, P. J. M. Greenslade, personal communication). By contrast the number in Australian tropical rain forest is normally under 100, even at the extreme tip of Cape York in lowland tropical rain forest (Taylor, 1972). Lowland tropical rain forest in Australia is impoverished in terms of species, perhaps because of the distance from the source area in south-east Asia, its small area, climatic fluctuations and, in much of Queensland where it occurs, temperatures are limiting being lower than nearer the equator. It is mainly colonized by stage I species (*sensu* Wilson, 1961) expanding from New Guinea (McArthur and Wilson, 1967). By contrast only about 20 species of ants are found in the cool temperate rain forests of Tasmania and these are largely confined to logs. This type of forest also appears to be depauperate in herbivorous insect species (McQuillan, in preparation).

The conservation of Australia's insect fauna therefore presents a continuum of problems. At one extreme there are biologically important individual species of extremely restricted distribution and at the other highly diverse radiations often covering the whole continent in their range.

IV. CURRENT THREATS TO INSECTS

A. Vegetation Clearance and Degradation of Habitat

Vast tracts of land have been cleared for pasture and arable agriculture since European settlement and the extent of change in 13 vegetation types over 200 years is shown in Fig. 6. Forest and woodland have decreased considerably in area with the exception of open woodland which has increased. The most depleted vegetation types of all are those on land suitable for growing wheat in subhumid semiarid climates, and in Western Australia and parts of South Australia only small remnants of the original vegetation now remain. In Victoria nearly three-quarters of the forest has been greatly modified since the advent of Europeans (Fig. 7). In addition a high proportion of coastal vegetation has disappeared as a result of residential subdivision and, recently, tourist developments on the eastern seaboard.

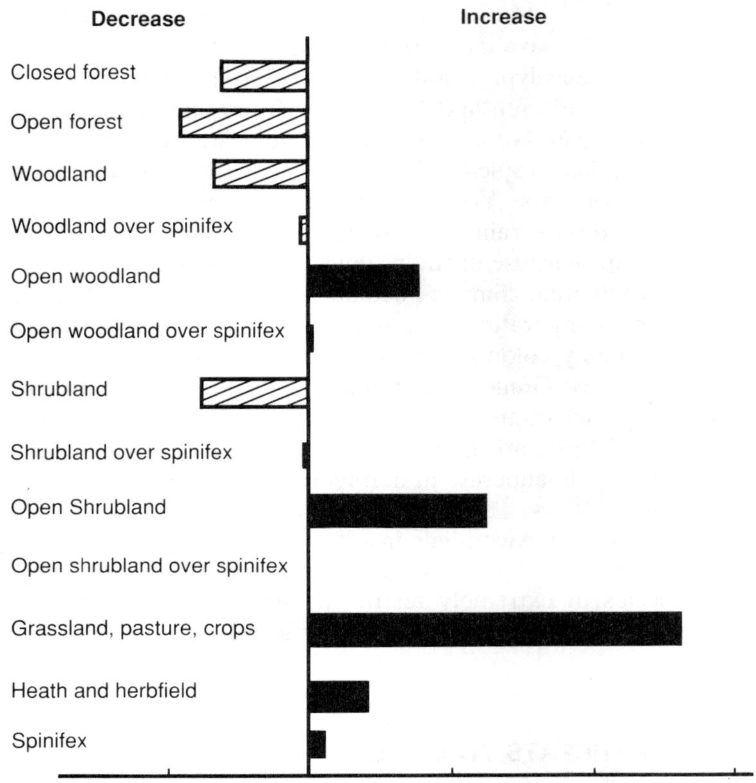

Fig. 6. Percentage increase (■ to right of vertical) or decrease (▨ to left of vertical) in 13 vegetation types in Australia over the last 200 years (after Atlas of Australian Resources, 1990).

There are severe land degradation problems in Australia, largely caused by a combination of clearing, overcultivation and overstocking (particularly during periods of drought), by feral mammals and by irrigation (Ive and Cocks, 1989). Topsoil has been lost together with its associated insects and it is estimated that as much as 50% of the continent may be affected to some extent (Fig. 8).

Even forests that have not been cleared are often subject to a policy of burning every 4–7 years with the object of removing low shrubs and leaf litter to reduce the hazard of major forest loss from naturally occurring bushfires. It is becoming apparent that there is a decline in the ability of forest to persist in its original form under burning regimes which are more frequent than those under which the forest originally evolved (Noble,

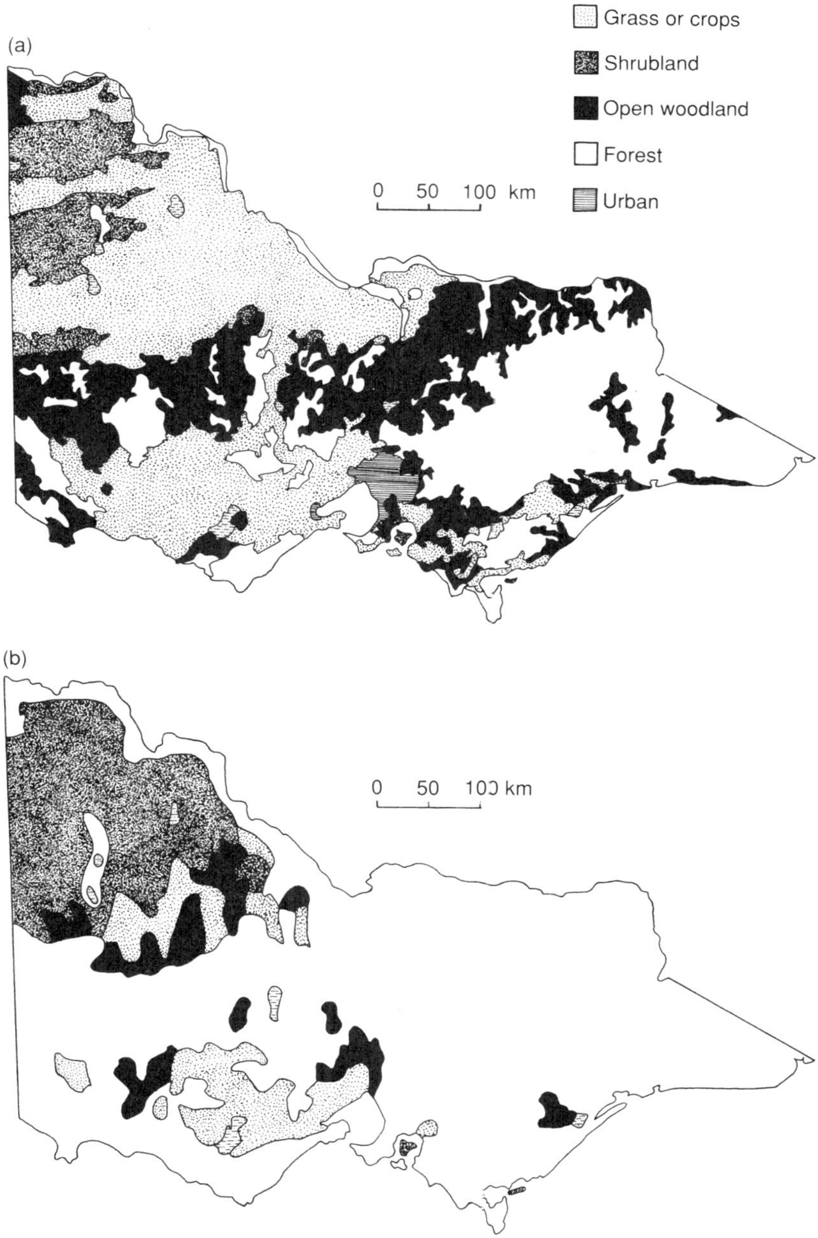

Fig. 7. Victoria, showing changes in tree cover since European settlement: **(a)** 1980, **(b)** before settlement. Maps based on those of Forests Commission, Victoria (*Common Ground*, June/July 1989).

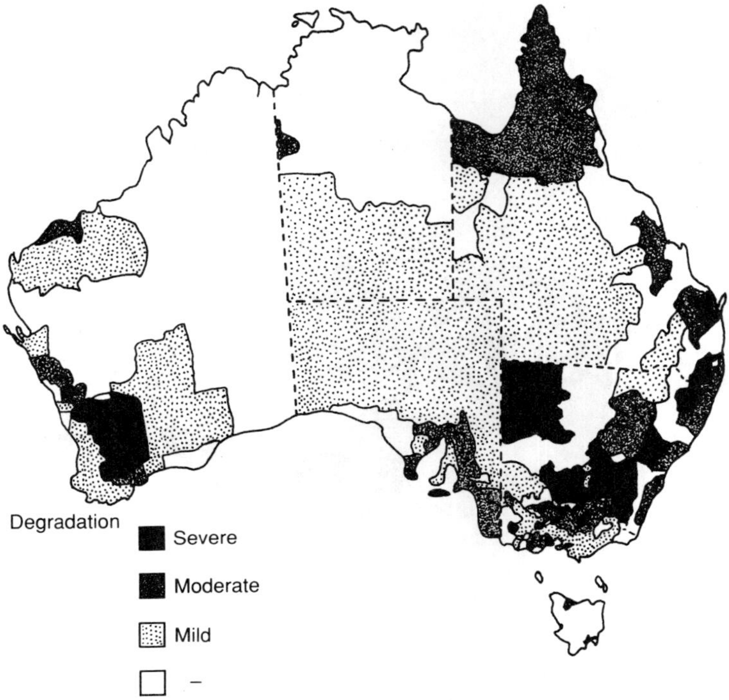

Fig. 8. Soil degradation severity index in Australia (after Ive and Cocks, 1989).

1982). This "fuel reduction burning" is practised not in summer when populations are largely dormant, but in the cooler periods of spring or autumn when insects are active and are at their most vulnerable. Main (1987b) found that one relic araneomorph family of spiders (the Anapidae) only occurred on sites which had not been burnt for over 50 years. Other families of small-bodied spiders (1.0–5.0 mm long) were best represented on the "older" sites, i.e. those which had remained unburnt for 17–30 years. A deep litter layer is vital for survival of many litter-dwelling spiders in eucalypt forest and coastal heaths in the south-west of Western Australia.

Karst areas can each carry a suite of endemic species of cave insects, but they are vulnerable to modification of the terrestrial environment that can alter inputs underground. Mining and hydroelectric schemes may cause severe but more localized problems affecting primarily aquatic habitats (Michaelis, 1984). The flooding of Lake Pedder in Tasmania in the early 1970s for instance, caused many changes in populations of aquatic invertebrates (Lake *et al.*, 1978).

B. Other Threats

Deliberately or accidentally, Europeans have been responsible for the introduction of large numbers of exotic organisms to Australia. Most introduced arthropod species may spread rapidly through the "culture steppe" (*sensu* Matthews, 1978) but are only rarely able to colonize native vegetation. When they do, they compete with and may deleteriously affect the native species. Apart from rabbits and stock which degrade vegetation, there are others which affect native insects more directly as predators. The black rat (*Rattus rattus*) has been implicated in the extinction of the endemic Lord Howe Island stick insect *Dryococelus australis*, and the cane toad (*Bufo marinus*), introduced into Queensland to control sugar cane pests and gradually spreading over much of the continent, is a voracious predator of insects. The brown trout (*Salmo trutta*), introduced to Tasmania for recreational purposes, is believed to be responsible for the elimination of much of the State's native aquatic fauna from certain water bodies (Wells *et al.*, 1983), and there are similar problems with the deliberate release of surplus aquarium fish into streams and lakes on the mainland. The European wasp (*Vespula germanica*) which was accidentally introduced, is both predator and competitor. It can overwinter in many parts of Australia and may form huge perennial colonies containing over a million cells (Spradbery, 1973). Evidence is accumulating that the honey bee competes with native insect pollinators for nectar (Blyth, 1987) and can form feral colonies in most parts of southern Australia where there is adequate water.

The decomposer system may be directly affected by other introduced species. A Collembolan, *Hypogastrura purpurescens*, has spread along waterways in Tasmania and colonized the ground layer of remote areas of rain forest by this means (P. Greenslade, unpublished data). In the winter rainfall zone of southern Australia, the introduced Portuguese millipede (*Ommatoiulus moreletii*) has surges of abundance (Baker, 1985) and must also be regarded as a competitor of native detritivores.

Non-specific biological control agents are a threat to native species. However the current programme to enhance toxicity in a non-virulent virus infecting the native noctuid pests *Helicoverpa armigera* and *Helicoverpa punctigera* (Gould, 1988) is unlikely to be a threat to related non-pest species because the virus cannot reproduce in adults. Controls on the introduction of genetically engineered organisms to Australia do not seem to be adequate at present. Application must be made to the Genetic Manipulation Advisory Committee for permission to import or experiment with transgenic animals or plants but the committee has no legal powers. Illegal importation and release would be difficult to police. In addition, any organism destined for biological control must be approved by the federal government under the Biological Control Act but this legislation was enacted to protect the importer and economic plants rather than the environment.

The literature on accidental introductions to Australia is now considerable and strict quarantine laws are administered by the federal government to reduce the probability of further introductions. Total protection is not possible and new pest species, such as *Chrysomya bezziana*, the screw worm fly, regularly arrive (Rajapaksa and Spradbery, 1989). So far this species had failed to become established. Pest plants can also be so invasive that they can eliminate large areas of native vegetation. Over the past 10 years in the Northern Territory, *Mimosa pigra*, forming an almost monospecific tall shrubland, has replaced native sedgeland, riparian, aquatic, paperbark and monsoon forest communities (Braithwaite *et al.*, 1989). These authors state that "a massive loss of animal and plant species may well occur if the spread of this aggressive weed is not halted".

Australia does not yet have an acid rain problem and pollutants are mainly relatively local and restricted to soil and aquatic habitats. Chemical pollution by fertilizers and pesticides has rendered much agricultural land unsuitable as insect habitat, and native vegetation may be sprayed for locust and mosquito control. Plantations of native tree species for the forestry industry can be particularly subject to heavy insect attack as the trend towards monocultures of even-aged cloned seedlings increases. Control by aerial spraying, as sometimes adopted in Tasmania, renders plantations unsuitable for insect conservation.

C. Greenhouse Effects

Predictions for the outcome of greenhouse effects in Australia vary, but the majority suggest that some of the vegetation types which are rarest and carry the highest proportion of endemic species, such as alpine environments, are likely to become even more restricted in future (Nix, 1982; Busby 1987, 1988). The mallee also is likely to be markedly reduced in area in southern Australia if the area under arid climatic regimes extends further south and winter rainfall, mediterranean climates disappear (Greenwood and Boardman, 1988). Rising sea levels may eradicate some coastal ecosystems. As yet, the question of conservation in the face of possible very rapid change in climate has hardly been considered in Australia.

V. THE HISTORY AND STATUS OF INSECT CONSERVATION IN AUSTRALIA

A. History

Interest in conservation came early to Australia and the first parks were proclaimed before the turn of the century in New South Wales and South Australia. Until very recently however, interest in conservation of animals has been restricted to vertebrates. An early book (McKeown, 1949), which dealt in part with insects, was in advance of its time and it was not until 30 years later that the subject of insect conservation was again seriously addressed in a small publication by Key (1978). During the 1970s several states adopted legislation to protect individual species (some papilionid butterflies in Queensland, cave insects in Tasmania and buprestid beetles and the ant *Nothomyrmecia* in Western Australia). The legislation was enacted to control collecting, but it is doubtful that it has been of benefit in the conservation of the insects concerned and it may even have resulted in increased exploitation of the butterfly species by drawing attention to their putative desirability. In no case was such species-orientated legislation accompanied by provision for active research, habitat reservation or management of the taxa concerned. Consequently many Australian entomologists became disenchanted with the protection of individual species. Furthermore, lack of information on distributions makes it difficult to prepare well-documented lists of those under threat. Efforts by the Australian Entomological Society's Conservation Committee during these years concentrated on habitat protection by preparing submissions to land enquiries (Marks, 1972; New, 1976; Taylor, 1976). This was the state of insect conservation up to the early 1980s when New (1984) reviewed the subject in Australia and the IUCN's Invertebrate Red Data Book (Wells *et al.*, 1983) listed 20 threatened invertebrates (plus all onychophorans) including nine species of insects from Australia (Table III).

In the past 5 years there has been an increase in activity in insect conservation, by government bodies, specialist groups and the public. Public attitudes to insect conservation are gradually changing, although general natural history and conservation societies are still almost exclusively concerned with the problems of plants and vertebrates. The despoilatory attitude predominant among lessees and owners of land is being replaced by a more ecological approach, largely because problems of soil degradation have encouraged farmers to consider "sustainable agriculture". Even among biologists, the belief that if examples of all vegetation communities are conserved this will adequately reserve invertebrates, is becoming modified. With an order of magnitude more insects than vascular plants, many insect distributions are on a finer scale than those of plants and commonly influenced by more subtle ecological factors. As a start Yen (1987), for example, has shown that in the Victorian mallee, Coleoptera reflect plant communities more accurately than vertebrates but samples from eight vegetation types represented 11 distinct beetle communities.

Some species have attracted considerable media coverage because they have either been rediscoveries of "living fossils" previously thought to be extinct or because it was believed that they were restricted to small areas scheduled for development (Table III). It is notable that despite this, for instance in the cases of the Eltham copper and *Nothomyrmecia*, there have been only small or no increases in habitat reservation as a result, and their role as umbrella species has been negligible. They have succeeded, however, in focusing the attention of the public on conservation of organisms other than plants and vertebrates, and the Eltham copper became a significant election issue at state level in early 1989. At a broader scale taxonomically, some states, notably Tasmania (Greenslade, 1985) and New South Wales (Nadolny, 1987), have commissioned reports on the conservation status of their invertebrate faunas, and several lists of supposedly endangered species have been published (Michaelis, 1984; Kennedy and Burton, 1986; Hill and Michaelis, 1988). The lists did not aim at being comprehensive and the first concentrated on freshwater invertebrates, the second on marine mollusca and the third listed 250 species of which a high proportion were either Lepidoptera or Drosophilidae.

Other aspects of insect conservation are at an early stage in Australia. For instance there has been no progress in the mechanisms whereby habitat reservation of the listed species can be addressed, although suggestions were made at an Entomological Society's symposium on insect conservation (Majer, 1987; New, 1987). Strategic research into management of reserves for invertebrate conservation and into habitat creation have not been attempted in Australia. Only limited captive breeding programmes have been carried out in connection with two butterfly houses, one at the Melbourne Zoo and the other in Queensland.

B. Legislation

A factor in understanding the legislative process in Australia as regards land management is that legal responsibility for developments lies with the state and not with the federal government. The latter is without power to overrule decisions made by a state government. However, it may under its powers for external affairs, refuse to grant licences to export natural products resulting from any development. Two recent cases in which states challenged the right of the federal government to nominate areas for World Heritage status* were referred to the Supreme Court and were won by the federal government, but this does not affect its powers as far as within-state developments are concerned. Only small parcels of land are under its direct control, a few off-shore islands (for example Christmas Island, Ashmolean Reef, Heard Island) and some national parks considered to be of world significance (Kakadu, Uluru). Although there is federal legislation which requires an environmental impact statement (EIS) to be prepared before any major development can take place, this can be by-passed. Invertebrates are included in the legislation but under the general term "wildlife". As they are not specified as such, it is rare for them to be considered in any EIS. However federal government intervention has resulted in invertebrates playing a major role in two recent enquiries, that of the Lemonthyme and Southern Forests of Tasmania (Helsham *et al.*, 1988) and the north Queensland rain forests. In both cases the relevant state government contested both the right of the federal government to nominate the World Heritage boundaries, and the boundaries themselves. The first case was won by the federal government in the High Court but in neither state are the World Heritage boundaries finally decided (1989). In both enquiries it was recognized that the major part of faunal diversity lay in the invertebrates. As with any legal case of this nature, one major difficulty is in the lack of congruence between the proferred scientific probabilities and the legal verities required. World Heritage status offers the best protection to threatened sites, since State parks and other reserves are subject to changes in status at the political whim of state or federal government.

---

* World Heritage sites are listed under the "Convention Concerning the Protection of the World Cultural and National Heritage", to which 108 nations are party. The convention, administered by Unesco in Paris, is a mechanism for protecting sites and monuments of such exceptional interest and universal interest that they are the responsibility of all mankind.

Legislation that allows nomination of sites for the "heritage value" of the invertebrates that they contain exists under the Australian Heritage Act of 1977. This act established the Register of the National Estate administered by the Australian Heritage Commission which is a statutory body. The Commission has to be notified if any disturbance is planned to a site by federal government. The act has no powers to prevent activities such as logging in national estate forests, although some compromise is usually achieved as to the extent of clear felling. So far no sites have been nominated for listing for invertebrates, but there is nothing to prevent such applications being made in the future. Sites such as the Lockerbie scrub in Cape York, the type locality of 340 species of Lepidoptera and Coleoptera alone (Monteith, 1974), could well be nominated. In the Australian Antarctic Territory, under the treaty regulations, a few "specially protected areas" have been established which have much stronger protection than state reserves. A permit issued by the appropriate minister is required before they can be visited and permits are not issued to casual visitors (Bonner, 1989).

Other national legislation relevant to insect conservation prohibits the import and export of the fauna without permit. This is supported by the Convention in International Trade in Endangered Species of Wild Fauna and Flora (CITES), to which Australia is party. However there appears to be little illegal trading in Australian insects because, apart from the Queensland stag beetle *Phalacrognathus muelleri* (Table III), few species command high prices overseas.

C. Federal and State Initiatives

The federal body with responsibility for insect conservation is the Australian National Parks and Wildlife Service, but so far this department has given priority to vertebrates. In 1989 an endangered species unit was set up which appeared initially concerned with vertebrates and plants. Its terms of reference, however, state that it is responsible for endangered habitats and any species in danger of becoming extinct. A workshop to discuss the role of this unit took place in late 1989 and invertebrates were among the topics considered.

Recently there has also been a federal commitment to conservation using a habitat approach with the establishment of the National Rainforest Conservation Programme. Funds have been directed through the states for research in return for guarantees of forest protection. Invertebrate projects are in progress in four states, Tasmania, New South Wales, Queensland and Western Australia (in the north-west).

Each state government differs in its attention to the conservation of insects, ranging from virtually nil in South Australia to considerable in Victoria and Tasmania. A number of invertebrate projects are being supported in these two states and there is a conviction in the relevant departments that invertebrates deserve equal if not more attention than vertebrates because of the larger number of species involved and the poorer information base. A very recent development (July 1989) has been the adoption of an Australian version of the Council of Europe's "Charter on Invertebrates" by the Council of State Nature Conservation Ministers. It is too soon for any improvements in insect conservation to have resulted from this charter but we can predict that invertebrate conservation will now have to receive more serious attention at state level.

Although the policy in South Australia is to concentrate on vertebrates and plants because of lack of resources, it has a good record in conservation generally and two schemes have been pioneered which benefit insects indirectly. These are heritage agreements for native vegetation on private land, and rural parks (McPhail and Dendy, 1989). Federal subsidies for clearing native vegetation were virtually abolished in 1973, but clearing continued. Legislation was enacted in 1980 to help private land holders willing to retain native vegetation on their properties. It provided reimbursement of rates, and subsidy for fencing native vegetation, to landowners who enter into heritage agreements with the state government to manage these areas in perpetuity. In the first 2 years of the scheme, protection agreements covered 170 sites totalling 15 000 ha. The small size yet large number of these reserves means that they are a significant addition to insect conservation by providing a widely dispersed collection of remnant habitats. Other states may introduce similar schemes.

The concept of rural parks was established in South Australia in 1989 with the Innamincka Regional Reserve. This is the first controlled multiple-use reserve within a conservation framework in Australia (McPhail and Dendy, 1989). One drawback seems to be inadequate provision of areas where stock grazing is completely prohibited.

Tasmania, in contrast, has supported projects on temperate rain forest invertebrates and endemic butterflies and, under the World Heritage Directed Grants, funded field work on Lepidoptera (two projects), Psocoptera, Chironomidae, Trichoptera and Odonata, as well as a number of studies on non-insect invertebrates. Although it does not mention invertebrates as such, the State's Forest Practices Code is designed to provide some protection for wildlife during forestry operations. It has no power to enforce adherence to its recommendations, however.

A recent development in Tasmania has led to one of the most significant advances in national invertebrate conservation. In order to protect two undescribed species of an unknown genus of onychophorans, commonly named the blind velvet worm and the giant velvet worm, of restricted distribution, nearly 2000 ha in five areas of state forest in north-east Tasmania have been designated Wildlife Priority Areas by the Tasmanian Forestry Commission. Although this offers no legal protection, it does mean that velvet worm conservation has priority over forestry operations in gullies and their habitat will be protected from fire. It has taken the combination of an amateur collector, two professional taxonomists, two state wildlife officers, and in the region of $10 000 in federal grant money for three research projects to collect and map the distributions of these two onychophorans in detail (Tait and Briscoe, 1987; Mesibov, 1988). Although there has been some media coverage of these species, it is minimal compared with that given to the Eltham copper or to *Nothomyrmecia*, no public or political pressure was involved and the cost was relatively small. What it did take was close collaboration between state officials and academics and a few small, carefully directed, research projects. Although this could be called a single species approach, at least here the onychophorans are true "umbrella" species as the area concerned is large and the habitat in which they occur (well-rotted logs in *Eucalyptus* forest) also contain a range of endemic Tasmanian insects.

Because of a small group of dedicated and active insect conservationists, Victoria has progressed more than any other state in its legislation by enacting in 1988 the Flora and Fauna Guarantee Act which specifically includes invertebrates and has the stated intention of ensuring that no animal or plant species becomes extinct. A list of threatened species and habitats is to be compiled to which names can be submitted for consideration by an advisory committee. Included in the legislation is the power to invoke an Interim Conservation Order which gives temporary protection to a critical habitat or listed species by applying a moratorium on the actions of any landowner which might threaten the "survival, abundance or evolutionary development of native species or communities of flora or fauna". Once a species (taxon) is listed, the onus falls on the state government to investigate its status and, if necessary, to formulate a management plan. Because this legislation was only enacted recently it is not yet possible to report any benefits to insect conservation.

In Victoria too a project to document insect distributions on a scale similar to that possible in Europe is being attempted. The best documented group, the butterflies, are being mapped on a ten by ten minute grid for the whole of Victoria (Entomological Society of Victoria, 1986). The state is similar in size to the British Isles and about 120 species are involved. From the small number of squares from which any records were known prior to 1985 (Fig. 9), it is clear once again that the detailed distributional data that are available for species in Europe do not exist for Australia. A conservation policy for all Victoria's invertebrates is in preparation (Vaughan, 1989).

In other states there has been no similar effort to incorporate invertebrates in the brief of official nature conservation agencies. In New South Wales for instance, they are not even included in the definition of the "wildlife" which is protected in national parks and reserves. New South Wales is the only state where much of the rain forest, although fragmented, is fully protected as a World Heritage area. Nevertheless clear-felling tracts of the southern eucalypt forests for wood chips is proceeding and is attracting much conservation interest at present. Although Queensland has been notable for its lack of action at state level in matters of conservation, the presence of a rare mangrove butterfly, *Acrodipsas illidgei*, at Redland Bay near Burleigh Heads has saved 160 ha of land from development as a resort (Samson, 1989). A judicial appeal by the developers against the local council's decision not to approve the project was

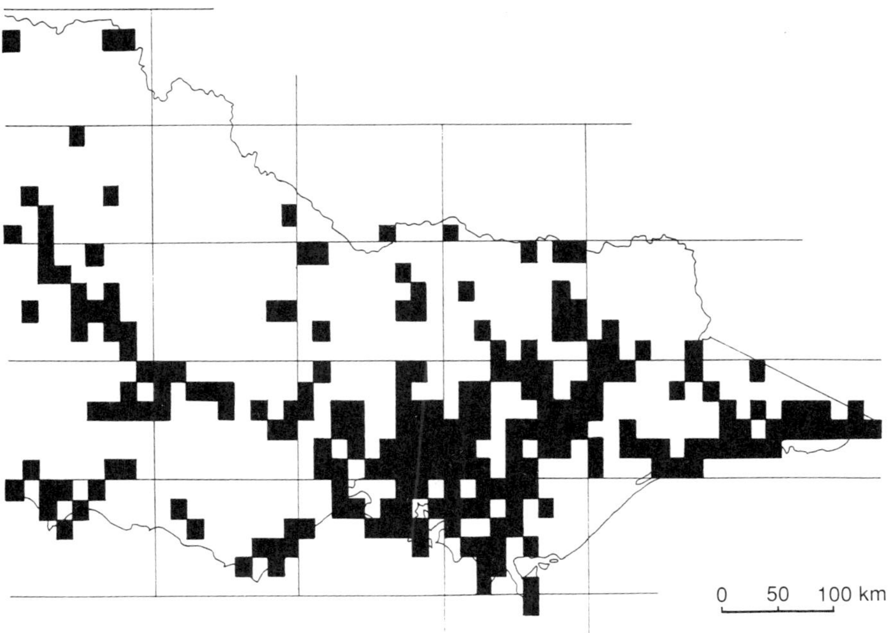

Fig. 9. Distribution of ten minute quadrats from which butterflies had been recorded in Victoria up to 1985 (Entomological Society of Victoria, 1986)

unsuccessful. Western Australia has legislation to protect roadside vegetation (Saunders and Hobbs, 1985) which, like the South Australia Heritage Agreements, indirectly benefits insect conservation. Conservation problems in the sparsely populated Northern Territory have centred on mining and tourist developments in national parks, control of feral vertebrates such as the buffalo, and the effects of fire. The insect fauna here is probably the least known of any state.

D. The Future

The future for insect conservation in Australia seems better than at any other time in the history of the European influence on the continent. The current political climate is strongly supportive of environmental protection and the recently announced Statement on Environment (Anon., 1989) included mention of the importance of biodiversity. It proposed initiatives in the compilation of faunal inventories and documentation of endangered habitats. This should facilitate the inclusion of insects in conservation programmes and accelerate the reservation of critical habitats. However such ideals will clearly demand augmentation of available taxonomic skills in most states, a trend opposite to that which is occurring at present.

What then are the immediate requirements for more rapid progress in insect conservation? Above all, politicians, land managers and the public need to be convinced to take insect conservation seriously and to appreciate the vital need for regeneration of natural habitat for invertebrates. Since powers concerning decisions on land in Australia reside with the states, this has led to a confusing and disparate pattern of conservation effort across the country. Changes in governmental responsibility for land decisions can only be resolved by national referendum. Introductions for biological control need to be more rigorously screened than has been the case in the past. Some deliberate introductions have had unexpected and damaging effects in Australia as well as elsewhere (Howarth, 1983; Howarth and Ramsay, Chapter 4). Whilst Australia was the first country to practice regular fumigation of aircraft arriving from overseas (Russel *et al.*, 1984), surveys reveal live insects still arriving by this means. The mechanisms by which information from taxonomists and ecologists is transferred to land managers has improved recently, largely because biologists have become more involved in strategically directed projects, and because of the development of nationwide data bases, but there is room for further improvement. Rapid yet rigorous survey methods using "priority groups" of insects to detect degradation and loss of habitat still need to be developed (New, 1987), and although some efforts have been made in this direction much remains to be done.

Obviously more information is required on the taxonomy of the fauna, together with its distribution and ecology, but it is unlikely that the rate at which this is being accumulated will increase substantially as there is no prospect of any increase in employment of entomologists. Encouragement of a larger pool of amateurs may be the only possibility here but this must be a relatively long-term process.

One partial solution to the lack of taxonomic and distributional data would be to develop a model to predict the level of endemism and species richness of sites as a practical aid to their assessment and ranking. One such model was attempted in Tasmania by calculating the percentage of endemic species found in 14 higher taxa and plotting the habitats in which they were predominantly found on the habitat templet of Southwood (1977) and Greenslade (1983) (Greenslade P., 1985) (Fig. 10). A summary of the data (Table IV) shows that the highest levels of endemism tend to be found in the most severe habitats, that is those which are predictably unfavourable and in which **A**- or adversity selection for conservation of adaptation operates. Correlates of **A**-selection include long life histories, low reproductive rates, few offspring, low dispersal ability and conservative genetic systems, all characteristics that would tend to make a species most susceptible to disturbance. This gives some indications for conservation priorities. Endemism is lowest in temporary and disturbed habitats and as agricultural land where *r*- or exploitation selection for productivity is dominant and the fauna contains a large proportion of highly dispersive, often exotic, species. This was shown at generic level for Collembola in Tasmania (Fig. 4). Diversity is greatest in relatively stable, favourable habitats such as lowland eucalypt forest and woodland under *K*- or interaction selection for persistance in crowded, competitive, species-rich environments. Using Collembola this model has been shown to have value in measuring the level of disturbance through grazing intensity in improved pastures (Greenslade and Greenslade, 1987). Whether it could be of wider value in Australia needs to be investigated.

VI. CONCLUSION

Australia is the only developed large country with a reasonably high proportion of its vegetation still relatively undisturbed and with the ability to protect what remains. But because of the long time needed to collect accurate data to adequately satisfy land managers, and the need to support the immediate protection of ecosystems which are being destroyed at an ever increasing rate, conservationists are in a difficult position. A moratorium on further alienation of land carrying native vegetation until sufficient information is available is not practical. But, for example, we

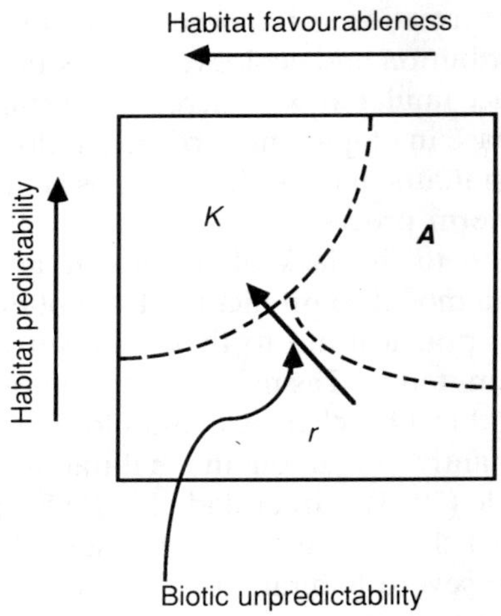

Fig. 10. Habitat templet of Greenslade (1983).

TABLE IV. Classification of some Tasmanian taxa according to level of endemism at species level, associated habitat and predominant selection type operating in that habitat.

Selection type	Usual habitat	Taxon	% endemism
A[a]	Cold torrents	Psephenidae	100
	Cold fast streams	Plecoptera	89
	Caves	Trechini	93
	Burrows	Parastacidae	88
	Cold alpine lakes and streams	Anaspididae	80
	Deep soil	Oligochaeta (terrestrial)	92
	Logs (late decay)	Lucanidae	83
			91
*k***A**[b]	Cold slow streams	Tipulidae	75
	Lakes	Turbellaria (terrestrial)	69
		Trichoptera	70
		Oligochaeta (aquatic)	60
K	*Eucalyptus* forest	Coccinellidae	30
r	Pasture	Collembola	2–4

[a] Overall % endemism for **A** taxa = 91 (150/165 species, excluding Trechini).
[b] Overall % endemism for *k***A** taxa = 69 (167/241 species)

are regularly asked by state forestry commissions before they give approval to fell a coupe for information of any insects that occur there which might suggest that retention is desirable. National Estate committees also are asked to comment on whether there is any entomological information on the many small parcels of land proposed for listing. In very few of these cases is there any and their role as reservoirs for native invertebrates is only now being appreciated.

Much of the action to date on insect conservation has been of the "crisis management" type exemplified by single species such as the Eltham Copper. The single species approach can be of value in reserving the small parcels of land which may be adequate for many relict Gondwanan and older taxa. It is of no value in protecting the diverse Australian radiations characteristic of eucalypt and other dominant vegetation which could perhaps be protected in rural regional parks. Safeguarding a wide range of habitats is clearly vital, even though many of these cannot, at present, be managed specifically to conserve invertebrates. The habitat approach, although avoiding the problems caused by the lack of taxonomic and distributional data, is hampered by difficulty in identifying all insect habitats and in assessing priorities.

Another solution might be to reverse the normal procedure whereby protection is afforded to those areas about which most is known. At present, "protection" is based on solid documentation. Perhaps it is those habitats about which we know least which should be accorded priority for protection. Many Australian vegetation associations are small and are clearly under-represented in formal reserves (Specht *et al.*, 1974). Of interest also is the development of an ecological philosophy which has its roots in ethical rather than biological tenets (Johnson, 1991). The concept that an ecosystem has a right to exist independently rather than in terms of human perception or for anthropomorphic reasons avoids the requirement to comprehensively document faunas and distributions before they can be afforded protection. A combination of heuristic model, "negative gearing" of sites, and ethical principles would appear at first sight to be a last resort but might ultimately prove to be the most constructive approach to insect conservation.

Acknowledgements

We are grateful to the many people, too numerous to list by name, for information used in producing this manuscript. In particular we thank P. J. M. Greenslade for his contributions to the text.

REFERENCES

Andersen, A. N. (1983). Species diversity and temporal distribution of ants in the semi-arid mallee region of northwestern Victoria. *Aust. J. Ecol.* **8**, 127–137.

Andersen, A. N. and Yen, A. L. (1985). Immediate effects of fire on ants in the semi-arid mallee region of north-western Victoria. *Aust. J. Ecol.* **10**, 25–30.

Anonymous (1989). "Statement on the Environment". Australian Government Publishing Service, Canberra.

Atlas of Australian Resources (1990). "Vegetation" volume 6, 3rd series. Canberra, Australian Surveying and Land Information group.

Audley-Charles, M. G. (1987). Dispersal of Gondwanaland: relevance to evolution *In* "Biogeographical Evolution of the Malay Archipelago" (T. C. Whitmore, ed.), pp. 5–25. Clarendon Press, Oxford.

Baker, G. H. (1985). The distribution and abundance of the Portuguese millipede *Ommatoiulus moreletti* (Diplopoda: Iulidae) in Australia. *Aust. J. Ecol.* **10**, 249–259.

Barker, W. R. and Greenslade, P. J. M. (eds) (1982). "Evolution of the Flora and Fauna of Arid Australia". Peacock Press, Adelaide.

Beadle, N. C. W. (1981). "The Vegetation of Australia". Cambridge University Press, Melbourne.

Braithwaite, R. W., Lonsdale, W. M. and Estbergs, J. A. (1989). Alien vegetation and native biota in Tropical Australia: the impact of *Mimosa pigra*. *Biol. Conserv.* **48**, 189–210.

Blyth, J. D. (ed.) (1987). "Beekeeping and Land Management". Department of Conservation and Land Management, Perth.

Bonner, W. N. (1989). Environmental assessment in the Antarctic. *Ambio* **18**(1), 83–89.

Bridgewater, P. B. (1987). The present Australian environment – terrestrial and freshwater. *In* "Fauna of Australia. Volume 1A. General Articles" (G. R. Dyne and D. W. Walton, eds), pp. 69–100. Australian Government Publishing Service, Canberra.

Busby, J. R. (1987). Floristic communities and bioclimates of *Nothofagus cunninghamii* rainforest in south-eastern Australia. *In* "The Rainforest Legacy. Australian National Rainforests Study – The Nature, Distribution and Status of Rainforest Types" Vol. 1 pp. 23–31. Special Australian Heritage Publication Series No. 7(1). Australian Government Publishing Service, Canberra.

Busby, J. R. (1988). Potential implications of climate change on Australia's flora and fauna *In* "Greenhouse. Planning for Climate Change" (G. I. Pearman ed.), pp. 387–398. Brill and CSIRO, Australia.

De Deckker, P. and Williams, W. D. (eds) (1986). "Limnology in Australia". Melbourne/ W. Junk, Dordrecht.

Delfosse, E. S. and Cullen, J. M. (1985). CSIRO Division of Entomology submission to the inquiries into biological control of *Echium plantagineum* L., Paterson's curse/Salvation Jane. *Plant Protect. Q.* **1**(1), 24–40.

Dyne, G. W. and Walton, D. W. (eds) (1987). "Fauna of Australia. Volume 1A. General Articles". Australian Government Publishing Service, Canberra.

Entomological Society of Victoria (1986). "Preliminary Distribution Maps of Butterflies in Victoria". Melbourne.

Fox, L. R. (1981). Defense and dynamics in plant-herbivore systems. *Am. Zool.* **21**, 853–864.

Frakes, L. A., McGowran, B. and Bowler, J. M. (1987). Evolution of Australian Environments. *In* "Fauna of Australia. Volume 1A. General Articles" (G. R. Dyne and D. W. Walton, eds), pp. 1–16. Australian Government Publishing Service, Canberra.

Gilbert, N. (1988). Control of fecundity in *Pieris rapae*. V. Comparisons between populations. *J. Anim. Ecol.* **57**, 395–410.
Gould, F. (1988). Genetic engineering, integrated pest management and the evolution of pests. *Trends Biotechnol.* **6**(4), S15–18.
Greenslade, P. (1985). "The conservation status of Tasmanian non-marine invertebrates". Unpublished report, Tasmanian Parks and Wildlife Service, Hobart.
Greenslade, P. (1987). Generic biogeography of Tasmanian Collembola. *In* "Soil Fauna and Soil Fertility" (B. R. Striganova, ed.), pp. 23–31. Proceedings of the 9th International Colloquium on Soil Zoology, Moscow, August 1985.
Greenslade, P. and Greenslade, P. J. M. (1987). A new approach to environmental assessment. *In* "Soil Fauna and Soil Fertility" (B. R. Striganova, ed.), pp. 245–252. Proceedings of the Ninth Colloquium of Soil Zoology, Moscow, 1985.
Greenslade, P. J. M. (1979). "A Guide to Ants of South Australia". South Australian Museum, Adelaide.
Greenslade, P. J. M. (1983). Adversity selection and the habitat temple. *Am. Nat.* **122**(3). 352–365.
Greenslade, P. J. M. (1985) Pterygote insects and the soil: their diversity, their effects on soils and the problem of species identification. *Quaest. Entomol.* **21**, 571–585.
Greenslade, P. J. M. and Greenslade, P. (1989). Ground layer invertebrate fauna. *In* "Mediterranean Landscapes in Australia. Mallee Ecosystems and their Management" (J. C. Noble and R. A. Bradstock, eds), pp. 266–284. CSIRO, Australia.
Greenslade, P. J. M. and Halliday, R. B. (1983). Colony dispersion and relationships of meat ants, *Iridomyrmex purpureus* and allies, in an arid locality in South Australia. *Insectes Sociaux* **30**, 82–9.
Greenwood, G. and Boardman, R. (1988). Climatic change and some possible effects upon the terrestrial ecology of South Australia. *In* "Greenhouse '88: Planning for Climate Change" (T. Dendy, ed.), pp. 112–132. Department of Environment and Planning, Adelaide, South Australia.
Heath, J. (1973). "Provisional Atlas of the Insects of the British Isles. II". Biological Records Centre, Huntingdon.
Heatwole, H. (1987). Major components and distribution of the terrestrial fauna. *In* "Fauna of Australia. Volume 1A. General Articles" (G. R. Dyne and D. W. Walton, eds), pp. 101–135. Australian Government Publishing Service, Canberra.
Helsham, M. M., Hitchcock, P. P. and Wallace, R. H. (1988). "Report of the Commission of Inquiry into the Lemonthyme and Southern Forests", Volumes 1 and 2. Australian Government Publishing Service, Canberra.
Hill, L. and Michaelis, F. B. (1988). "Conservation of Insects and Related Wildlife". Australian National Parks and Wildlife Service, Canberra.
Howarth, F. G. (1983). Classical biocontrol: panacea or Pandora's box? *Proc. Hawaiian Entomol. Soc.* **25**, 239–244.
Ive, J. and Cocks, D. (1989). Rural land degradation in Australia. The role of trees in ecological solutions. *Aust. Conserv. Farmer* **1**(3), 18–22.
Jarman, S. J., Brown, M. J. and Kantvilas, G. (1984). "Rainforest in Tasmania". National Parks and Wildlife Service, Tasmania.
Jarman, J., Brown, M. J. and Kantvilas, G. (1987). The classification, distribution and conservation status of Tasmanian rainforests. *In* "The Rainforest Legacy. Australian National Rainforests Study. Vol. 1 — The Nature, Distribution and Status of Rainforest Types", pp. 10–21. Australian Government Publishing Service, Canberra.
Johnson, L. (1991) "A Morally Deep World." Cambridge University Press, Cambridge, in press.

Kennedy M. and Burton, R. A. (eds) (1986). "Threatened Species Conservation Strategy for Australia. Policies for the Future". Ecofund, Australia.

Key, K. H. L. (1978). "The Conservation Status of Australia's Insect Fauna". Australian National Parks and Wildlife Service, Canberra.

Kitching R. L. (1981). The geography of the Australian Papilionoidea. *In* "Ecological Biogeography of Australia" (A. Keast, ed.) pp. 979–1005. W. Junk, The Hague.

Lake, P. S., Swain, R., Richardson, A. M. and Coleman, D. (1978). Freshwater environments and their fauna. *In* "The South West Book" (H. Gee and J. Fenton, eds). Australian Conservation Foundation, Melbourne.

MacArthur, R. H. and Wilson, E. O. (1967). "The Theory of Island Biogeography". Princeton University Press, Princeton, NJ.

Main, B. Y. (1987a). Persistence of invertebrates in small areas: Case studies of trapdoor spiders in Western Australia. *In* "Nature Conservation: the role of Remnants of Native Vegetation" (D. A. Saunders, G. W. Arnold, A. A. Burbidge and A. J. M. Hopkins, eds), pp. 29–39. Surrey Beatty, Chipping Norton, New South Wales.

Main, B. Y. (1987b). Ecological disturbance and conservation of spiders: implications for biogeographic relics in southwestern Australia. *In* "The Role of Invertebrates in Conservation and Biological Survey" (J. D. Majer, ed.), pp. 89–97. Western Australian Department of Conservation and Land Management Report, Perth, WA.

Majer, J. D. (ed.). (1987). "The Role of Invertebrates in Conservation and Biological Survey". Department of Conservation and Land Management, Perth.

Margules, C. R. (1987). "The Wog Wog Habitat Patch Experiment: Background, Objectives, Experimental Design and Sample Strategy". Technical Memorandum 85/18 (Revised July 1987), CSIRO Institute of Biological Resources, Canberra.

Marks, E. N. (1972). "A Submission to the House of Representatives Select Committee on Wildlife Conservation". Transcript of Evidence X, 6383–6404.

Matthews, E. (1978). "Insect Ecology". University of Queensland Press, Brisbane.

McKeown, K. C. (1949). "Nature in Australia". Angus and Robertson, Sydney.

McPhail, I. R. and Dendy, T. (1989). Biological conservation in South Australia. *Aust. Biol.* **2**(2), 18–26.

McQuillan, P. B. (1991). *Nothofagus* (Fagaceae) and its insect fauna — an overview and preliminary synthesis, in preparation.

Mesibov, R. (1988). "Tasmanian Onychophora. A report for the Department of Lands, Parks and Wildlife". Unpublished report, Tasmanian Parks, Wildlife and Heritage, Hobart, Tasmania.

Michaelis, F. B. (1984). Possible effects of forestry on inland waters of Tasmania: a review — *Env. Conserv.* **11**, 331–343.

Monteith, G. B. (1974). "Focus on Cape York". Report from the Fauna Subcommittee of the Entomological Society of Queensland, Brisbane.

Nadolny, C. (1987). "Rainforest butterflies in New South Wales: their Ecology, Distribution and Conservation". New South Wales National Parks and Wildlife Service, Sydney.

New T. R. (for the Australian Entomological Society) (1976). "Submission to the Inquiry into the Impact on the Australian Environment of the Current Woodchip Industry Program". Senate Hansard: transcripts of reports to the Australian Standing Committee on Science and the Environment, Canberra.

New, T. R. (1983). Systematics and ecology: reflections from the interface. *In* "Australian Systematic Entomology: a Bicentenary Perspective" (E. Highley and R. W. Taylor, eds), pp. 50–79. CSIRO, Melbourne.

New, T. R. (1984). "Insect Conservation — an Australian Perspective". W. Junk, Dordrecht.

New, T. R. (1985). A revision of the Australian Myrmeleontidae (Insecta: Neuroptera) 1. Introduction, Myrmeleontini, Protoplectrini. *Aust. J. Zool. Suppl.* 104.

New, T. R. (1987). Insect conservation in Australia: towards rational ecological priorities. *In* "The Role of Invertebrates in Conservation and Biological Survey" (J. D. Majer, ed.), pp. 5–20. Department of Conservation and Land Management, Perth.

New, T. R. (1988). "Associations between Insects and Plants". New South Wales University Press, Kensington, New South Wales.

Nix, H. (1982). Environmental determination of biogeography and evolution in Terra Australis. *In* "Evolution of the Flora and Fauna of Arid Australia" (W. R. Barker and P. J. M. Greenslade, eds), pp. 47–66. Peacock Press, Adelaide.

Noble, J. C. (1982). The significance of fire in the biology and evolutionary ecology of mallee *Eucalyptus* populations. *In* "Evolution of the Flora and Fauna of Arid Australia" (W. R. Barker and P. J. M. Greenslade, eds), pp. 153–159. Peacock Publications, Adelaide.

O'Farrell, A. F. (1970). Odonata. *In* "Insects of Australia". Melbourne University Press, Melbourne.

Rajapaksa, N. and Spradbery, J. P. (1989). Occurrence of the Old World screw-worm fly *Chrysomya bezziana* on livestock vessels and commercial aircraft. *Aust. Vet. J.* **66**(3), 94–96.

Russel, R. C., Rajapaksa, N., Whelan, P. I. and Langsford, W. A. (1984). Mosquito and other insect introductions to Australia aboard international aircraft, and the monitoring of disinsection procedures. *In* "Commerce and the Spread of Pests and Disease Vectors" (M. Laird, ed.), pp. 109–141. Praeger, New York.

Samson, P. R. (1989). Morphology and biology of *Acrodipsas illidgei* (Waterhouse and Lyell), a myrmecophagous lycaenid (Lepidoptera: Lycaenidae: Theclinae). *J. Aust. Entomol. Soc.* **28**(3), 161–168.

Sant, G. J. and New, T. R. (1988). "The Biology and Conservation of *Hemiphlebia mirabilis* Selys (Odonata, Hemiphlebiidae) in Southern Victoria". Technical Report 82, Arthur Rylah Institute for Environmental Research Melbourne.

Saunders, D. and Hobbs, R. (1989). Corridors for conservation. *New Sci.* 28 January 1989, 63–68.

Smith-White S. (1982). Summary and re-integration. *In* "Evolution of the Flora and Fauna of Arid Australia" (W. R. Barker and P. J. M. Greenslade, eds), pp. 371–379. Peacock Press, Adelaide.

Southwood T. R. E. (1977). Habitat, the templet for ecological strategies? *J. Anim. Ecol.* **46**, 337–365.

Specht, R., Roe, E. M. and Broughton, V. H. (1974). Conservation of major plant communities in Australia and Papua New Guinea. *Aust. J. Bot. Suppl. Ser.* **7**, 1–667.

Spradbery, P. J. (1973). "Wasps. An Account of the Biology and Natural History of Solitary and Social Wasps with Particular Reference to those of the British Isles". Sidgwick and Jackson, London.

Tait, N. N. and Briscoe, D. A. (1987). Report on WHA directed research contract to Dr N. N. Tait and Dr D. A. Briscoe: Onychophora in Tasmania. Unpublished report, Tasmanian Lands Parks and Wildlife Service, Hobart.

Taylor, R. W. (1972). Biogeography of insects of New Guinea and Cape York Peninsula. *In* "Bridge and Barrier: the Natural and Cultural History of Torres Strait" (D. Walker, ed.), pp. 213–230. Australian National University, Canberra.

Taylor, R. W. (1976). "Submission to the Inquiry into the Impact on the Australian Environment of the Current Woodchip Industry Program". Senate Hansard: transcripts of reports to the Australian Standing Committee on Science and Environment, Canberra.

Vaughan, P. J. (1991). "A Draft Conservation Strategy for the Non-Marine Invertebrates and Non-vascular Plants of Victoria" in press.

Wells, S. M., Pyle, R. M. and Collins, N. M. (1983). "The IUCN Invertebrate Red Data Book". IUCN, Gland.

Wilson, E. O. (1961). The nature of the taxon cycle in the Melanesian ant fauna. *Am. Naturalist,* **95**, 169–193.

Working Group on Rainforest Conservation (1985). Rainforest Conservation in Australia. A report to the Hon. Barry Cohen, M. P., Minster for Arts, Heritage and Environment.

Yen, A. L. (1987). A preliminary assessment of the correlation between plant, vertebrate and Coleoptera communities in the Victorian mallee. *In* "The Role of Invertebrates in Conservation and Biological Survey" (J. D. Majer, ed.), pp. 73–88. Department of Conservation and Land Management, Perth.

4

The Conservation of Island Insects and their Habitats

F. G. HOWARTH AND G. W. RAMSAY

I. Introduction ... 71
II. Islands ... 72
A. Hawaii ... 73
B. New Zealand ... 74
C. Characteristics of Island Biology ... 74
III. Island Conservation ... 77
A. Impacts of Invasive Alien Species ... 80
B. Alteration of Habitats by Land Use Changes ... 85
C. The Biodiversity Crisis ... 86
D. Ecological Myths ... 87
E. Cultural Biases ... 91
IV. Recommendations and Solutions ... 93
A. Research ... 93
B. Education ... 94
C. Reserve Management ... 95
D. Legal and Legislative Actions ... 97
E. Controlling the Introduction of Alien Species ... 99
F. Pest Control Programmes ... 100
G. The Future ... 100
References ... 101

I. INTRODUCTION

To assess the status of island insects and their habitats is a daunting task. Not only are there thousands of islands in the world's oceans, with their native insect faunas totalling millions of species, but also, broadly defined, nearly all organisms live within island-like habitats that are surrounded by inhospitable "seas". Nevertheless, oceanic islands have long been recognized as special places rich in unique species and ideal natural laboratories for deciphering concepts in evolution and ecology (Simon and Sugden, 1987). It is also appropriate in a review of conservation biology to focus on islands because their faunas appear to be more sensitive to environmental changes, and more prone to extinction than continental ones (Vitousek, 1988; Loope *et al.*, 1988).

Our thesis is that islands are in fact microcosms of continents, and that ecological and evolutionary processes are identical on both. Only the relative area (size) of ecosystems, taxonomic diversity, endemism, isolation from potential colonists and historical development (including geological, biological and human history) are different. However, even for these characteristics, island-like situations occur frequently on continents. For example, near Chillagoe, Queensland, Australia, four small, old limestone towers occur within a 2 km^2 area. Several cave-adapted arthropods are endemic to these towers; some even to single towers within the group, being unable to cross the inhospitable surface terrain between them (Hoch and Howarth, 1989).

What is happening now on islands is a preview of coming attractions for the continents. This review focuses on island insects, but includes information on other terrestrial invertebrates when relevant.

II. ISLANDS

Islands may be broadly divided according to whether they are located in mid-ocean or fringing the continents, entirely low lying or with raised areas. True oceanic islands differ from continents and fringing continental islands in three respects: geographical isolation, youthfulness and small size with requisite small population sizes.

Low islands have little physiographic and climatic diversity. Most have been disturbed by humans since prehistoric times and are now largely inhabited by strand, vagile and human-dispersed species. A few small islands, most notably poorly known Henderson in the South Pacific, have partly escaped the impacts of humans and provide a glimpse of what once existed on low islands (Fosberg *et al.*, 1983).

High islands intercept prevailing winds, creating extreme rainfall gradients. Annual rainfall can vary between 25 cm on leeward coasts to over 1000 cm at mid-elevation windward sites. Additionally, habitats are zoned by altitude above sea level. Geological processes such as volcanism and erosion can subdivide the habitat further, so that on high islands habitat diversity can be extreme.

We consider two disparate island groups in some detail to demonstrate unifying principles — one a young, extremely isolated mid-oceanic island chain, Hawaii, and the other a group of old, long isolated fringing continental islands, New Zealand.

A. Hawaii

The Hawaiian Islands are the most isolated group of high islands on earth, being nearly 4000 km from the nearest comparable land area. They are the summits of giant submarine volcanoes, and each island or group of islands has always been isolated from the others by deep straits 40 km or more wide (Dalrymple *et al.*, 1973). The islands are progressively older to the north-west, ranging from about 500 000 years for Hawaii Island to 30 million years for Kure Atoll about 2500 km away. Kauai Island at only 5.1 million years is the oldest of the eight main or high islands (Simon and Sugden, 1987). The total land area is 16 770 km^2, of which more than half (10 455 km^2) occurs on Hawaii Island.

The ancestors of the biota of the Hawaiian Islands had to disperse across vast oceanic distances and establish reproducing populations. Many arriving propagules undoubtedly perished. Wind currents, especially storms and the jet stream, carried most of the successful propagules to the islands (Holzapfel *et al.*, 1978, Carlquist, 1980, 1981). Biologists visiting the islands are at first struck by what is absent. Only 50% of the known orders and 15% of the known families of insects are represented in the native fauna. Only 350–400 separate colonizations can account for the total insect fauna now estimated to include over 10 000 endemic species (Zimmerman, 1948; Gagne, 1988) of which about 5000 species have been described. The pattern of a few colonizers giving rise to many closely related species is a characteristic of the Hawaiian biota (Zimmerman, 1948; Howarth *et al.*, 1988; Wagner *et al.*, 1990). Hawaiian *Drosophila* (more than 450 species (Hardy, 1965; Carson, 1987)) and *Hyposmocoma* (350–500 species (Zimmerman, 1978)) head a list of nine genera that each contain more than 100 native species. At least six genera will be added when (or if) taxonomic studies are completed.

By far the majority of native Hawaiian arthropods are found on only one island, and many are even restricted to a small part of a single island. Such a restricted distribution explains in part the high overall diversity and vulnerability of many species. In total, more than 180 distinct natural communities (i.e. discrete groups of interacting species occupying a common area) are found in the Hawaiian Islands, a number comparable to the number of communities found in continental areas (Mueller-Dombois *et al.*, 1981; Cuddihy, 1989; Daws, 1988; Gagne and Cuddihy, 1990).

B. New Zealand

New Zealand is a fragment of ancient Gondwanaland with a long and complex geological history consequent of its position on the boundary of two moving tectonic plates. Like New Caledonia, it severed its land links 80–60 million years ago as the Tasman Sea formed (Fleming, 1979; King, 1984) and it now lies 2200 km from the nearest land. Its isolation has been long and extreme like that of a true oceanic island.

The westerly wind belt of the Forties gives New Zealand a wet and temperate climate, with warm summers and no marked dry season — much like that of Western Europe (Coulter, 1973; Hurnard, 1978). The axial mountain ranges are high, with permanent snow and leeward rain shadow. New Zealand (including its offshore and subantarctic islands) has a high habitat diversity, with 235 ecological districts combined into 82 ecological regions (Simpson, 1982). The terrestrial arthropod fauna is estimated to comprise at least 20 000 species, of which about 12 000 have been named (Ramsay *et al.*, 1988). In contrast with Hawaii, it has had a resident biota continuously for 100 million years, but subsequent immigrants have complemented the biota. Throughout New Zealand's long isolation, the biota was subjected to intense earth movements, frequent changes in geography and orogenies, as well as the Pleistocene glaciations during which only remnants of the original biota survived (Dumbleton, 1970; Fleming, 1979). Speciation was fostered by the development of ecological niches on the isolated, changing archipelago.

C. Characteristics of Island Biology

1. Endemism

Due to their isolation and history, island biotas have a high degree of endemism. Well over 90% of the terrestrial arthropods (Kuschel, 1975) and 81% of the flora (Godley, 1975) are endemic to New Zealand. For Hawaii the level of endemism is higher, over 99% for terrestrial arthropods (Zimmerman, 1948) and 89% for the flora (Wagner *et al.*, 1990).

2. Taxonomic diversity

Many higher taxa are unexpectedly absent on islands, as few colonizers were successful. This is especially true for the vertebrates, but many arthropod taxa are also missing. Conversely there are endemic families in New Zealand and some other islands, and island faunas in general are enriched at the species level. The more vagile groups within the birds, insects, spiders and molluscs are disproportionately well represented.

3. Ecological diversity

The number of habitats is often comparable to continental areas, but the youthfulness and isolation limit the number of species occupying given niches.

4. Adaptive radiation

Many Hawaiian arthropods, like island species in general, have broadened their ecological roles (niches) or have become specialized to fulfil functions atypical for their continental relatives in a very brief time. For example, pug moth caterpillars in the genus *Eupithecia* (Geometridae) have radiated in Hawaii into over 20 species of ambush predators, the only known carnivorous Lepidoptera larvae able to capture active prey (Montgomery, 1983). There are terrestrial damselfly naiads, blind big-eyed hunting spiders, blind underground tree crickets, terrestrial and arboreal saldid water bugs, and a lygaeid seed bug that scavenges on moribund arthropods (Zimmerman, 1948; Cobben, 1980; Howarth, 1987). These represent examples of macroevolution on microcontinents in minitime. In New Zealand too there are many examples, such as the rodent niche being occupied by stenopelmatid wetas, which are called "invertebrate mice" (Ramsay, 1978a). There is a trend for aquatic groups to become terrestrial and *vice versa* (McLellan, 1975; Dugdale, 1975; Zimmerman, 1948). New Zealand has an exceptionally diverse litter fauna, both in number of taxa and in degree of sympatry. Solem *et al.* (1981) found a fivefold greater number of land snail species present than is usual elsewhere. Coleoptera and Lepidoptera are major components (Watt, 1975; Moeed and Meads, 1986; Dugdale, 1975). The virtual absence of native termites and ants may partly explain the lushness and persistence of the litter, its great faunal diversity and the degree of flightlessness. The introduction of ants and termites may spell disaster for elements of this fauna.

5. Flightlessness

A conspicuous theme among island insects is the evolution of flightlessness, and the Hawaiian and New Zealand faunas contain some of the finest examples. Of the 11 orders of insects that established in Hawaii with wings, only one, the Odonata, has not evolved flightless species. Since most of the dominant flightless continental groups, such as ants and termites, were poor dispersers to islands, alate natives have evolved to take their place. Often flightlessness has evolved within each group independently from separate ancestral species on each island in turn (Zimmerman, 1957; Hardy and Delfinado 1974), showing that the evolution of flightlessness is a general process rather than an exceptional phenomenon. What makes this process exciting in Hawaii is that in many cases, flighted and flightless sibling species often live side by side, making it possible to study the process. (See Dempster, Chapter 7, for another view of the loss of vagility.)

6. *Pollination*

In contrast to the fact that inbreeding plants are more likely to establish on islands, the proportion of outbreeding plant species among the native floras is much higher than in continental areas (Carlquist, 1980). The major pollinating groups are insects and birds. Mutualistic associations are likely, but pollination biology on islands is very little studied and very vulnerable to perturbation.

7. *Seasonality*

In contrast to most continental areas, seasonality is greatly reduced on islands. Winter diapause is even rare among New Zealand arthropods living under the extreme subalpine and alpine biotopes, where it would be expected to be prevalent. Island climate, moderated by the surrounding ocean, tends to be relatively uniform and plays a less important role in arthropod activity and population dynamics than do biotic associations. Therefore alien species may have a much greater detrimental effect on native populations than otherwise would be the case.

8. *Longevity*

Island species often have increased longevity. The large and spectacular carnivorous New Zealand land snails (Rhytididae: *Powelliphanta*) have been reported to live for 40 years (Meads *et al.*, 1984). Adults of the large molytine weevil, *Karocolens pittospori* (Fig. 1), lived for more than 3 years in captivity, and larval development took approximately 1 year (Bennett, 1987). The Wellington speargrass weevil, *Lyperobius huttoni* (Fig. 2), is also long lived (Bull, 1967). The wetas take 2 years to reach maturity, and the adults may live for several years (Ramsay, 1955) (Fig. 3). Hawaiian achatinellid snails are also long lived, which is considered a major factor in their vulnerability to predation (Hadfield, 1986).

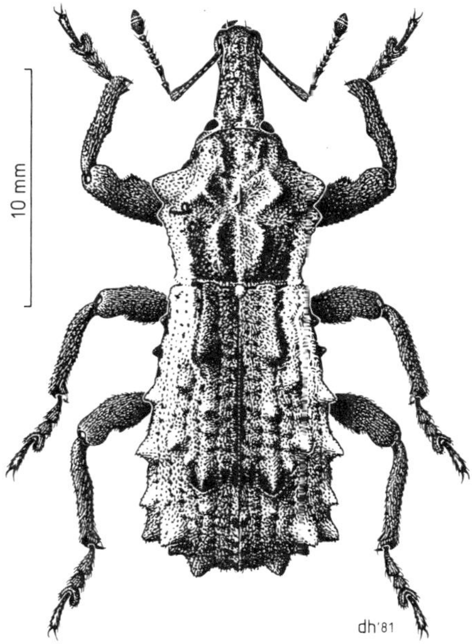

Fig. 1. *Karocolens pittospori* Kuschel, a long-lived large weevil recently discovered on a rodent-free offshore island in New Zealand, it is unusual in being associated with Pittosporaceae. Conservation status: vulnerable. Credit: D. H. Helmore.

III. ISLAND CONSERVATION

The major factors causing decline of native insect faunas are impacts of invasive alien organisms, including plants, vertebrates, invertebrates and microorganisms, and the alteration of habitats and communities resulting from the impact of human use and abuse of natural resources (Pyle *et al.*, 1981; New, 1984; Crosby, 1986). Wodzicki (1981) recognized similar problems for vertebrate conservation on the islands of the South Pacific. These perturbations are greatly exacerbated by the increasing ability of humans to exploit the global ecosystem, in concert with our decreasing appreciation of the natural world. Additionally, taxonomic and ecologic ignorance, myths, cultural biases and human greed block finding and implementing appropriate solutions to these problems. The severity of the injury experienced by any particular island or continental community results from different histories of human impact rather than from any inherent fragility or instability within particular natural ecosystems.

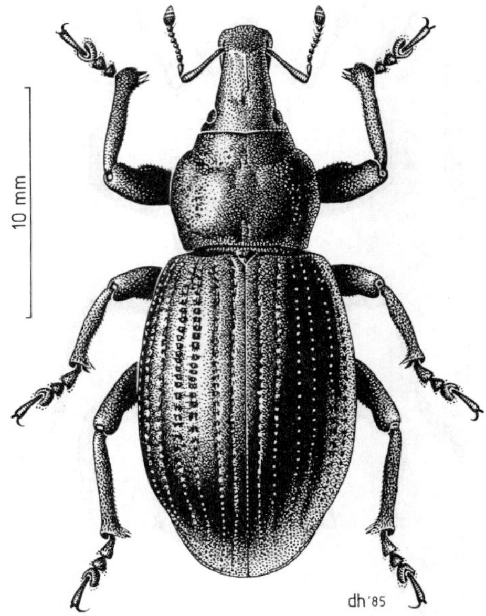

Fig. 2. *Lyperobius huttoni* Pascoe (Wellington speargrass weevil). Occurs in subalpine areas in Canterbury and Marlborough with several lowland coastal populations at Sinclair Head, Wellington. The latter populations are seriously depleted by habitat destruction and rat predation. A special reserve has been established and measures taken to rehabilitate the habitat. Conservation status: Rare/Endangered. Credit: D. H. Helmore.

Polynesians colonized Hawaii nearly 2000 years ago and New Zealand about 1000 years ago causing the loss of unique species by introduction of animals (e.g. Polynesian rat and dog into New Zealand and the rat, dog and pig into Hawaii) and by modification of the environment. Fire replaced climate in dictating vegetation type, and 50–75% of New Zealand has been burned during the last 1000 years (Molloy, 1970; McGlone, 1989). The severity of Polynesian impacts on Hawaii have only recently been recognized with the discovery of an extinct avifauna (Kirch, 1982; Olson and James, 1984). After western contact, Hawaii was subjected to more than a century of abuse from feral ungulate populations and the nearly complete abandonment of the traditional island culture in favour of anything alien before biological surveys began. Currently, over 75% of the land area has been converted to human use (Gagne, 1988). Gagne and Howarth (1985) listed six major perturbations, not mutually exclusive, that were important in the extinction of 27 species of Hawaiian macrolepidoptera. They are, in order of importance, biological control intro-

Fig. 3. *Deinacrida rugosa* Buller (Stephens Island weta). Confined to Stephens and Mana Islands, Cook Strait, where the populations continue to florish despite the presence of mice (on Mana Island only). Introduced to Maud Island in 1978 where it has increased and expanded its distribution. Was probably once widespread on mainland New Zealand and other offshore islands. Conservation status: Rare. Credit: D. H. Helmore.

ductions, habitat loss, alien mammals, host loss, alien arthropods and hybridization with an invading alien relative.

A sign of the times is the emergence of posthumous taxonomy for modern species described after their extinction. In New Zealand remains of three large weevil species (Fig. 4) were recently discovered, which were probably annihilated by the Polynesian rat (Kuschel, 1987). In 1977 Tobias' caddisfly (*Hydropsyche tobiasi*) was described from specimens collected on the River Rhine, France, in the 1920s. None have been seen since (Collins, 1987a). In Hawaii three earwigs, *Anisolabis* spp., were described from museum specimens and apparently are now extinct perhaps from the impacts of introduced species (Brindle, 1980), and two noctuid moths, *Heliothis* (*Heliocoverpa*) spp., were annihilated by rabbits on Lizianski and possibly by sterile crosses with introduced relatives or by introduced parasites on the main islands (Hardwick, 1965). In addition, Cobben (1980) described *Saldula longicornis* from Oahu on the basis of 50–100-year-old specimens. Regrettably, posthumous taxonomy will

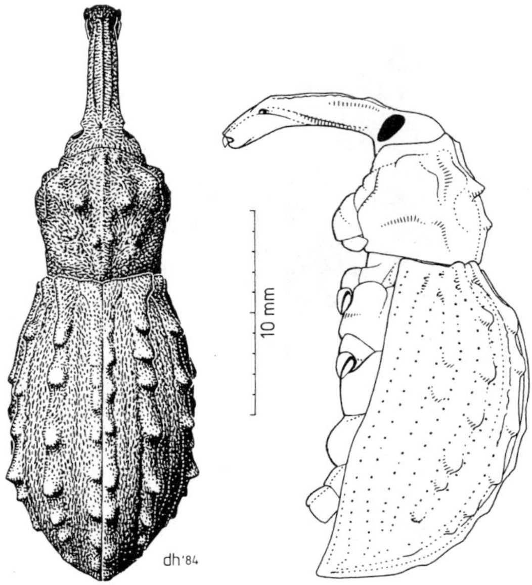

Fig. 4. *Tymbopiptus valeas* Kuschel. A large extinct weevil, probably eliminated by the Polynesian rat. Specimens have been recovered from limestone cave sinkholes and from remains of forests buried under volcanic ash approximately 1680 years ago, in the central region of North Island, New Zealand. Credit: D. H. Helmore.

expand as collections from altered or lost ecosystems are identified (Collins, 1987a).

A. Impacts of Invasive Alien Species

Biological pollution (the effects of invading alien species) is the most insidious and pervasive environmental threat. Unlike other human-caused perturbations, it is virtually irreversible and has the potential to undo all other conservation programmes. Alien species do not respect human boundaries but invade all suitable habitats to which they can disperse. The invaders include nearly all classes of plants, animals and microorganisms, and in concert they compete with, parasitize, prey on, spread diseases among or destroy the habitats of native organisms. Many of their effects and impacts may go unrecognized in natural areas and unmonitored habitats, if not specifically sought. It may not be until an alien is well entrenched that attempts are made to deduce in hindsight what major factors were involved in the decline or disappearance of native species.

The world's biota is being homogenized, as increasing world transport has broken the splendid isolation that allowed the evolution of native island species (Atkinson, 1985; Drake *et al.*, 1989). The destruction caused by ungulates, rodents, rabbits, mongooses, ferrets, birds, mosquito fishes and certain other vertebrates on islands is well known (Decker, 1973; Wodzicki, 1981; King 1984; Tomich, 1986; Loope *et al.*, 1988). Feral and domestic ungulates continue to catalyse the world's biota and pasturize its forests! The cascading demise of the native biota of Laysan following the introduction of rabbits remains one of the best-known examples of the detrimental effects of alien species in an island ecosystem (Bryan, 1926; Tomich, 1986).

Similarly in New Zealand, more than 48 species of mammals and 130 species of birds have been introduced (excluding zoo animals) by human agency (Thompson, 1922). Fortunately, not all established, but those that did have had a devastating effect, both directly and indirectly, on the terrestrial invertebrate fauna. Furthermore, the effects are continuing as a steady state has not yet been reached. Although it has been known for some time that rodents have a deleterious effect on island ecosystems, the profoundness of their impact has only recently been realized (Ramsay, 1978b; Bremner *et al.*, 1984). They have eliminated an almost unknown fauna, the remnants of which now persist only on a few rodent-free offshore islands and in alpine habitats above the bush-line. The impact of rodents on the biota (including invertebrates) of Lord Howe Island, where they went ashore from a wrecked yacht in 1918, is well documented (King, 1984). Before 1964 rats gained a foothold on Big South Cape Island, New Zealand, using the mooring line of a fishing boat and devastated the biota, exterminating a large flightless weevil (Fig. 5) (Watt, 1975) and possibly other insect species. The impact of rodents on the British fauna has never been considered to our knowledge, but undoubtedly there must have been effects when the Romans introduced the house mouse, the crusaders brought back the ship rat in the twelfth century and the Norway rat arrived during the eighteenth century (King, 1984).

Invertebrates also can be terribly destructive. Over 2500 alien invertebrates are believed to be established in Hawaii. At least 25% were deliberately introduced for supposedly useful purposes (Swezey, 1931; Funasaki *et al.*, 1988). The others were presumably inadvertently introduced as stowaways. Some, both intentionally and unintentionally introduced, have invaded virtually all habitats from sea level to the summits of the highest mountains, have become pests, and threaten native species (Howarth, 1983, 1985; Gagne and Howarth, 1985).

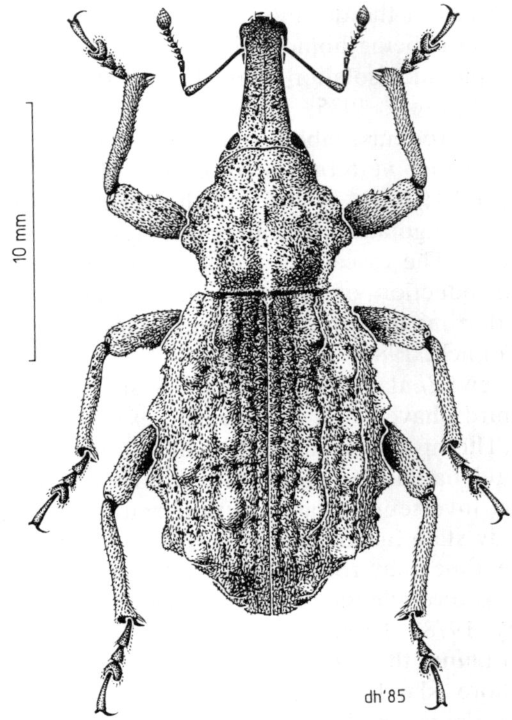

Fig. 5. *Hadramphus stilbocarpae* Kuschel (Stilbocarpa weevil). Confined to small rat-free offshore islands in Foveaux Strait, and off Big South Cape Island, and to the Snare Islands. Exterminated on Big South Cape Island by invading rats. A distinct population may still persist near Puysegur Point and islets in Breaksea Sound, Fiordland. Conservation Status: Rare.

Many alien invertebrates feed directly on native plants, affecting their survival and that of associated native faunas. Many alien predators threaten native prey and compete with native predators. Some are toxic to natives. Aliens spread diseases among native species, e.g. mosquitoes vectoring bird pathogens, ambrosia beetles and true bugs transmitting plant diseases. Earthworms and termites drastically alter soil structure and nutrient cycling and probably adversely affect regeneration of native plants. Introduced organisms for mosquito and snail control and escapees from the aquarium trade disrupt freshwater aquatic communities. Even apparently innocuous aliens may be food for alien predators, reservoirs for diseases, pollinators for alien weeds or dispersal agents for aliens thereby keeping these associated alien populations higher, which may inflict greater injury on native ecosystems. Control procedures, including chemical, biological and cultural methods that target alien pests, may be detrimental to native species (Howarth, 1985). A current scheme to eradicate four alien tephritid fruit flies from Hawaii poses serious risks to the native arthropods and their associated biota. If consessions are made in the promise of success, the programme may become the single most detrimental project to hit the Hawaiian biota.

The number of alien terrestrial invertebrates in New Zealand is unknown but is unlikely to be as high as that of Hawaii. More and more aliens are being recognized as the fauna becomes better studied, however. Some terrestrial invertebrates have reached New Zealand inadvertently through human trade and travel activities, and yet others, including serious agricultural pests, have managed to cross the Tasman Sea on their own. Deliberate introductions for biological control purposes number 351 importations and 225 releases, of which 70 have become established (Cameron *et al.*, 1989). No known extinctions have resulted from these introductions. Some aliens have been destructive to the insect fauna (Ramsay *et al.*, 1988). New Zealand is fortunate in that no native plant has been seriously affected by introduced alien invertebrate phytophages apart from ephemeral epidemics of the manuka blight (Hoy, 1954).

Social and colonial insects historically have had far greater adverse effects than most other invertebrates (Howarth 1985). These include the social termites, ants, wasps and bees and the colonial aphids, mealybugs and their relatives. Ants, especially those with large aggressive colonies, have been strongly implicated in the extinction or extirpation of many native species (Zimmerman, 1948; Solem, 1976; Lubin, 1984; Medeiros *et al.*, 1986). Several species of dangerous ants, including *Pheidole megacephala*, have become established in New Zealand, and their possible spread is a potential time bomb for leaf-litter invertebrates.

In 1977 an aggressive race of the yellow jacket, *Vespula pensylvanica*, became established in Hawaii and quickly spread throughout suitable habitats on all the main islands. Its phenomenonal population explosion and spread corresponded with an alarming decline in many native arthropods and may even have affected native bird numbers (Gambino *et al.*, 1987). Similarly, *Vespula vulgaris* and *V. germanica* have invaded New Zealand and are almost certainly having a severe but unmonitored impact not only on terrestrial arthropods through predation but also on the ecosystem as a whole by pre-empting the honeydew resource of the southern beech forests (Moller *et al.*, 1987).

Many insect parasites and predators purposely introduced for biological control of pest species have expanded their diets to include native species and even alien herbivores introduced to control weeds. Biological control introductions were implicated in the extinctions of at least 15 species of native Hawaiian moths, five of which were actually the targets of the control programme (Gagne and Howarth, 1985).

The oft cited (e.g. see DeBach, 1974) classic example of successful biological control, that of the levuana moth, *Levuana iridescens*, on Fiji by the purposeful introduction of the tachinid parasite, *Bessa remota*, from Malaysia in 1925, was described in great detail by Tothill *et al.* (1930). The last authentic specimen of this endemic monotypic genus of Zygaenidae was collected in 1929 (Robinson, 1975). The species declined from being a widespread local pest to endangered in only 2 years, and its demise is probably the best documented study of extinction among the insects. Another unrelated zygaenid, *Heteropan dolens*, was extirpated from Fiji at the same time (Robinson, 1975). Whether the fly, which remains common on other hosts in Fiji, caused the extinction of other native Fijian Lepidoptera is not known, because pre-release surveys were not conducted. Other documented extinctions caused by biological control introductions include land snails by *Euglandina rosea* (Murray *et al.*, 1988; Hadfield, 1986), aquatic organisms by mosquito fish (Minckley and Deacon, 1968; Moore and Gagne, 1982; Murdoch and Bence, 1987) and vertebrates by mongooses (Honegger, 1981). Even though purposeful introductions under the aegis of classical biological control have been attempted against only a small percentage of the world's pest species, the accumulating evidence indicates that classical biological control introductions have been a major factor in species extinctions, and the environmental damage is at least comparable to the much more widely used chemical control programmes (Howarth, 1983, 1985). Ironically, little consideration is being given to the impacts on non-target organisms by biological control practitioners. Entomology students, administrators and the public are still being told that classical biological control is risk free to the environment (DeBach, 1974; Anon., 1988).

Alien plants have also altered habitats in Hawaii. Agricultural conversion, including croplands and forestry plantations, have virtually replaced native ecosystems in many places. More insidious, if not more serious for conservation is the penetration of the remaining native habitats and reserves by alien plant species (Wodzicki, 1981; Smith, 1985). These plants are having a profound effect on the native biota by competing with native host species, changing the leaf-litter communities, and supporting invasive alien animals, so that in many habitats most of the ecological processes are being carried out by alien species (Howarth, 1985). In New Zealand there are as many alien plant species established in the wild as there are natives (Webb *et al.*, 1988), and in Hawaii nearly as many (Wagner *et al.*, 1990). In contrast, although Nihoa was subjected to intensive utilization by Polynesians off and on for 700 years (Cleghorn, 1988), few alien species were introduced, and a diverse endemic biota survives.

B. Alteration of Habitats by Land Use Changes

Natural processes, such as vulcanism, erosion, subsidence and sea level changes all affect native island species, but human activities have hastened the pace of change by modifying the land for agricultural, horticultural, forestry, industrial (including mining and mineral exploitation), urban, recreational and military purposes and by means of fire, drainage of wetlands, impoundments, land reclamation, channellization of rivers, pollution from waste disposal, use of pesticides, fertilizers, and so on.

Conversion of land to agriculture is a continuing process. The current land tax structure in Hawaii favours clearing native forests for pasture, plantations, wood chipping and other commercial ventures. Some prime native habitats have recently been cleared for questionable economic reasons. Recent experience in New Zealand has been the same — cheap loans have enabled farmers to clear and develop land, much of it under native vegetation. After such land is developed once, it is commonly abandoned as uneconomic. However, changed government policies, especially the removal of subsidies, have altered this even to the extent that the Minister of Finance has been described as a great conservationist! In Hawaii too recent changes in the local government portend more enlightened environmental policies.

Many Hawaiian insects are extremely host specific. In some groups related sympatric species even divide up the resources of a single host species, e.g. *Drosophila* (Montgomery, 1975), Cerambycidae (Gressitt, 1978) and many Lepidoptera and Homoptera (Swezey, 1954). As their host plants become rarer and more scattered, these host-specific species become extremely vulnerable to extinction. Most native leafhoppers (Cicadellidae) and planthoppers (Cixiidae and Delphacidae) remain undescribed, and many may be going extinct without any documentation. Some members of the endemic moth genus *Mapsidius* are of special concern because their only known hosts (*Charpentiera* species) are now extremely rare. Host specificity is also high in New Zealand (more than 60% in some insect families) (Dugdale, 1975), but, although some are extremely rare because of scarcity of their host, none are known to have become extinct through loss of host.

The desertification of Kahoolawe shows how loss of habitat can lead to cascading extinctions and loss of benefit to humans. Kahoolawe is the eighth largest Hawaiian island (116 km^2) and until the last century supported human settlements. Introduction of ungulates, especially sheep, destroyed the forest cover and rain, instead of percolating, washed the soil into the sea. Springs and permanent streams dried up, and the island no longer supports either a human population or much of its native biota. Other Pacific islands are precariously close to similar fates as watersheds are compromised by land clearing and other impacts. Goats and other herbivores, including the Australian brush-tailed possum in New Zealand, are particularly serious in this respect (King, 1984; Loope *et al.*, 1988; Stone and Scott, 1985).

Fire, wood chipping for biomass energy and for pressed board manufacture, silviculture, mining, pollution, military manoeuvres, including practice bombing and resultant fires, and powerline construction also take their toll. Freshwater aquatic habitats are being modified by impoundments, water diversion, stream channellization, land reclamation, drainage, watershed destruction, pollution and alien species.

The military and multinational corporations are responsible for much of the destruction. Until recently the clear-felling of large areas of prime indigenous forest for the Japanese wood-chipping industry in several parts of New Zealand proved difficult to prevent as so much of the forest is privately owned.

C. The Biodiversity Crisis

A critical problem facing conservation biologists is the taxonomic impediment (Wilson, 1985; Ramsay, 1986), which results from both the lack of knowledge concerning invertebrates and the lack of trained personnel to obtain that knowledge. For both Hawaii and New Zealand only 50% of the native insect fauna is estimated to have been described, but the level of knowledge is uneven among the various taxa. Other islands are probably much less well known, e.g. Henderson Island, for which less than 5% of the native insects are known to science (Fosberg *et al.*, 1983). This deficiency hinders the development of effective conservation strategies. Conservation programmes can only be as good as the systematics research upon which they are based (Wilson, 1985). Whether a population under study is alien or endemic, or whether it consists of one or 10 or 100 closely related species all bear on management recommendations.

The status of major portions of the Hawaiian macrolepidoptera and indeed much of the native arthropod fauna is impossible to assess because of the incomplete state of the biological survey (Gagne and Howarth

1985). Some Hawaiian groups are so poorly known systematically that they cannot be determined at this time, and their management is largely ignored. For example, the three major terrestrial crustacean groups, crabs, amphipods and isopods, were only recognized as components of the native fauna within the past two decades, and they remain largely unstudied. Unfortunately the terrestrial crabs are all now extinct. Among the insects, the native leaf bugs (Miridae) total almost 50 described species, but a revision of only one genus (*Nesiomiris*) by Gagne (in preparation) will double that number. At least 200 new species wait in collections, and the group is undoubtably the largest native hemipteran family in the islands. In virtually every native group from Hawaii to New Zealand that has been studied using modern methods, new species are recognized.

There is an assumption that species are discrete units in nature, but the more critical our studies become the less support such a thesis engenders. Current reserve management, with its emphasis on rare and distinct species, may mean our understanding of evolution will be an artificial result of biased management rather than good science, as the resources to prove the myth wrong could be lost. The lesson for conservation biology is that the more populations of endemic species we can preserve for future studies, the greater their combined value is to science and the better our understanding will be of evolutionary ecology and conservation biology.

D. Ecological Myths

Several biases and myths prevent biologists from asking the right questions in research programmes and thereby prevent them from finding the appropriate answers. These ecological myths stem from our ignorance of ecological processes, the parochial development of biology in man-disturbed north temperate continental biomes, and our religious and cultural biases. These myths include beliefs concerning the concept of discrete species, the relative unimportance of invertebrates in ecosystem functioning, ecosystem stability and vulnerability to disturbances, alien invasiveness, and community ecology and ecosystem functioning in temperate versus tropical areas and on islands versus continents. Also, if the responsible group believes no problem exists, very likely no action will be taken. Lack of taxonomic and ecological information is not unique to tropical islands. There was insufficient information available to save the large blue butterfly (*Maculinea arion*) from extirpation from Britain even after 200 years of observation and study (Wells *et al.*, 1983).

The role of invertebrates in ecosystem functioning is much less well understood than is the role of vertebrates. In part, this results from our anthropocentric bias by which we are able to recognize the activities of larger organisms, especially those most familiar to us, while smaller, less familiar creatures go unnoticed until the damage is done and the causes obscure. Given their dominance in terms of number of species and biomass, the insects must play a profound role in ecosystem functioning (Wilson, 1987), especially on isolated islands where many vertebrate groups are absent and where a few invertebrate groups are disproportionately well represented.

Although community ecology and ecosystem functioning operate under the same rules in both temperate and tropical areas, different geologic, climatic and biotic histories may make conclusions from studies in one region inappropriate for others. For example, a myth still perpetuated in biology texts is that insects are rare in marine habitats. That may be true in north temperate regions where these books are prepared, but beaches and shallow marine habitats in the tropics sometimes teem with insects. In fact, nearly half of the orders are represented; yet marine biologists do not study insects, and entomologists generally conduct their studies above the high tide mark. Thus, the myth persists, and the marine insect fauna is largely ignored in conservation programmes! There was also a myth that "barren" habitats, e.g. lava flows, mountain tops and caves, on young islands are too ephemeral and too young to support native communities. All of these habitats are now known to harbour native arthropod communities on even the youngest Hawaiian Island (Howarth, 1987). Phreatic habitats also have been recently shown to support diverse communities in New Zealand and elsewhere. Clearly, one must assume nothing in evolutionary biology but must, as the early philosophers admonished, "open the horse's mouth and count the number of teeth".

The ability to resist perturbations has been viewed as an inherent characteristic of the ecosystem (Vitousek, 1988). Evolutionary theory predicts that native organisms are admirably adapted to their environment. Extinctions result from populations being unable to adapt to changes or disturbances to which they have had no prior experience, that is, a novel perturbation (Howarth, 1985; Perry *et al.*, 1989). Populations that experience and survive specific repeated perturbations are under strong natural selection pressures to cope with the perturbation or go extinct. One must define the perturbation as it affects the community under study not as it affects human values or past experience of the researcher. For example, fire can be either a severely detrimental novel perturbation in some ecosystems or a necessity for maintaining a fire-climax community. The absence of a regular seasonal flood or the occurrence of a minor flood at an unusual season may be more damaging to a natural flood plain community and some cave communities than would be an unusually severe one at the appropriate time. Other ecosystems characterized by natural disturbances include those that regularly experience lava flows, land slides, high surf, frosts, storms and other climatic phenomena. These naturally perturbed sites on islands are largely still recolonized by native pioneers.

When native Hawaiian forests are severely disturbed, they are often replaced by alien species. Outwardly, it appears that island ecosystems are inherently fragile and cannot compete with alien species. However, the replacement results from the interplay of many factors. Four major factors are the following:

1. A novel perturbation starts the process.
2. The novel perturbation often continues to act, such as goat or cattle grazing, which favours alien pioneers.
3. Aggressive pioneer species are more likely to be introduced than others, as these have adapted to human-disturbed ecosystems and travel well.
4. Ecological processes and mutualistic associations (such as soil mycorrhiza) may be disrupted, making recovery a long successional process (Perry *et al.*, 1989), and, where the ecosystems are small, a seed source may not be available, following a large disturbance.

Research under the International Biological Programme demonstrated that native island species were not inherently "fragile" compared to invading aliens and that native communities could recover under protective management when the novel perturbations were removed or mitigated (Mueller-Dombois *et al.*, 1981; Stone and Scott, 1985). However, small population sizes and the resultant small number of refuges, the severity and frequency of the novel perturbations and the close interrelationships within native groups act in concert to increase their vulnerability to extinction (Gagne, 1988). A larger area or greater habitat complexity often allows more species to coexist since at least some individuals within each population are able to find refuges from which they can continually recolonize the main habitat. This may explain in part extinctions in freshwater aquatic habitats and on islands since in both, with their smaller populations and limited immigration rates, predator–prey encounters can run their course to extinction (Murdock and Bence, 1987). Human activities, agroecosystems, land use changes and the establishment of alien, human-associated biota represent a plethora of novel perturbations, especially on islands where humans only recently arrived (in evolutionary time). These factors are largely responsible for the perceived fragility of island ecosystems.

Another myth concerns the concept of empty niches in relation to invading alien species. "Niche" is an ecological concept, i.e. a functional role in an ecosystem, and, therefore, there are few if any empty niches in natural functioning ecosystems. In fact, the evidence from Hawaii suggests that newly formed ecological niches fill in ecological time, like succession (Mueller-Dombois and Howarth, 1981; Howarth, 1987), rather than in evolutionary time as is commonly assumed. At the same time few if any niches can be completely full. Species packing (i.e. the number of species that can share a given resource) is a complex, dynamic phenomenon and unrelated to the colonizing ability of an alien species. In contrast to the long accepted theory, the invasive ability of an alien is dictated by how close its genetic repertoire matches the new environment rather than to any inherent fragility or to the presence of "empty" niches within the invaded community. With the establishment of each new alien, the chance of its associated biota becoming established becomes greater. In fact, as more species become established, it becomes more probable that additional species will find a niche; thus the chance of an alien species becoming established is directly proportional to the number of species already present (Mueller-Dombois and Howarth, 1981; Howarth, 1985) rather than inversely proportional, as is usually assumed.

Also well entrenched is the myth that island ecosystems are inherently fragile. The extreme view that the native Hawaiian biota was going extinct, and nothing could be done to save it (Perkins, 1897), hampered conservation efforts for more than half a century and still is expressed even within some conservation groups. As a result, well-meaning public and private agencies introduced alien species to replace the doomed native forests, an example of a self-fulfilling prophecy. However, in the wake of even the most severe perturbation, there will be some native species capable of surviving and exploiting resources in the altered ecosystem. Even in downtown Honolulu, outwardly like other urbanized areas, endemic Hawaiian insects still persist, and a few are successful, e.g. the biting midge *Forcipomyia hardyi* and the pyralid moths, *Mestolobes miniscula* and *Omiodes blackburni*. Some native Hawaiian insects have become agricultural pests, most notably the pyralid moths, *Omiodes* spp., which attack sugar cane and coconut palms. In New Zealand two indigenous insects, the scarabaeid grass grub and the hepialid caterpillar, porina, have become serious pests of pasture. Virtually any novel perturbation in a community will enhance some resident species while annihilating or reducing others.

An alternative myth states that no insect is rare if it is sought after in the right place at the right time. This is an assumption made by many collectors and pest control strategists to justify their activities. If only it were true! Many immigrant pest species on islands have been extirpated for the same reasons that the native species have declined. There is a temptation to collect rare and unusual species, and unfortunately, at least in New Zealand where there is a tendency for the interesting species to be long lived, the taking of specimens may seriously affect their populations. To date there have been few known problems caused by human collecting, except for certain *Powelliphanta* snail populations in New Zealand (Powell, 1946; Meads *et al.*, 1984). However, the potential for problems to occur is very real, where only a few remnant populations exist or where the demand for specimens develops as a consequence of increased knowledge and interest in the fauna. In Hawaii the native tree snails were subjected to intense collecting pressure in the nineteenth and early twentieth centuries, which Hadfield (1986) considered a significant factor in their decline and extinction. On the other hand, populations of the Society Island tree snails, *Partula* spp., remain robust despite commercial exploitation for jewelry; yet on Moorea and Tahiti they face extinction from the purposely imported predatory snail, *Euglandina rosea* (Murray *et al.*, 1988).

E. Cultural Biases

Many cultural biases severely hinder development and implementation of conservation action programmes. These biases include lack of appreciation for invertebrates, native species and the natural world in general, and human greed, materialism, a belief in unlimited resources and reliance on quick, short-term technological fixes for solving environmental problems. Other problems are public apathy, lack of political and financial support, antagonism from resource exploiters, and a general lack of appreciation that sustainable use of natural resources is really the only long-term option we have if our civilization is to survive.

The public's low esteem for invertebrates and fear of "creepy crawlies" is exploited by the advertising industry's (Fleet Street and Madison Avenue) message that the only good bug is a dead bug. Their message that all insects (equals "invertebrates" to the public) are harmful and should be killed "dead" is so persuasive that even many applied entomology textbooks imply that all predators of insects are "beneficial" regardless of their prey. Our dependence on invertebrates for survival is mostly ignored (Wilson, 1987).

Western culture bestows little appreciation for the natural world. Many islands, including Hawaii, lack an environmentally aware constituency able to act on behalf of natural systems (Gagne, 1988). This low esteem for native species versus alien species is reflected in the term "exotic", which has come to mean both something foreign, non-native, as well as something wondrous. Western culture is often entranced by the exotic. The problem is greatly magnified in Hawaii because of Hawaii's great ethnic diversity. Each ethnic group has introduced its own culturally important organisms, i.e. those considered "useful", exotic or aesthetic. Some of these introductions may be harmful to the native biota, and other harmful organisms may ride as hitchhikers or find suitable hosts when they finally arrive by other means.

Conservationists and land managers sometimes appear overwhelmed by the tasks of understanding, assessing the status of, and conserving invertebrates, and they pray that a sufficient number of interesting invertebrates will survive in reserves established for other reasons. However, many native invertebrates may not reside in such reserves, and, unless their status is monitored, even the invertebrate species included in reserves may be lost. There is an unfortunate political view that once a species is successfully established in captivity, there is no longer a need to save it in the wild.

The belief that anyone can exploit any natural resource without having to account for future consequences is based, in part, on the assumptions that natural resources are infinite and there for our use. This materialistic point of view sometimes leads to antagonism between exploiters and conservationists. That economic development must always be in conflict with conservation goals is, of course, a myth, if the goal is to optimize the returns.

Governments often base land use decisions on biased economic analyses favouring the few most visible resources that are important in national and international trade, i.e. short-term profits and commercial activities, rather than a more complete and accurate analysis of the value to sustaining the local human economy. Often the economic value of a forest is higher under sustainable development of its resources than its value under exploitation for its commercial products (Peters *et al.*, 1989).

Some biases also affect political and legal actions. As a consequence of its cultural heritage, the United States government can control populations of larger vertebrates on private lands but has little such legal authority for plants and other organisms. Quarantine regulations are often based on the misguided assumption that all plants are useful.

There is an almost child-like belief in the ability of science and technology to solve the world's problems. This belief reinforces the reliance on quick, short-term fixes for environmental problems. The loss of credibility when it is realized that this is not so could have serious deleterious repercussions. We choose not to notice that we have already destroyed much of the world's living resources.

IV. RECOMMENDATIONS AND SOLUTIONS

Hawaii, New Zealand and many other oceanic islands have lost so much of their native biota since the arrival of humans that it may seem that any conservation action is too little and too late. However, recent surveys confirm that there is an abundance of native species left to be saved and that islands are important natural laboratories for studies in evolution, ecology and conservation biology (Simon and Sugden 1987; Ramsay *et al.*, 1988; Daws, 1988). Recent studies on islands have also demonstrated that the native biotas do recover under protective management (Mueller-Dombois *et al.*, 1981). Solutions include the development of more effective research programmes, education programmes, reserve management, legal and legislative actions, quarantine programmes, review procedures to curtail the introduction of alien species, environmentally sound pest-control programmes, and monitoring programmes.

A. Research

Extinction studies have been done in hindsight after the game is lost. On islands we still have the opportunity to study declining populations in the field, but the need is urgent. Several native Hawaiian groups recently disappeared alarmingly fast on some islands: i.e., *Megalagrion* damselflies, *Dyscritomyia* flies and *Achatinella* snails on Oahu, Collembola, *Odynerus*, Sphecidae and *Hylaeus* (*Nesoprosopis*) on most islands. Some still have good populations on one or more islands and present good opportunities for conservation biology. In New Zealand the few offshore islands that remain rodent- and possum-free also provide this opportunity. By long-term monitoring surviving populations there is an excellent chance to recognize the beginning of decline and its causes and to develop ameliorating strategies before the problem spreads.

Long-term ecological research is needed both to distinguish short-term population fluctuations from the longer term irreversible changes in ecosystems, and to revise several myths in ecology concerning ecosystem stability and vulnerability to invasions and other perturbations. Conservation problems often proceed more quickly on islands than in continental areas, as well as proceed sequentially within island groups. These circumstances provide the opportunity to conduct comparative research on the effects of different management strategies. Ecologists will be better able to solve the analagous problems on the continents by using solutions first worked out on islands.

The correct ecological questions must be investigated in order to understand ecological processes (Perry *et al.*, 1989). For example, biological control practitioners have long assumed that purposely introduced species posed no threat to the environment (Anon., 1988), and therefore environmental impacts were not addressed or recorded in studies of the efficacy of the introduction (Howarth, 1983, 1985; Murdock and Bence, 1987). Most of the evidence of environmental damage has been gathered serendipitously during unrelated studies (e.g. see Murray *et al.*, 1988). However, there is a growing awareness of the potential for ecosystem injury, and public review procedures are being strengthened. Recently in New Zealand the proposed introduction of a mite to control gorse was the subject of an environmental impact report (EIR) and much public discussion (Hill, 1987).

Biosystematic surveys and research are needed to identify all habitats in every ecological area so that appropriate best remaining viable examples of each habitat and ecosystem can be protected. The inventory of biological diversity is woefully incomplete. Only 10% to perhaps 50% of insect species of various islands have been described and are known to science, making it difficult or impossible to formulate effective conservation strategies. All biological information should be backed up with well-documented voucher specimens, which should be deposited and cared for in appropriate dedicated permanent collections. These collections will allow reassessment of the results and allow valid comparisons of changes in the fauna over time. This in turn will help identify the causes of the change (Danks, 1988). We have to demystify taxonomy and focus on the task of determining the biological diversity of the world. Species on islands and in the tropics may be going extinct faster than they are being collected and described, and faster than management plans are being developed to save them.

B. Education

In order to effect long-term solutions, we need to develop strong environmental education programmes that overcome the public's fear of the perplexing array of strange invertebrates, instil an appreciation of the value, aesthetics, right to life and importance of insects and other invertebrates for human welfare, and inspire a higher esteem for native rather than for alien ("exotic") species. Public education programmes are often most effective when reinforced by action, e.g. strict quarantines not only intercept many potential pests but also reduce introductions by impressing upon the public the risks associated with alien species. Education will also enhance the political status of conservation, as well as increasing public awareness of the plight of invertebrates.

There is a need for educational programmes that counter the perceived need to control false pests, e.g. many horticultural problems and cosmetic or trivial crop damage. A good advertising agency could be employed to counter the cultural biases against invertebrates. Currently, the popular advertising message reinforces the cultural fear of invertebrates.

Schools and training colleges need resource management curricula enabling teachers and students to form an environmentally aware constituency, that would act on behalf of native species. Ultimately each school would have the resources to take responsibility for monitoring the environment within its area (e.g. see Brown, Chapter 14).

C. Reserve Management

Preserving biological diversity means protecting in perpetuity as many living populations as possible. The most effective way to do this is to assure that good viable examples of every natural community and ecosystem is represented in reserves. Currently in Hawaii, nearly half (88) of the 180 recognized native communities are not represented in reserves and are globally endangered (Gagne, 1988). Although it is desirable to keep management to a minimum, ecosystems are dynamic and change is continual and universal. This change is often the result of human activity and affects not only human environments but also neighbouring natural systems, e.g. acid precipitation, global warming, the creation of culture steppe and especially alien species introductions. Natural changes that may have serious implications for conservation also occur (Gerrish, 1989). Thus some management of nature reserves is essential, and the condition of the reserve should be continually monitored. Ideally, management should create conditions by which a community can take care of itself. Conservationists should continue to exploit the human bias towards saving rare and unusual species. However, for both scientific reasons and conservation goals, it is just as important to save the numerous closely related populations of widespread variable native species. In contrast to much of Europe and parts of the north temperate region, where natural communities have long been modified by humans, many relatively intact natural communities still exist on Pacific islands and elsewhere. The aim on islands is to protect remaining examples of these nearly pristine habitats.

Conservation programmes must first identify the problem, then work out a solution that is geared to the area and situation. For example, reforestation programmes should use local native species; efforts to protect one species should not globally endanger others (e.g. translocating endangered species). The saddle-back, an endangered native New Zealand bird, has been introduced to Little Barrier Island, where the only surviving population of the protected giant weta, *Deinacrida heterocantha* occurs. The bird preys upon this and other insects. In the past the national wildlife refuges in Hawaii were managed "strictly for the birds" sometimes to the detriment of other rare native species. Soil conservation programmes, land use changes and recreational programmes in parks and reserves should not put the survival of the indigenous ecosystem and its included native species at risk. Recognition that the threats to natural communities result from novel impacts rather than some inherent fragility of native organisms should improve our understanding of ecology and should allow better ability to predict the effects of invading organisms and mitigative actions in natural systems and to the development of better management programmes.

We also need to rehabilitate endangered ecosystems. Management actions must mitigate not only the perturbation but also the injury or damage. The latter may include disrupted associations and processes, which must be identified and re-established (Perry *et al.*, 1989). New Zealand has had considerable success with eliminating introduced vertebrates from islands (Merton, 1978), particularly rodents, for which new techniques have been developed (Taylor and Thomas, 1989).

Many offshore islets have thus far escaped some perturbations that affect native populations on neighbouring islands and harbour relict populations of insects no longer found on the main islands (Bremner *et al.*, 1984). These populations give ecologists a second chance to save them, as well as provide baseline information for assessing change in the rapidly disappearing lowland communities. They also provide opportunities for translocation of threatened species. For example, a rodent-free island in New Zealand was used successfully to establish a third population of the protected giant weta *Deinacrida rugosa* (Fig. 3) (Meads, 1988). Even though Kahoolawe has undergone profound changes, many unique insect species still survive there. Thus offshore islands can play important roles in the conservation of island species. In many island groups, biological surveys of these islets need to be completed and conservation plans implemented.

To solve the taxonomic impediment Ramsay (1986) suggested that a few appropriate invertebrate indicator species be determined for each kind of habitat. These indicator species should be chosen on the basis of ease of sampling and recognition as well as their vulnerability to disturbance (New, 1984). Aids to their identification can be prepared so that managers will be able to appreciate and monitor the health of the invertebrate populations within the reserves. For example, the high altitude aeolian ecosystem on Mauna Kea on Hawaii Island is controlled by substrate characteristics. The most vulnerable substrate is composed of loose accumulations of volcanic cinders, and the wekiu bug, *Nysius wekiuicola*, is a conspicuous resident of this habitat (Howarth, 1987). Monitoring both the status of the cinder habitat and the population of the bug would enable diagnosis of the health of the whole community.

Realistic collecting policies for invertebrates need to be implemented. With the need for invertebrate surveys and biosystematics studies remaining among the most pressing conservation priorities in most habitats, regulated specimen collecting must be encouraged in all but a few cases.

Captive breeding programmes can be developed for invertebrates where necessary and appropriate. Rearing methods for invertebrates lag far behind those for vertebrates, but such programmes can provide useful biological information for determining some of the factors involved in their decline and managing related species in the wild, as well as providing opportunities for public education concerning the conservation of invertebrates (Collins, 1990).

Although we dislike to have to face triage and would recommend that conservationists tackle whichever projects they feel they can win, some campaigns seem more important than others in the larger endeavour to protect biological diversity. The amount of resources sometimes given to token projects to save a species no longer living in the wild or retaining its genetic diversity, for example the Hawaiian crow, would, in part at least, be better spent on invertebrate surveys and habitat monitoring programmes.

D. Legal and Legislative Actions

Governments have a responsibility to the international community to protect their native biota and must ensure that there is effective coordination among public agencies empowered to regulate the use of natural resources (Collins, 1987b). For example, conflicting mandates and inadequate budgets restrict the effectiveness of the US Fish and Wildlife Service and the National Park Service in carrying out their important function to protect native species. In New Zealand, some of the conflicting mandates have recently been resolved with a reconstitution of environmental responsibilities into a single Department of Conservation.

All decisions on resource and land use changes should be based on sustaining the living resources. Government agencies must base economic analyses and land use decisions on optimum benefit to sustaining the local economy, rather than on maximum profits from the few most visible short-term national and international values, as is currently done (Peters *et al.*, 1989). For example, the experience from Kahoolawe, where short-term benefits from sheep grazing destroyed the island's capacity to support both human populations and many of its native species, illustrates that for islands in particular, in addition to their biodiversity, the greatest value of the remaining native forests may be as watersheds. On Oahu Island, public access to the forests above Honolulu is restricted to protect these important watersheds from direct human disturbance, but, ironically, feral pigs roam freely. In essence, Honolulu's watershed is currently being "managed" by pigs. Sport hunting is generally in conflict with species conservation on islands because nearly all game animals are alien species, and environmental damage to native communities occurs at any population level (Gagne, 1988).

The establishment of legally protected species may give strong administrative advantages. However, legal protection for island invertebrates lags far behind efforts devoted to vertebrates and some plants. In New Zealand, 26 species of terrestrial arthropods and at least three genera of terrestrial molluscs are officially recognized as "animals" and thereby granted protection. In Hawaii, all members of the snail genus *Achatinella*, which is endemic to Oahu Island and which contained some 40 species (of which less than 19 survive), are the only Hawaiian invertebrates officially listed as endangered and protected under the US Endangered Species Act. The no-eyed, big-eyed hunting spider, *Adelocosa anops*, and the blind cave amphipod, *Spelaeorchestia koloana*, have been proposed as endangered and threatened, respectively, but action has been indefinitely delayed. Another 300 species are currently under review (Anon., 1989). Legally protected species provide flagships for research and management actions as well as some funds for habitat conservation, and additional species under threat should be listed. Also threatened habitats should be identified and given legal protection. More appropriate regulations on invertebrate conservation need to be developed and enforced. Administrative deadlines for planning and accomplishing conservation goals should be established and adhered to.

New Zealand is at present reviewing its legislation relating to resource use (Resource Management Law Reform (RMLR)). Conservation, including the concept of sustainability, is being given proper priority. In Hawaii too, there are some hopeful signs that conservation priorities are changing (King *et al.*, 1989).

E. Controlling the Introduction of Alien Species

The most pressing conservation problem facing island insects, and eventually continental species also, is the continual introduction of non-native species of plants and animals. Thus, more effective quarantine measures and improved review and regulatory procedures for proposed alien introductions, including those for biological control, zoological and botanic gardens, private collections, pet shops, plant nursery businesses and commercial ventures, are desperately needed. Society must discourage alien introductions in principle. Persons proposing to introduce alien species undertake a serious burden, and those responsible must demonstrate convincingly in an objective public forum that the alien species poses miniscule potential risk to native species, human health and the economy. Long-term monitoring studies are needed for all purposefully introduced species to determine the true fate of these species and how they affect their ecosystems.

Zoos and botanic gardens, both public and private, have a definite beneficial role in conservation biology, but they also impose risks:

1. Escapees from both public and private botanic gardens remain one of the major conservation problems on islands (Gagne, 1986). Botanic gardens remain full of major potential weeds, yet few horticulturalists realize that some of their ornamental flowers have the capacity to escape cultivation and establish in the wild.
2. Captive breeding programmes for plants and vertebrates have yet to protect any of the invertebrates associated with the captive species. Where local native plant and vertebrate species are being propagated within their historic range, their associated invertebrates should be included in the captive breeding programme.
3. Conservation biologists cannot put all their eggs in one basket. We must counter the expedient view that once a species is successfully established in captivity, there is no need to save it in the wild.
4. Botanic and zoologic gardens are not long-term solutions, as institutions often shift priorities with changing times and staff. Captive breeding is at best a stop gap measure. For example, Wahiawa Botanic Garden on Oahu Island once contained a number of rare native plants, some now nearly extinct in the wild, but several staff changes and reduced funding have changed the garden to a public park with alien ornamental plants and concurrent little regard for native species. Professional labels, marking some of the spots where one could see the last individual of a native tree, still attest to the past aspirations of the garden. Such gardens should have a charter or legal responsibility that binds them to conservation.

F. Pest Control Programmes

Oversight procedures for determining pest status should separate false pests from true pest species, as well as minimize conflicts among different interests. Conflicts may arise between both what control methods are used, such as impacts on non-target organisms and chemical pollution (Jepson, 1989), and what constitutes a pest, such as lantana and melastomes used as ornamentals while being considered noxious weeds by others. Control methods must be appropriate, environmentally sound, and target only certified pests. Often both real and perceived pest species are controlled by whatever arsenal seems expedient, with detrimental effects on the native biota. There are no panaceas in pest control; any action to kill or limit one species will also adversely affect associated species. By their very nature eradication proposals pose serious threats to the environment and should be treated with caution.

Although often attractively economical in the short term, classical biological control, which is largely irreversible, unpredictable and self-dispersing, and which has serious environmental risks, should be used only for certified pests where sufficient information is available to minimize the risks (Howarth 1983, 1985; Roberts, 1986). Protocols for biological control introductions should include ecological studies on both the target and the agent to prove efficacy and minimize impacts on non-target organisms. A public review process to minimize conflicts of interest and the monitoring of the introduced organisms should be undertaken in a long-term effort to determine the efficacy of the agents as well as their effects on the adopted ecosystems. Classical biological control should not target native species or their close relatives. Classical biological control is now being expanded into tropical biomes in earnest, yet the survey of biodiversity there is still so incomplete that many species of native insects and other invertebrates may go extinct without ever being recognized or catalogued.

G. The Future

Even though the problems concerning invertebrate conservation seem at times to be hopeless, we must be wary of being hoisted by our own petard. Doomsday attitudes among conservation strategists often convey a sense of hopelessness to the public, which then hinders effective action and support. Furthermore, when a predicted crash does not happen, conservationists may lose credibility.

Our society is based on growth and the exploitation of natural resources rather than sustainability of the global ecosystem. It is inevitable that this will eventually reach a sustainable level as far as mankind is concerned, which can be either implemented by planned management or catastrophically imposed. The pressure for maximum and immediate profit at the expense of resources should not be allowed to jeopardize the future of any ecosystem. The choice is ours. We are in the predicament of the Lorax (Seuss, 1971), watching the world being exploited, destroyed and polluted as a consequence of human greed and materialism. The credibility of science and technology is being eroded as they fail to solve the ecological crisis. Who will take on the role of the Lorax and speak for the insects?

Acknowledgements

We thank Drs N. M. Collins and J. A. Thomas for organizing this forum to promote insect conservation, and The Royal Entomological Society of London and sponsors of the meeting for providing support for our presentation. We also thank Drs N. Evenhuis, H. Hoch, C. P. Stone and N. C. Howarth, and H. A. Ramsay for their expert review and helpful comments on the manuscript and Mr D. H. Helmore, DSIR Mt Albert Research Centre, Auckland, for drawing the figures.

REFERENCES

Anonymous (1988). Entomology research initiatives. Council of Entomology Department Administrators, December 1988, pp. 1–11.

Anonymous (1989). Endangered and threatened wildlife and plants; animal notice of review. *Fed. Reg.* **54**(4), 554–579.

Atkinson, I. A. E. (1985). The spread of commensal species of *Rattus* to oceanic islands and their effects on island avifaunas. *ICBP Tech. Publ.* **3**, 35–81.

Bennett, B. G. (1987). Observations on rearing *Karocolens pittospori* (Coleoptera: Curculionidae). *N. Z. Entomol.* **9**, 34–37.

Bremner, A. G., Butcher, C. F. and Patterson, G. B. (1984). The density of indigenous invertebrates on three islands in Breaksea Sound, Fiordland, in relation to the distribution of introduced mammals. *J. R. Soc. N. Z.* **14**, 379–386.

Brindle, A. (1980). The cavernicolous fauna of the Hawaiian Islands 12. A new species of blind troglobitic earwig (Dermaptera: Carcinophoridae), with a revision of the related surface-living earwigs of the Hawaiian Islands. *Pacific Insects* **21**, 261–274.

Bryan, E. H. Jr (compiler) (1926). "Insects of Hawaii, Johnston Island and Wake Island". Bernice P. Bishop Museum Bulletin 31, 94 pp.

Bull, R. M. (1967). A study of the large New Zealand weevil, *Lyperobius huttoni* Pascoe 1876 (Coleoptera: Curculionidae: Molytinae). Unpublished MSc Thesis, Victoria University of Wellington, Wellington, 125 pp.

Cameron, P. J., Hill, R. L., Bain, J. and Thomas, W. P. Eds, "A Review of Biological Control of Invertebrate Pests and Weeds in New Zealand 1874–1987." CAB International Institute of Biological Control (CIBC) Technical Communication 10, xiii + 426 pp.

Carlquist, S. (1980). "Hawaii, a Natural History". Pacific Tropical Botanic Garden, Lawai, Kauai, Hawaii.

Carlquist, S. (1981). Chance dispersal. *Am. Sci.* **69**, 509–516.

Carson, H. L. (1987). Tracing ancestry with chromosomal sequences. *Trends Ecol. Evol.* **2**, 203–207.

Cleghorn, P. L. (1988). The settlement and abandonment of two Hawaiian outposts: Nihoa and Necker Islands. *Bishop Museum Occ. Pap.* **28**, 35–49.

Cobben, R. H. (1980). The Saldidae of the Hawaiian Archipelago (Hemiptera: Heteroptera). *Pacific Insects* **22**, 1–34.

Collins, N. M. (1987a). Invertebrate diversity. *IUCN Spec. Rep. Bull.* **18**(7/9), 3–4.

Collins, N. M. (1987b). "Legislation to conserve insects in Europe." Amateur Entomologists Society Pamphlet No. 13, pp. 1–30.

Collins, N. M. (ed.) (1990). "The Management and Welfare of Invertebrates in Captivity". National Zoo Federation, London.

Coulter, J. D. (1973). Ecological aspects of the climate. *In* "The Natural History of New Zealand. An Ecological Survey" (G. R. Williams, ed.), pp. 28–60. Reed, Wellington, xviii + 434 pp.

Crosby, A. W. (1986). "Ecological Imperialism. The Biological Expansion of Europe, 900–1900". Cambridge University Press, Cambridge, xvi + 368 pp.

Cuddihy, L. W. (1989). Vegetation zones of the Hawaiian Islands. *In* "Conservation Biology in Hawai'i" (C. P. Stone and D. B. Stone, eds), pp. 27–37. University of Hawaii Press, Honolulu.

Dalrymple, G. B., Silver E. A., and Jackson, E. D. (1973). Origin of the Hawaiian Islands. *Am. Sci.* **61**, 294–308.

Danks, H. V. (1988). Systematics in support of entomology. *Ann. Rev. Entomol.* **33**, 271–296.

Daws, G. (1988). "Hawaii The Islands of Life". The Nature Conservancy of Hawaii. Signature Publication, Honolulu, 156 pp.

DeBach, P. (1974). "Biological Control by Natural Enemies". Cambridge University Press, London, 323 pp.

Decker, B. G. (1973). Unique dry-island biota under official protection in northwestern Marquesas islands. *Biol. Conserv.* **5**, 66–67.

Drake, J. A., Mooney, H. A., di Castri, F., Groves, R. H., Kruger, F. J., Rejmanek, M. and Williamson, M. (eds). (1989). "Scope 37, Biological Invasions A Global Perspective". Wiley, Chichester, xxiv + 525 pp.

Dugdale, J. S. (1975). The insects in relation to plants. *In* "Biogeography and Ecology in New Zealand" (G. Kuschell, ed.), Monographie Biologicae Vol. 27, pp. 561–589, W. Junk, The Hague.

Dumbleton, L. J. (1970). Pleistocene climates and insect distribution. *N. Z. Entomol.* **4**, 3–23.

Fleming, C. A. (1979). "The Geological History of New Zealand and its Life". Auckland University Press, Auckland, 141 pp.

Fosberg, F. R., Sachet, M. H. and Stoddart, D. R. (1983). Henderson Island (Southeastern Polynesia): Summary of current knowledge. *Atoll Res. Bull.* **272**.

Funasaki, G. Y., Lai, P.-Y., Nakahara, L. M., Beardsley, J. W. and Ota, A. K. (1988). A review of biological control introductions in Hawaii: 1890 to 1985. *Proc. Hawaii. Entomol. Soc.* **28**, 105–160.

Gagne, W. C. (1986). Hawaii's botanic gardens: panacea or Pandora's box in the conservation of Hawaii's native flora? *Newsletter Hawaii. Bot. Soc.* **25**, 7–10.

Gagne, W. C. (1988). Conservation priorities in Hawaiian natural systems. *Bioscience* **38**, 264–271.

Gagne, W. C. (1991). Insular Evolution and Speciation of the Genus *Nesiomiris* Kirkaldy (Heteroptera: Miridae) in the Hawaiian Islands. *Bishop Museum Bulletins in Entomology*, Bishop Museum Press, Honolulu, in preparation posthumously by colleagues.

Gagne, W. C. and Cuddihy, L. (1990). Vegetation. *In* "Manual of the Flowering Plants of Hawaii" (W. L. Wagner, D. R. Herbst and S. H. Sohmer, eds), pp. 42–92. University of Hawaii Press and Bishop Museum Press, Honolulu.

Gagne, W. C. and Howarth, F. G. (1985). Conservation status of endemic Hawaiian Lepidoptera. "Proceedings of the 3rd Congress European Lepidopterologists, Cambridge 1982", pp. 74–84. Societus Europaea Lepidopterologica, Karlsruhe, 211 pp.

Gambino, P., Medeiros, A. C. and Loope, L. L. (1987). Introduced *Paravespula pensyvanica* (Saussure) yellowjackets prey on Maui's endemic arthropod fauna. *J. Trop. Ecol.* **3**, 169–170.

Gerrish, G. (1989). 'Ohi'a dieback and forest life cycles. *In* "Conservation Biology in Hawai'i" (C. P. Stone and D. B. Stone, eds), pp. 27–37. University of Hawaii Press, Honolulu.

Godley, E. J. (1975). Flora and vegetation. *In* "Biogeography and Ecology in New Zealand" (G. Kuschell, ed.), Monographie Biologicae Vol. 27, pp. 177–229. W. Junk, The Hague.

Gressitt, J. L. (1978). Evolution of the endemic Hawaiian cerambycid beetles. *Pacific Insects* **18**, 137–167.

Hadfield, M. G. (1986). Extinction in Hawaiian Achatinelline snails. *Malacologia* **27**, 67–81.

Hardwick, D. F. (1965). The corn earworm complex. *Mem. Entomol. Soc. Can.* **40**, 1–247.

Hardy, D. E. (1965). "Insects of Hawaii. Vol. 12. Diptera: Cyclorrhapha II, Series Schizophora, Section Acalypterae I, Family Drosophilidae". University of Hawaii Press, Honolulu.

Hardy, D. E. and Delfinado, M. D. (1974). Flightless Dolichopodidae (Diptera) in Hawaii. *Proc. Hawaii. Entomol. Soc.* **21**, 365–371.

Hill, R. (1987). "The Biological Control of Gorse (*Ulex europaeus* L.) in New Zealand: an Environmental Impact Assessment". D.S.I.R., Auckland. 56 pp + 7 appendices.

Hoch, H. and Howarth, F. G. (1989), Six new cavernicolous cixiid planthoppers in the genus *Solonaima* from Australia (Homoptera: Fulgoroidea). *Syst. Entomol.* **14**, 377–402.

Holzapfel, E. P., Clagg, H. B. and Goff, M. L. (1978). Trapping of air-borne insects on ships on the Pacific, part 9. *Pacific Insects* **19**, 65–90.

Honegger, R. E. (1981). List of amphibians and reptiles known or thought to have become extinct since 1600. *Biol. Conserv.* **19**, 141–158.

Howarth, F. G. (1983). Biological control: panacea or Pandora's box? *Proc. Hawaii. Entomol. Soc. 1980* **24**, 239–244.

Howarth, F. G. (1985). Impacts of alien land arthropods and molluscs on native plants and animals in Hawai'i. *In* "Hawai'i's Terrestrial Ecosystems: Preservation and Management" (C. P. Stone and J. M. Scott, eds), pp. 149–179. University of Hawaii Press, Honolulu.

Howarth, F. G. (1987). Evolutionary ecology of aeolian and subterranean habitats in Hawaii. *Trends Ecol. Evol.* **2**, 220–223.

Howarth, F. G., Sohmer, S. H. and Duckworth, W. D. (1988). Hawaiian natural history and conservation efforts: what's left is worth saving. *BioScience* **38**, 232–237.

Hoy, J. M. (1954). Scale insects associated with manuka species in New Zealand. *N. Z. J. Agric.* **89**(6), 601, 603–604.

Hurnard, S. M. (1978). Climatic factors in the seasonality of New Zealand insects: a meteorological viewpoint. *N. Z. Entomol.* **6**, 337–343.

Jepson, P. C. (ed.) (1989). "Pesticides and Non-target Invertebrates". Intercept Ltd, Wimborne, Dorset, xiv + 240 pp.

King, C. (1984) "Immigrant Killers. Introduced predators and the conservation of birds in New Zealand". Auckland. Oxford University Press, Oxford, 224 pp.

King, L. B., Campbell, F., Edelson, D. and Miller, S. (1989). 'Extinction in Paradise: Protecting our Hawaiian species". Mimeo Report, Natural Resources Defense Council, Washington.

Kirch, P. V. (1982). The impact of the prehistoric Polynesians on the Hawaiian ecosystem. *Pacific Science* **36**, 1–14.

Kuschel, G. (1975). Introduction. *In* "Biogeography and Ecology in New Zealand" (G. Kuschel, ed.), Monographie Biologicae Vol. 27, pp. xv–xvi. W. Junk, The Hague.

Kuschel, G. (1987). The subfamily Molytinae (Coleoptera: Curculionidae): general notes and descriptions of new taxa from New Zealand and Chile. *N. Z. Entomol.* **9**, 11–29.

Loope, L. L., Hamann, O. and Stone, C. P. (1988). Comparative conservation biology of oceanic archipelagoes: Hawaii and the Galapagos. *BioScience* **38**, 272–282.

Lubin, Y. D. (1984). Changes in the native fauna of the Galapagos Islands following invasion by the little red fire ant, *Wasmannia auropunctata. Biol. J. Linnean Soc.* **21**, 243–251.

McGlone, M. S. (1989). The polynesian settlement of New Zealand in relation to environmental and biotic changes. *In* "Moas, Mammals and Climate in the Ecological History of New Zealand" (M. R. Rudge, ed.). *N. Z. J. Ecol.* **12** (Suppl.), 115–129.

McLellan, I. D. (1975). The freshwater insects. *In* "Biogeography and Ecology in New Zealand" (G. Kuschel, ed.), Monographie Biologicae Vol. 27, pp. 537–559. W. Junk, The Haque.

Meads, M. J. (1988). "A brief review of the current status of the giant wetas and a strategy for their conservation". Ecology Division (DSIR, Private Bag, Lower Hutt, New Zealand) Report No. 11, 17 pp.

Meads, M. J., Walker, K. J. and Elliott, G. P. (1984). Status, conservation, and management of the land snails of the genus *Powelliphanta* (Mollusca: Pulmonata). *N. Z. J. Zool.* **11**, 277–306.

Medeiros, A. C., Loope, L. L. and Cole, F. R. (1986). Distribution of ants and their effects on endemic biota of Haleakala and Hawaii Volcanoes National Parks: a preliminary assessment. *In* "Proceedings of the 6th Conference in Natural Sciences, Hawaii Volcanoes National Park" (C. W. Smith and C. P. Stone, eds), pp. 39–51. Cooperative National Park Resources Studies Unit, University of Hawaii, Honolulu.

Merton, D. V. (1978). Controlling introduced predators and competitors on islands. *In* "Endangered Birds" (A. S. Temple, ed.), Ch. 15, pp. 121–128. Croom Helm, London.

Minckley, W. L. and Deacon, J. E. (1968). Southwestern fishes and the enigma of endangered species. *Science* **159**, 1424–1431.

Moeed, A. and Meads, M. J. (1986). Seasonality of litter-inhabiting invertebrates in two native forest communities of Orongorongo Valley, New Zealand. *N. Z. J. Zool.* **13**, 45–63.

Moller, H., Clapperton, K., Gaze, P., Sandlant, G., Thomas, B. and Tilley, J. (1987). Honeydew, life blood of South Island forests. *Forest Bird* (246) **18**(4), 14–16.

Molloy, B. P. J. (1970). Evidence for post-glacial climate change in New Zealand. *J. Hydrol. N. Z.* **8**(2), 56–67.

Montgomery, S. L. (1975). Comparative breeding site ecology and the adaptive radiation of picture-winged Drosophila. *Proc. Hawaii. Entomol. Soc.* **22**, 65–102.

Montgomery, S. L. (1983). Biogeography of the moth genus *Eupithecia* in Oceania and the evolution of ambush predation in Hawaiian caterpillars. *Entomol. Gen.* **8**, 27–34.

Moore, N. W. and Gagne, W. C. (1982). "*Megalagrion pacificum* (McLachlan): A preliminary study of the conservation requirements of an endangered species". Report Odonata Specialist Group, International Union for Conservation of Nature and Natural Resources, No. 3. Utrecht, 5 pp.

Mueller-Dombois, D., Bridges, K. W. and Carson, H. L., (eds). (1981). "Island Ecosystems: Biological Organization in Selected Hawaiian Communities". US/IBP Synthesis Series No. 15. Hutchinson Ross, Pennsylvania, 583 pp.

Mueller-Dombois, D. and Howarth, F. G. (1981). Niche and life-form integration in island communities. *In* "Island Ecosystems: Biological Organization in Selected Hawaiian Communities" (D. Mueller-Dombois, K. W. Bridges and H. L. Carson, eds), Vol. 15, pp. 337–354. US/IBP Synthesis Series.

Murdock, W. W. and Bence, J. (1987). General predators and unstable prey populations. *In* "Predation" (W. C. Kerfoot and A. Sih, eds), pp. 17–30. University Press of New England, Hanover.

Murray, J., Murray, E., Johnson, M. S. and Clarke, B. (1988). The extinction of *Partula* on Moorea. *Pacific Science* **42**, 150–153.

New, T. R. (1984). "Insect Conservation — An Australian Perspective". W. Junk, Dordrecht, 184 pp.

Olson, S. L. and James, H. F. (1984). The role of Polynesians in the extinction of the avifauna of the Hawaiian Islands. *In* "Quaternary Extinctions: A Prehistoric Revolution" (P. S. Martin and R. G. Klein, eds), pp. 768–780. The University of Arizona Press, Tucson.

Perkins, R. C. L. (1897). The introduction of beneficial insects into the Hawaiian Islands. *Nature* **55**(1430), 499–500.

Perry, D. A., Amaranthus, M. P., Borchers, J. G., Borchers, S. L. and Brainerd, R. E. (1989). Bootstrapping in ecosystems. *BioScience* **39**, 230–237.

Peters, C. M., Gentry, A. H. and Mendelsohn, R. O. (1989). Valuation of an Amazonian rainforest. *Nature* **339**, 655–656.

Powell, A. W. B. (1946). The Paryphantidae of New Zealand. No. 5. Further new species of *Paryphanta*, *Wainuia*, and *Rhytida*. *Rec. Auckl. Inst. Mus.* **3**(2), 99–136.

Pyle, R., Bentzien, M. and Opler, P. (1981). Insect conservation. *Ann. Rev. Entomol.* **26**, 233–258.

Ramsay, G. W. (1955). The exoskeleton and musculature of the head and the life cycle of *Deinacrida rugosa* Buller 1870. Unpublished MSc thesis, Victoria University College, Wellington, New Zealand, 161 pp.

Ramsay, G. W. (1978a). Invertebrate mice. *N. Z. Entomol.* **6**, 400.

Ramsay, G. W. (1978b). A review of the effect of rodents on the New Zealand invertebrate fauna. *In* "The Ecology and Control of Rodents in New Zealand Nature Reserves" (P. R. Dingwall, I. A. E. Atkinson and C. Hay, eds), pp. 89–97. Department of Lands and Survey Information Series 4.

Ramsay, G. W. (1986). The taxonomic impediment to conservation. *The Weta* **9**(2), 60–62.

Ramsay, G. W., Meads, M. J., Sherley, G. H. and Gibbs, G. W. (1988). Wildlife Research Liaison Group — Research on terrestrial insects of New Zealand. *WRLG Res. Rev.* **10**, 1–49.

Roberts, L. I. N. (1986). The practice of biological control — implications for conservation, science and the community. *The Weta* **9**, 76–84.

Robinson, G. S. (1975). "Macrolepidoptera of Fiji and Rotuma". Classey Ltd, Oxon, UK, 362 pp.

Seuss, Dr. [Geisel, A. S.] (1971). "The Lorax". Random House, New York. Unpaged.

Simon, C. M. and Sugden, A. (eds) 1987. [Hawaiian special issue.] *Trends Ecol. Evol.* **2**, 173–229.

Simpson, P. (1982). "Ecological Regions and Districts of New Zealand. A Natural Subdivision". Wellington, Biological Resources Centre publication ISSN-0111–8625, ii + 63 pp.

Smith, C. W. (1985). Impact of alien plants on Hawai'i's native biota. *In* "Hawai'i's Terrestrial Ecosystems: Preservation and Management" (C. P. Stone and J. M. Scott, eds), pp. 180–250. University of Hawaii Press, Honolulu, 584 pp.

Solem, A. (1976). "Endodontid Land Snails from Pacific Islands (Molusca: Pulmonata: Sigmurethra). Part I: Family Endodontidae". Field Museum Natural History Press, Chicago.

Solem, A., Climo, F. M. and Roscoe, D. J. (1981). Sympatric species diversity of New Zealand land snails. *N. Z. J. Zool.* **8**, 453–485.

Stone, C. P. and Scott, J. M. (eds) (1985). "Hawai'i's Terrestrial Ecosystems: Preservation and Management". University of Hawaii Press, Honolulu, 584 pp.

Swezey, O. H. (1931). Records of introduction of beneficial insects into the Hawaiian Islands. *In* "Handbook of Insects and Other Invertebrates of Hawaiian Sugar Cane Fields" (F. X. Williams, compiler), pp. 368–389. Hawaiian Sugar Planters Association, Honolulu.

Swezey, O. H. (1954). "Forest Entomology in Hawaii". Bishop Museum Special Publication No. 44, Honolulu, 266 pp.

Taylor, R. H. and Thomas, B. W. (1989). Eradication of Norway rats (*Rattus norvegieus*) from Hawea Island, Fiordland, using brodifacoum. *N. Z. J. Ecol.* **12**, 23–32.

Thompson, G. M. (1922). "The Naturalisation of Animals and Plants in New Zealand". Cambridge University Press, Cambridge, x + 607 pp.

Tomich, P. Q. (1986). "Mammals of Hawaii". Bishop Museum Special Publication No. 76, Honolulu, 375 pp.

Tothill, J. D., Taylor, T. H. C. and Paine, R. W. (1930). "The Coconut Moth in Fiji (A History of its Control by Means of Parasites)". Imperial Institute of Entomology, London, vii + 269 pp.

Vitousek, P. M. (1988). Diversity and biological invasions of oceanic islands. *In* "Biodiversity" (E. O. Wilson, ed.), pp. 181–189. National Academy Press, Washington, DC.

Wagner, W. L., Herbst, D. R. and Sohmer, S. H. (1990). "Manual of the Flowering Plants of Hawaii". University of Hawaii Press and Bishop Museum Press, Honolulu.

Watt, J. C. (1975). The terrestrial insects. *In* "Biogeography and Ecology in New Zealand" (G. Kuschell, ed.), Monographie Biologicae Vol. 27, pp. 507–535. W. Junk, The Hague.

Webb, C. J., Sykes, W. R. and Garnock-Jones, P. J., with contributions by Given, D. R., Brownsey, P. J., Macmillan, B. H. and Moore, L. B. (1988). "Flora of New Zealand. Volume IV. Naturalised Pteridophytes, Gymnosperms, Dicotyledons". DSIR Botany Division, Christchurch, New Zealand, xviii+ 1365 pp.

Wells, S. M., Pyle, R. M. and Collins, N. M. (compilers) (1983). "The IUCN Invertebrate Red Data Book". IUCN, Gland, Switzerland, 632 pp.

Wilson, E. O. (1985). The biological diversity crisis: A challenge to science. *Iss. Sci. Technol.* **2**, 20–29.

Wilson, E. O. (1987). Diversity: the little things that run the world (the importance and conservation of invertebrates). *Conserv. Biol.* **1**, 344–346.

Wodzicki, K. (1981). Some nature conservation problems in the South Pacific. *Biol. Conserv.* **21**, 5–18.

Zimmerman, E. C. (1948). "Insects of Hawaii. Vol. 1. Introduction". University of Hawaii Press, Honolulu, 206 pp.

Zimmerman, E. C. (1957). "Insects of Hawaii. Vol. 6 Ephemeroptera-Neuroptera-Trichoptera, and supplement of Volumes 1–5", University of Hawaii Press, Honolulu, ix + 209 pp.

Zimmerman, E. C. (1978). "Insects of Hawaii. Vol. 9. Microlepidoptera". University of Hawaii Press, Honolulu, 1903 pp. in 2 vols.

5

The Conservation of Insects and their Habitats in Northern and Eastern Europe

KAURI MIKKOLA

I. Introduction ... 109
II. The Red Data Lists of Northern and Eastern Europe ... 111
III. Protection of Insects and their Habitats in Northern and Eastern Europe ... 112
IV. Causes of Threat and the Similarity of Species Lists in Finland and Sweden ... 114
V. Conclusions ... 117
References ... 119

I. INTRODUCTION

In order to understand the faunal composition of Northern Europe, it is necessary to recall the biogeographic history of the area. As recently as 20 000 years ago the whole region was covered by a continental ice sheet, several kilometres thick, and no higher insects lived in the area. In postglacial times, i.e. in the last 10 000 years, the climatic development has been uneven, and included some warmer and drier periods than at present. In addition, tree species did not return to the area together or in the present composition, but in a certain order, spruce (*Picea*) being one of the last to arrive.

This glacial history has several consequences for the modern insect fauna of Fennoscandia:

1. There are virtually no endemic species in the area because the time span of about 10 000 years is too short for speciation.
2. The modern fauna is predominantly composed of widely distributed

species with wide ecological amplitude and a good ability for dispersal. This is seen, for instance, in the fauna of Noctuidae: about 70% of the local Finnish species are shared with those of Spain and the Far East (K. Mikkola, unpublished). Of the species in Lapland, about three-quarters are Holarctic in distribution, i.e. common with North America.
3. The alternation of advances and retreats of the glaciers caused a mixing of arctic and alpine elements, and the subsequent cool periods led to a wide distribution of boreal forests in Europe which brought about the spread of boreal and montane elements.
4. The warm dry postglacial periods have left relicts which are now seen as disjunct occurrences of thermophilous species in the southern parts of Fennoscandia.
5. Fennoscandia is bounded by the ocean in the north and west. The faunal reinvasion originated mainly in the east and south-east, or in the south and south-west.

The European part of the Soviet Union, Russia, belonged mainly to the extensive periglacial belt during the glaciation, where dry tundra and steppe-like habitats prevailed. Because the steppes and forest steppes are characteristic of southern Russia, although nowadays mainly transformed to arable land, the existence of endemic species is at least a theoretical possibility.

II. THE RED DATA LISTS OF NORTHERN AND EASTERN EUROPE

The first red data list of insects dates back to the mid-1970s, when the endangered animals and plants of Finland were listed (Borg and Malmström, 1975; Malmström, 1975). A total of 84 insect species was included, of which 44 were Lepidoptera and 26 Coleoptera. A specialists' committee for a Finnish red data book was established by the Ministry of Agriculture in 1983, and the committee finished its work under the Ministry of Environment in 1985 (Rassi *et al.*, 1986). This included 352 species of insects (Table I).

In Sweden, a book about the endangered invertebrates in forestry (Ehnström and Waldén, 1986) included 221 species of insects. A comprehensive list of endangered insects of Sweden was published a year later (Andersson *et al.*, 1987) and included 707 species of insects (Table I).

In Norway, a series of books about rare insects is in preparation. The first volume consists of those belonging to a variety of small insect groups

TABLE I. The number of insects species in the Red Data lists of Finland (F; Rassi *et al.*, 1986), Sweden (S; Andersson *et al.*, 1987) and the Soviet Union (SU; Borodin *et al.*, 1984), and in Volume I of the rare insects in Norway (N; Aagard and Hågvar, 1987; this treats only the groups indicated with a species number).

	F	S	SU	N
Diplura	–	–	1	
Ephemeroptera	3	–	1	9
Odonata	5	5	11	11
Plecoptera	5	–	–	4
Orthoptera (*sensu lato*)	2	5	16	7
Embioptera	–	–	1	
Dermaptera	–	–	1	1
Hemiptera	26	41	5	15 (aquatic)
Neuroptera	4	5	4	14
Coleoptera	141	303	33	
Hymenoptera	27	24	24	
Diptera	5	71	1	
Mecoptera	1	–	–	1
Trichoptera	5	2	–	46
Lepidoptera	128	251	123	
Total insects	352	707	221	

(Aagard and Hågvar, 1987; Table I). It is mainly concerned with distributions and does not classify the threatened species as the Finnish and Swedish lists do. I know of no comprehensive red data list of insects from Denmark.

The Council of Ministers of the Nordic Countries has published a book of threatened animals and plants at intervals of about 5 years. The insects will probably be included for the first time in the next volume, with data also from Norway and Denmark. The Nordic entomological societies have compiled a booklet about the conservation of insects in the different countries (Hågvar, 1988).

Separate red data books for some of the republics of the Soviet Union have been published, for instance for the Soviet Federal Republic in 1983 (Yeliseyev *et al.*, 1983), which included 34 species of insects. The comprehensive red data book of the Soviet Union was published a year later (Table I; Borodin *et al.*, 1984, first edition without insects in 1978). A total of 221 species of insects are included, of which only one-third, namely 66 species, occur in Europe. Most species are from the Caucasus or from the Far East.

Surprisingly, and contrary to western practice, Lepidoptera such as *Papilio machaon* L., *Saturnia pavonia* (L.) and *Catocala fraxini* (L). were included in the Soviet red data book, some even in category II. These are widely distributed species which are not particularly rare, and neither they nor their habitats are generally endangered. It seems likely that their colourful appearance has more influence on their inclusion than the actual threat. A new edition of the book is in preparation, and the principles for including species will be altered, so that species represent ecological or taxonomic groups (A. V. Sviridov, personal communication). However, there will probably be no radical change of the species list other than to include many additional species. In the 1980s, at least two popular books and several volumes with articles about endangered insects have been published in the Soviet Union, some of them criticizing present composition of the red data list.

The new constitution of the Soviet Union, dating from the early 1980s, prohibits the ownership and import or export of insect collections. It has been revealed, however, that members of the All-Union Entomological Society may collect and own collections. It has also been simple to transport insect collections, at least over the border of Finland and the Soviet Union, provided one has a testimony about the material from a Finnish or Soviet museum for the Soviet customs officers. Collecting or taking species listed in the red data book is prohibited under penalty of up to 500 roubles (in 1988 equal to some $700); such sentences have already been given.

III. PROTECTION OF INSECTS AND THEIR HABITATS IN NORTHERN AND EASTERN EUROPE

The first insect to be given legal protection in Fennoscandia was the beetle *Cerambyx cerdo* (L.), in 1918 on the Swedish island Öland. It is noteworthy that the 12 giant oaks on which it lives were also protected. Between 1938 and 1950, protection was extended to four more local insects in Sweden; the beetles *Carabus intricatus* (L.) and *Cucujus cinnaberinus* (Scop.), the wasp *Scolia unifasciata* (Cyr.) and the butterfly *Lopinga achine* (Scop.). Recently, voluntary protection was given to four butterfly species in Sweden, and to one butterfly species in Denmark. No protection is known for Norway.

In Finland, the first insects to be protected were the butterflies *Parnassius apollo* (L.) and *P. mnemosyne* (L.), in 1976. Another butterfly, *Lycaena dispar* (Hw.), was protected in 1983 following the discovery of one small population. In 1989, 22 more insects were protected, nine of them additional butterflies as well as one zygaenid, two sesiid, four microlepidopteran, three geometrid and one noctuid moths, and two coleopterans (*Cucujus cinnaberinus* and *Osmoderma eremita* (Scop.). Although it has been acknowledged that it is not essential to protect these species from collectors in order to conserve their populations, this was considered to be a necessary first step in the conservation of the habitats or, in some cases, their food plants. This is because it is only protected species that receive conservation plans, which include the possibility of protecting their habitats.

A network of nature parks and national parks in the Nordic countries plays an essential role in the conservation of the insect fauna. The total area of reserves in Norway is 1 000 000 ha (3% of the total area), in Sweden 2 000 000 (4.5%) and in Finland 1 000 000 ha (3%). Even if these were selected mainly because of their landscapes, birds, mammals and flora, they also conserve valuable insect faunas.

In a few cases, insects have played an important role in the establishment of a natural reserve. In Sweden three areas were long ago protected in this way, and one sand pit was protected as a habitat for the wasp *Philanthus triangulum* (Fabr.). In Denmark, a few oak forests and heaths containing a rich fauna of insects have long been protected on Jutland. In addition, several wells and creeks were protected because of their valuable fauna of Trichoptera and Plecoptera. In 1984, three localities on Jutland were protected on entomological grounds.

From southern Finland there are, in addition to some older reserves of entomological importance, six recent reserves of 25–85 ha where insects are considered the most important group. Of these, three are bogs, two are broad-leaved forests and one, the so-called Mäntsälä forest, a virgin coniferous forest. In the latter, the beetle *Leptura thoracica* Cr., once classified as extinct, was rediscovered in the 1980s. In 1988, two dolomitic fjeld slopes in north-west Lapland, of 165 and 175 ha and famous for their Lepidopteran fauna, were given legal protection. However, there are still some problems in restricting the grazing of reindeers in these reserves.

Several natural reserves for insects are planned in Finland. These include an area of sand dunes at the tip of the Hanko Peninsula, where the only stable population of the moth *Apamea anceps* (D. & S.) lives, and a 60-ha virgin mixed forest at Sipoo, where the beetle *Athous mutilatus* Rosenh was discovered for the first time in Finland in 50 years. (Both were legally protected in 1990.)

Some of the more general laws and programmes also have a great importance for insects. The Swedish silvicultural law (1985) also contains a rule considering fauna and flora, and of the conservation of the so-called noble tree forests from 1984 prevents their conversion to coniferous forests. The Danish law of nature protection conserves biotopes such as bogs, heaths and shore meadows. In Finland, programmes that conserve peatlands (1979–1981) and luxurious forests (1989) are saving many localities that are important for their insect fauna.

In all the Nordic countries, records and databases are now being compiled of localities containing threatened insect faunas. One example of ecological research work is the current programme of research in southern Sweden on the butterflies *Pyrgus armoricanus* (Oberth.), *Parnassius mnemosyne* (L.), *Maculinea alcon* (D. & S.) and the beetle *Pytho kolwensis* Sahlb. (Pythidae). In Sweden and Finland, provisional experiments have been started to study the food plant ecology and to reintroduce the species to its old haunts of the butterfly *Parnassius apollo* (L.) (Mikkola 1987; Bengtsson *et al.*, 1989).

IV. CAUSES OF THREAT AND THE SIMILARITY OF SPECIES LISTS IN FINLAND AND SWEDEN

In the Finnish and Swedish Red Data lists, the cause of threats has been subjectively estimated. These were scored using the Finnish list (Mikkola, 1987) by giving the first-mentioned (the most important) reason a value of 3, the second 2 and the rest one point (Fig. 1).

The diagrams of the seed plants (Magnoliophyta) and the Lepidoptera are very similar. In both, undergrazing or abandonment of meadows, changes in forestry, building activity, ditching in peatlands and the effects of collectors have similar values, although for plants, changing methods in silvicultural practice has had a greater effect than the change of tree species. Among other insects are many aquatic species; for these, water quality has been the most important factor, while for aquatic plants it is mainly water regulation activity.

For Coleoptera, by far the most important causes of threat are those connected with modern forestry practice, accounting for nearly 75% of all listings. The decrease of rotting wood in forests and the decline in area of virgin forests have had the greatest effect, while the change of tree species has been less significant. For example, in Sweden about 400 000 ha have been planted with the North American tree species *Pinus contorta*.

Statistics about Finnish biotopes explain why so many insects have suffered. For example, the area of forest that is older than 140 years decreased by 350 000 ha from 1950 to 1975, and is now less than 6% of the total, while the percentage of forest dominated by spruce (*Picea abies*) has increased since the 1920s from 28% to 42%. In 1936, cattle were pastured in 46% of the forests, but this has now virtually ceased. Populations of peatland insects have also declined; by 1980, 83% of the peatbogs in southern Finland had been drained as had 39% of the total in northern Finland.

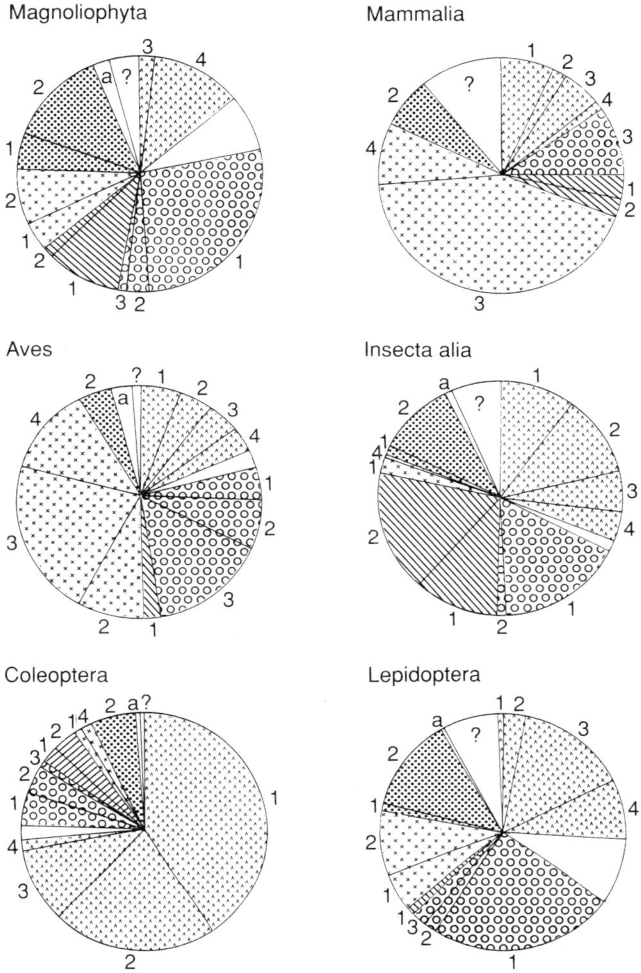

Fig. 1. Reasons for the decline of endangered species in Finland among flowering plants (Magnoliophyta), mammals, birds, butterflies and moths (Lepidoptera), beetles (Coleoptera) and other insects, displayed as pie diagrams (from Mikkola, 1987). The symbols and the numbers outside the diagrams are as follows. Symbols for coniferous forest: forestry and changes of forests; 1 = rotten wood, 2 = primeval forest, 3 = composition of tree species, 4 = methods of forestry. White: draining of peatbogs. Rings: agriculture; 1 = meadows grow over, 2 = changes of methods, 3 = chemical effects. Oblique lines: waters; 1 = building of waters, 2 = water quality. Crosses: direct effects of humans; 1 = wear, 2 = collecting, 3 = hunting, 4 = disturbance. Dots: indirect effects of humans; 1 = gravel pits, 2 = building, a = other reasons, ? = unknown reasons.

Annual forest fires in Finland now account for less than 1000 ha compared with 53 000 ha in 1924, resulting in the extinction or endangered status of the beetle *Agonum bogemanni* (Gyll.) and several species of bugs in the genus *Aradus*. In 1989, the first artincial forest fire was lit in a nature park in order to investigate the possibility of managing for such species.

Most of the luxurious forests of Finland were cleared for arable land centuries ago, as were the most fertile bogs. Oak forests were almost entirely destroyed for shipbuilding in the eighteenth century. The rest of these have been protected or are likely to be protected. But nowadays, only 8% of the original area is covered by deciduous forest. For insects, a particularly harmful recent change has been the widespread cessation of hay-making and grazing in meadows. Drainage, replanting with trees and the use of herbicides have also caused a dramatic loss of flowering meadows.

A comparison between the Finnish and Swedish Red Data lists (Table II) revealed a great different between Coleoptera and Lepidoptera. In both groups, about two-thirds of the endangered species of Finland also occur in Sweden, but of the species in common, nearly 85% of coleopterans but under 40% of the lepidopterans are also listed as endangered in Sweden. The greater consistency among beetles is probably due to their dependence on virgin forests, as already explained above, and these habitats are quite similar in both countries. The threatened butterflies and moths are mainly southern and south-eastern thermophilous species which live predominantly in anthropogenic habitats. These habitats are better represented on the Swedish islands Gotland and Öland and in southern Sweden than in Finland, which explains the discrepancy.

The consistency of the beetles on the two national lists prompted an examination of their threat classifications. Surprisingly, the different classes were close to a random distribution: in only 17 examples out of 39 (46.0%) was the class the same on both lists, compared with the 13 examples (33.3%) which would correspond to a random distribution.

TABLE II. A comparison of the Finnish (SF; Rassi *et al.*, 1986) and Swedish Red Data lists (S; Andersson *et al.*, 1987) for Coleoptera and Lepidoptera.

	Species in SF, classes 0–2	Occur in S	%	Listed by Andersson *et al.*	%
Coleoptera	72	46	63.9	39	84.8
Lepidoptera	38	26	68.4	10	38.5

V. CONCLUSIONS

Several biogeographic clines that are relevant for conservation run through Europe. For example, the ratio of natural to secondary habitats increases towards the north but the degree of endemism decreases. A somewhat similar trend runs from the west to the east. A purely European basis for conservation is inadequate, since for northern and eastern Europe at least, it is necessary to know and account for the status and ecological requirements of endangered species in Siberia. An extension of the taiga biome to Fennoscandia shows that many species are still common which are rare or endangered in central Europe.

One might expect that the existence of viable northern populations among species that are endangered in the south would offer good opportunities for autecological studies and be a source for reintroductions. There are, however, two probable reasons which mitigate against this: the species is probably acclimatized to the north and, when abundant, may also be more eurytopic, and have less critical habitat requirements than is the case with its disjunct southern populations. Further east, the species may still survive at about the same latitude as in Europe, and therefore occur in more similar ecological conditions which may offer better ground for conservation measures.

In general, it is difficult and controversial to compile an all-European list of endangered species, without any prefixes given for the species to restrict the area of threat to a part of Europe. The main reason for this is that some species that live in highly endangered biotopes in central Europe, such as the butterflies *Colias palaeno* (L.) and *Boloria aquilonaris* (Stichel) on peat bogs, are common in northern Europe and abundant throughout Siberia. To declare these, and some other species listed or proposed by the Bern Convention, as endangered in Europe may be controversial from the viewpoint of a northern European. This would greatly decrease the value of an All-European list. These species are nevertheless threatened in central Europe and are important indicators of endangered habitats. I therefore propose that consideration should be given to having prefixes on the list such as "excluding northern Europe", or, in addition to the main list, to have some groups of species with more restricted areas of threat.

The protection of insects and their habitats always involves protecting them from the adverse effects of human activities. For insects, the protection of habitats is of prime importance, and species protection is only occasionally effective. The problem may be that an administrator needs to have a species protected before he can protect its habitats. This is because biotopes containing endangered, and therefore protected, species may be monitored and managed more easily. This is the reason behind the recent protection of species in Finland. But entomologists now fear that accumulation of faunistic data on these species will stop, and that their conservation will be hampered rather than advanced. The same problem is faced in the Soviet Union with its high number of protected insect species.

The network of natural reserves in the Nordic countries is relatively representative, but inadequate. As a consequence, a fragmentation of habitats may cause problems with the dynamics and genetics of populations. Biotopes that are expected to become seriously fragmented are virgin forests and peatlands in southern Fennoscandia and the steppes in Russia. For certain ecological types of insects, management in the intervening areas may provide "stepping stones" which may secure the gene flow and facilitate recolonization after local extinctions. Examples of such action would include leaving dead and rotten trunks in commercial forests. Such aid is, of course, not possible where a special micro- and mesoclimate is peculiar to the habitat, as is the case with steppes. For some insect groups, such as caribid beetles, "a habitat mosaic at the regional scale seems to be a good conservation goal" (Niemelä *et al.*, 1988).

The fauna of northern Europe is unique not for its endemic species but for the combination of elements which spread to the area in postglacial times. There exist a great number of pairs of species of arctic or alpine, taigan or southern montane, or of central Asiatic or Mediterranean origin which live together only in Fennoscandia. These unique communities must be conserved as examples of the biodiversity created by the glacial history.

In a congress in Copenhagen in 1986, the Nordic entomological societies accepted a resolution where the following starting points for conservation of invertebrates were emphasized (Hågvar, 1988, translated from Norwegian):

1. The conservation of invertebrates, must have the same priority as that of vertebrates.
2. Endangered species survive only in special habitats which should be listed, conserved and, if need be, managed. Invertebrates alone may serve as an argument for establishing natural reserves.
3. The conservation of species richness and habitat diversity should be taken into account in forestry, agriculture etc. The recreation of lost habitats and the conservation of all habitat types is recommended.
4. Resources available for the registration of localities that are valuable for the conservation of insects should be increased in the Nordic countries, and a common Nordic plan for the conservation of insects should be devised.

Acknowledgements

I am grateful to Mr A. V. Sviridov (Moscow, USSR) for notes on the protection of insects in the USSR and Dr B. Ehnström (Stockholm, Sweden) for useful comments on the manuscript.

REFERENCES

Aagard, K. and Hågvar, S. (1987). Sjeldne insektarter i Norge. *Økoforsk utredning* **6**, 1–81.

Andersson, H., Coulianos, C.-C., Ehnström, B., Hammarstedt, O., Imby, L., Janson, L.-Å, Lindelöw, Å. and Waldén, H. W. (1987). Hotade evertebrater i Sverige. [Threatened invertebrates in Sweden.] *Entomol. Tidskr.* **108**, 65–75.

Bengtsson, B.-E., Elmquist, H. and Nyholm, E. (1989). Några rön kring apollofjärilen i Sverige samt försök att förklara dess tillbakagång. [On the Swedish apollo butterfly with an attempt to explain its decline.] *Entomol. Tidskr.* **110**, 31–37.

Borg, P. and Malmström, K. K. (1975). Suomen uhanalaiset eläin- ja kasvilajit. [Endangered animals and plants in Finland.] *Luonnon Tutkija* **79**, 33–43.

Borodin, A. M., Bannikov, A. G., Sokolov, V. E. *et al.* (1984). "Krasnaya kniga SSSR". Tom 1. Lesnaya Promyshlennost', Moskva, 390 pp.

Ehnström, B. and Waldén, H. H. (1986). "Faunavård i skogsbruket". 2. Denlägre faunen Skogsstyrelsen, Jönköping, 351 pp.

Hågvar, S. (ed.) (1988). "Vern av insektfaunaen i Norden". Stencil. Norsk Institut for Skogsforskning, Ås., 47 pp.

Malmström, K. K. (1975). Suomen uhanalaiset eläin- ja kasvilajit luetteloitu. [Endangered animal and plant species of Finland listed.] *Suomen Luonto* **34**, 81–84.

Mikkola, K. (1987). Förändringar av fjärilsfaunan i Finland i relation till biotopförändringar efter år 1950. [Changes in the Finnish lepidopteran fauna since 1950 in relation to environmental changes.] *Entomol. Medd.* **55**, 107–113.

Niemelä, J., Haila, Y., Halme, E., Lahiti, J., Pajunen, T. and Punttila, P. (1988) The distribution of carabid beetles in fragments of old coniferous taiga and adjacent managed forest. *Annales Zoologici Fennici,* **25**, 107–119.

Rassi, P. Alanen, A., Kemppainen, E., Vickholm, M. and Väisänen, R. (1986). "Uhanalaisten eläinten ja kasvien suojelutoimikunnan mietintö. II. Valtion painatuskeskus, Helsinki, 466 pp. [English summary: Threatened animals and plants in Finland. Helsinki 1987.]

Yeliseyev, H. B., Zabrodin, V. A., Fertikov, V. I., Kolosov, A. M. and Skarlato, O. A. (1983). "Krasnaya kniga RSFSR. Zhivotnye". Rosselkhozizdat, Moskva, 455 pp.

6

Mediterranean Insect Conservation

EMILIO BALLETTO AND ACHILLE CASALE

I. Introduction ... 121
II. The Mediterranean Insect Fauna ... 123
A. The Physical Background ... 123
B. The Origin and Composition of the Mediterranean Insect Fauna ... 123
III. Pleistocene Refugia ... 128
IV. Insect Conservation in the Mediterranean Area ... 129
A. Direct Conservation Measures ... 129
B. Indirect Conservation Measures ... 131
V. Threatened Environments of the Mediterranean Area ... 133
A. Vulnerable Environments ... 133
B. Threatened Environments ... 135
C. Endangered Environments ... 136
References ... 140

I. INTRODUCTION

The abundance, diversity and the high level of endemicity of the Mediterranean insect fauna are far too well known to need stressing here. All zoologists having personal experience of this area are not only familiar with the pleasure of frequent unexpected discoveries, but also with the occasionally excessively collection-orientated enthusiasm that every year drives many, particularly central and north European entomologists, to the temperate, varied and changing environments of the Mediterranean shores.

Conservation is a long-neglected issue in the Mediterranean countries. This not only applies to natural resources, but to many other broadly speaking cultural values as well, be they artistic, archaeological or other. Among the many factors contributing to this situation, one is certainly the fact that many Mediterranean countries have been lulled into the delusion that their riches were innumerable and, as such, everlasting. Local conditions of economic and cultural underdevelopment, high population density or even the extremes of endemic internal or external warfare have long prevented conservation-related problems from ranking very high in the scale of public priorities, particularly in the economically less developed parts of the region.

In some Latin countries, such as Italy, the long-rooted notion that culture is principally humanistic has proved particularly persistent. Accordingly, for law makers nature always come second to artistic or architectural values, and has long been treated as a more or less negligible part of the landscape. This is also why it may be difficult, even nowadays, to make a section of the general public understand that, say, the Mediterranean monk-seal (*Monachus monachus*), the wolf (*Canis lupus*) or the bear (*Ursus arctos*) can be worth protecting. However, luckily enough this part of the population is declining. Yet, bearing in mind that many people associate insects with something obnoxious, or even outrightly disgusting, the discovery that insect conservation has little appeal to the general public is not very surprising.

For a number of Mediterranean countries, perhaps more than for many parts of Europe, insect conservation is first of all a matter of general education (Guilbot, 1987; Collins and Wells, 1987). No local and limited intervention, or any other coercive action will ever succeed until the general population is brought to realize how great the ecological and economic importance of insects is for terrestrial ecosystems, and even to appreciate their function from a cultural or aesthetic point of view (Tassi, 1969; Pyle *et al.*, 1981; Pavan, 1986).

The reason why in at least some Mediterranean countries nature remains in a relatively healthy state when compared with the rest of Europe (Balletto and Kudrna, 1987), has in fact little to do with past or present conservation practices. Even in the most advanced of these regions, this is more a consequence of the concurrence of geographic factors and current models of development than of a real conscience of ecological values (Balletto, 1989). In these countries, in fact, large portions of the land are too steep and hilly to be suitable for most agricultural or industrial uses and are not very attractive even for residential or touristic purposes.

II. THE MEDITERRANEAN INSECT FAUNA

A. The Physical Background

As is well known, the Mediterranean is a closed sea, washing the shores of southern Europe, west Asia and north Africa. It connects with the Atlantic Ocean, the Black Sea and the Red Sea through the Strait of Gibraltar, the Bosphorous and the Suez canal, respectively. Its surface is 2 966 000 km^2, its width 3860 km from east to west, its maximum depth 4404 m and its coastline 12 700 km long.

The Mediterranean-type climate, which occurs in various parts of the world, is characterized by average yearly rainfalls of 250–1200 (locally 1500) mm and mainly concentrated in the winter and spring seasons. Minimum monthly temperatures range from 8 and 12°C (locally with some snowfalls or frosty nights), and maxima between 18 and 30°C (locally 35°C). This climate is affected by the geographical position of the areas involved, and by the presence of relatively shallow seas or oceanic currents. It is found in five widely separated regions: south-western and western parts of Western Australia, southern California and the west coast of the United States, central Chile and the west coast of South America and the whole Mediterranean basin, extending as far east as the Persian Gulf and the Caspian Sea (Dell *et al.*, 1986). As a consequence, the Mediterranean types of ecosystems are particularly heterogeneous, both in vegetation and in composition of their vegetal and animal communities (Castri, 1973). The Mediterranean basin fauna is highly characterized and different in many respects from that of any other region of the world. At least in this particular case, however, the summer heat and drought are often more important to the fauna than the winter cold, and for many animal species aestivation periods may be longer or more important than hibernation.

B. The Origin and Composition of the Mediterranean Insect Fauna

Assessing the overall importance of the Mediterranean fauna is a complex problem. We attempted to estimate the total insect fauna of the Mediterranean basin and produced a rather conservative estimate of 150 000 species, 70% of which have been already described. Even though this figure will be bound to increase as scores of new species are described every year, at a first glance it does not seem particularly high when viewed on a world-wide basis. As the recent studies on tropical rain forests made by Erwin and his research group would place the total number of living arthropod species at 30 million, including perhaps 24 million insects (Erwin, 1982; see also May, 1988), the Mediterranean fauna can only account for a very limited fraction (0.6%) of the world's insects. Bearing in mind that the great majority of all animal species live in the tropical forests, this fauna assumes somewhat more significance when compared to the small dimensions of the area of the countries of the Mediterranean Basin (about 1 000 000 km^2, i.e. 0.94% of the total world non-frosted area, calculated to be 106 343 000 km^2). Three-quarters of all European species live on the Mediterranean shores and many of them are confined to this area, which is characterized by a particularly high rate of endemicity (refuge areas). In some isolated mountain groups and major Mediterranean islands, endemics account for 15–20% of the total insect fauna, while this figure escalates to a staggering 90% of steno-endemics in some caves.

To consider the Mediterranean fauna just as a "transitional fauna" in the sense of Darlington (1957) would be, therefore, to downgrade its complexity considerably. Even though the Mediterranean basin is located almost on the edge of the Palaearctic region, between Europe, Asia and Africa, its fauna is not merely a result of geographical position, but is rather due to the interaction of many historical (palaeo-geographical and palaeo-climatic) and ecological factors. Among these, the following are particularly significant:

1. The highly complicated structural evolution of the palaeo-Mediterranean basin, at least during the whole Tertiary, only recently elucidated to some extent (see e.g. Alvarez *et al.*, 1974, Hsü *et al.*, 1977; Rögl and Steininger, 1983). Current interpretations involve complex microplate drift, particularly in the western part of the basin, the orogeny of the Alpine, Pyrenean, Balkano-Anatolian and Apennine chains, the late Miocene salinity crisis and subsequent isolation of the Para-Tethys basins, the opening of the Strait of Gibraltar, up to the final formation of the modern Mediterranean Sea, in the course of the Pliocene.
2. The progressive deterioration of the climate which began in the late Tertiary, starting with the early ice phases of the post-Miocene (Carraro *et al.*, 1975), Pliocene and Pleistocene, which brought about many profound modifications both in the general morphology of a number of mountain groups (glacial erosion) and coastlines (eustatic movements of the Basin), and in the composition and structure of all terrestrial ecosystems.
3. The post-glacial ecological modifications, either brought about by

further variations of the climate, such as the many different phases of Mediterranean afforestation (Beng, 1967), or more or less directly connected to human activity, such as the desertification of the north Sahara in the last 5000 years (Quezel, 1957).

This complex of factors has locally determined either a number of situations of vicariance, highly favourable for the start of cladogenetic events, or a series of dispersal processes, involving many pre-existent faunal (and floral) elements, or both. This complex of processes is what has brought about the patch-work of distributional patterns, often interpreted in terms of "chorological categories" (La Greca, 1964), that make the Mediterranean area so very remarkable from a biogeographical point of view.

In an attempt to sum up in dynamic terms the most important lines (Fig. 1), it can be said that the following principal components can be recognized within the present-day insect fauna of the Mediterranean area.

1. A more ancient component, certainly of a pre-Quaternary origin, comprising;

a. some "extra-palaearctic" elements representing the last remnants of the original early tropical faunas which all but disappeared from the area (see Larsson, 1978). Some well-known examples are *Graellsia isabelae* or *Charaxes jasius* among the Lepidoptera, but also the Carabid beetles of the genus *Drypta*, *Odacantha* and *Aptinus*, or even the Diptera Psychodidae of the genus *Atrichobrunettia* (Brunettini) and one undescribed *Sycorax* (Sycoracinae).

b. Other pre-Quaternary elements with affinities either in the Gondwanian or in the Angarian area. These ancient forest dwellers were probably tied, initially, to the subtropical or temperate woodlands (Laurisilva: *Lauro-Prunetea*). Currently these elements are well represented in the soils and forest litter in some upper montane refugia, as well as on some Mediterranean and Macaronesian islands, and among the cave-dwelling fauna. Among insects, they represent probably the most significant stock, and that with the highest percentage of endemics. Among the many available examples among the Coleoptera, it is worth recalling here some members of the genus *Carabus*, many Trechid beetles confined to the high mountains or caves, many Staphylinoidea, Buprestidae, Chrysomelidae and Scarabaeidae, and the fabulous Euchirine *Propomacrus bimucronatus* (currently confined to Anatolia and the east Balkans, but belonging to a group represented only in east and south-east Asia). Among the Lepidoptera, it is certainly worth recalling here the celebrated *Acanthobrahmaea europaea*.

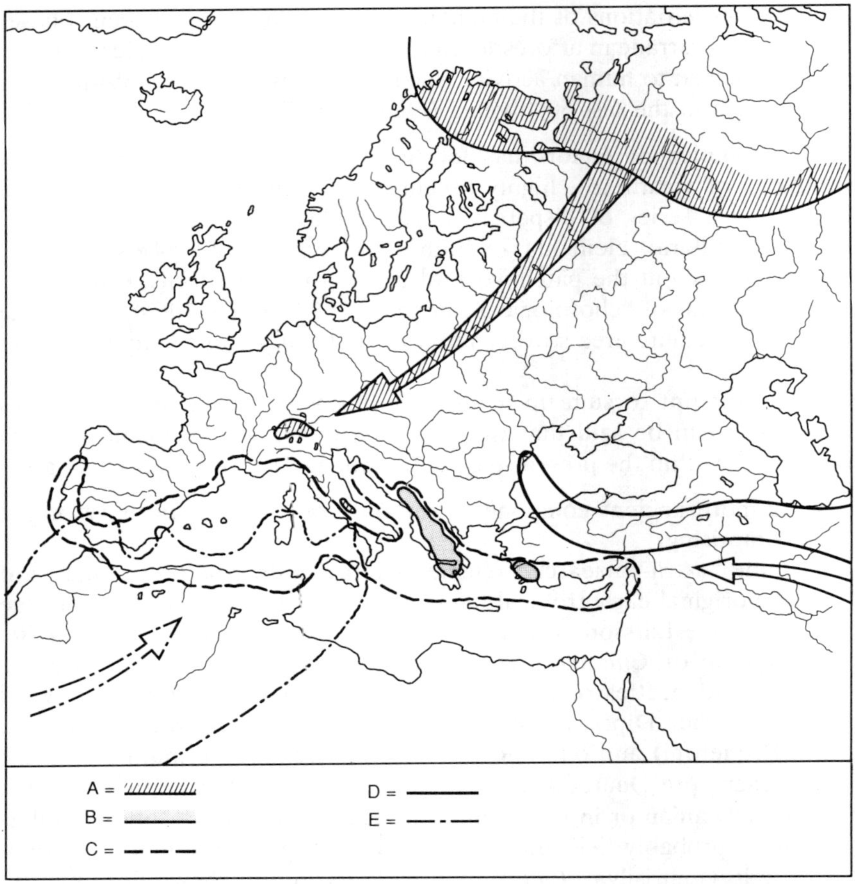

Fig. 1. Some distribution types in the Mediterranean area: (A) Coleoptera, Staphylinidae: genus *Mannerheimia*; (B) Coleoptera, Carabidae: tribe *Omphreini* (palaeo-endemic); (C) Lepidoptera, Nymphalidae: *Charaxes j. jasius*; (D) Neuroptera: *Wesmaelius mongolicus*; (E) Odonata: *Orthetrum nitidinerve*.

2. A more recent set of elements, having a northern (boreal) or north-eastern (Siberian or central Asiatic) origin, tied to the temperate cold forest types (tundra), or originally from the steppes. These elements colonized the Mediterranean area during the glacial periods of the Plio-Pleistocene, partially substituting the pre-existent fauna, or overlapping it.

a. Many examples can be quoted among the boreal–alpine or boreal–orophilous faunal elements, some of which were the object of more or less recent studies by several biogeographers (e.g. Holdhaus and Lindroth, 1939; Focarile, 1974). Among the many possible examples we can quote here the Mecoptera of the genus *Boreus*, the Diptera of the genus *Chionea*, the Coleoptera Staphylinidae of the genus *Mannerheimia* (and some *Atheta*), together with the many well-known mountain-dwelling (or "orophilous") elements now characterized by a Euro-central Asiatic or Euro-Siberian distribution. Among "classical" examples we can also recall some butterflies such as those belonging to the genus *Parnassius, Pieris callidice, Colias palaeno, Albulina orbitulus*, members of the genus *Agriades, Vacciniina optilete* etc. This same category also includes many forest-dwelling (soil or litter), generally phytophagous insects, which have accompanied the more recent phases of afforestation in the course of the last 15 000 years. Noteworthy in this connection is that the beech tree (*Fagus sylvatica*), with a cohort of exclusive species (*Rosalia alpina* etc. among the Coleoptera, *Aglia tau* among the Lepidoptera etc.) only succeeded in colonizing the Alps and some other mountains of southern Europe in these comparatively recent times, together with the fir tree and the red fir (Kral, 1972).

b. Within this category both the hygrophilous elements of the plains (fens etc.), and the xerothermophilous "recent" elements (some of which originating from the temperate steppes), must also be recalled. The former comprise a very large number of species of northern origin (Coleoptera: Carabidae and Dryopidae, Odonata, etc.) now characterized by relic distributions in the few residual wetlands (peat bogs, oxbow lakes, river mouths, marshes, retrodunal lakes etc.). The latter profited from the favourable opportunities that allowed their dispersal during the ipsothermic phases of the interglacial and post-glacial periods, occasionally differentiating *in situ*. Among them, some striking examples can be found among the Carabid, Curculionid, Scarabaeid and Tenebrionid beetles (genus *Asida*), as well as among the Lepidoptera, Orthoptera and Heteroptera. These insects either colonized some particularly xeric longitudinal valleys of the Alps (Meregalli and Osella, 1978 and literature cited therein), or the mountains of the Anatolic, the Balkan, the Italian or the Iberian Peninsulas, or even the Maghreb (Satyrid butterflies of the genus *Pseudochazara*, *Polyommatini* blues of the subgenus *Agrodiaetus*, and many others).

3. Last but not least we would like to mention here a group of species whose range of distribution has recently undergone a sudden increase as a consequence of human activities, or that were introduced outright into the area by humans, either in an active or passive way. This includes several pests, such as the Colorado beetle (*Doriphora*), but also a score of attractive large-sized Lepidoptera, such as *Phylosamia cynthia, Antheraea pernyi* and *Antheraea yamamay*, which became naturalized in many more or less restricted areas of the Mediterranean basin (Rougeot, 1971; Casale, 1973).

III. PLEISTOCENE REFUGIA

What has been said about the origin of the Mediterranean fauna should make clear that at least in so far as the whole area is concerned, the concept of Pleistocene refugia should no longer be employed with the (geographically) restrictive meaning that was so cherished by the biogeographers of the times of Jeannel (1942) and Holdhaus (1954). Taken together, in fact, the lands of the Mediterranean Basin represent a single broad refuge area. Several subunits, however, can be recognized both in the four major peninsulas (Iberian, Italian, Balkan and Anatolian), and in the bigger or smaller islands, and also in the Mediterranean shores and lowlands. At the other end of the spectrum lie the high plateaus and the many mountain chains of the Mediterranean region, several of which rise well over 3000 m above sea level.

In greater detail, it must be stressed that the following have survived in the Mediterranean area:

1. Many ancient thermophilous faunal elements that were already present in the area, where they sought refuge against the deterioration of the Pleistocene climate.
2. Other more recent microthermophilous elements that thermal events drove from the north into the Mediterranean area during the Pleistocene, and are now represented by relic populations in the coldest parts of the area.

Classic refuge areas for the former group have been, and still are, first of all the more ancient mountain chains (the Kabylie Mountains of Algeria, the Atlas and Rif Mountains of Morocco, the Betic massifs and the Central Cordilleras of Iberia, the Pyrenees, the Alps, the Balkans, the Taurus Mountains, the Cèdres Mountains of Lebanon, etc.), but also the pre-Alpine belts, the mountain groups of some of the major islands, down to the tiniest biotopes, the forest soils and the caves, among others.

Refuge areas for members of the second group (but to a degree also for some of the first) were the wetlands of the plains, the intermorainic basins, the steppe areas, the heaths and the fens, and again the highest mountains.

Many of the environments that allowed members of both categories to survive have recently disappeared, mostly because of human action, while others are now heavily threatened. The conservation of the Mediterranean insect fauna also depends upon the conservation of their remnants.

IV. INSECT CONSERVATION IN THE MEDITERRANEAN AREA

Like any other form of natural protection, insect conservation can be achieved by two different strategies:

1. Direct conservation measures to assist individual species deemed to be threatened with extinction.
2. Indirect measures aimed at the conservation of the environments and biotopes where these species live.

A. Direct Conservation Measures

As to the first approach, several European and other Mediterranean countries (Leduc, 1987; Malausa, 1987; Montoya, 1987) have now adopted conservation measures, generally banning or reducing the collection and/or sale of many rare or threatened species. This was achieved either within the framework of an international convention (CITES, Bern Convention etc., see Collins, 1987, for a review), or also independently. In the second case, the approach normally involves more species together, and the scenario can be rather varied. The range varies from instances where all collection is in practice entirely forbidden (Algeria), to others like that of Turkey where there is no law on collecting *per se* but doing research requires a normally unobtainable permit from the government, to yet different cases, as can happen in Spain, where there is no general law on insect collection (Montoya, 1987), but collectors may be required to obtain a permit from the local Alcade. Otherwise, collection bans concern a score of species only (France).

It would certainly be a mistake to underrate the effect of these kinds of measures, as on the one hand they can certainly discourage improper collection (at least in some cases), while on the other, they represent something of a symbolic example. Yet, measures of this type cannot stand alone, and also have at least two practical drawbacks:

1. In the first place, species are often chosen more on the basis of aesthetic or "sentimental" criteria than on purely scientific grounds. In other words, those species which are bigger, more colourful and notorious for being the most sought after and localized are also more liable to become the object of a conservation programme. In contrast, other species that are less well known but more threatened with extinction are easily neglected. For example, while rather considerable efforts have been devoted in Italy and France (Malausa and Drescher, 1987; Caddy *et al.*, 1987) to the conservation of the beautiful and "rare" (localized) *Carabus olympiae*, nothing has been done for the rather drab *Carabus italicus, C. clathratus* and *C. alysidotus*. Yet it is the second group rather than the former species that has undergone the most serious decline in recent years, as a consequence of habitat alterations in the densely populated Italian plains.
2. Another, and unfortunately more serious, problem is with law enforcement. This might apply to all legislation more or less in the same way, but it is a sad fact that among the many countries that have ratified the Bern Convention, only a score are really doing anything to preserve the habitats of species listed in its Appendix B, as required. The facts are even worse where measures against collection and trade are concerned, and it is a sad experience that occasionally such regulations have even proved detrimental. Particularly among the Coleoptera, species enclosed in the several European "red data lists" (*Carabus olympiae* is just one of many similar cases) have all but disappeared from the scores of more or less unofficial trade lists. The only noticeable change, in fact, has been in their price (now considerably higher), as there has been no apparent change in their being exchanged, traded, or indeed collected, normally by highly destructive and invisible "dormant" traps. Recent price hikes, on the other hand, have prompted many local collectors to search actively for this renewed (from the early 1900s) source of income, under the pressure of the many dealers who have rediscovered these forbidden treasures, often on the wave of the media. Other, less prodigal European collectors will keep on visiting the mountains of central Sardinia and Corsica during the months when the larvae of *Papilio hospiton* are well grown (to avoid collecting too many parasitized specimens), even though they know that they are now risking more than previously. The same applies to *Papilio alexanor* in France, to *Graellsia isabelae* in France and Spain, etc.

B. Indirect Conservation Measures

Coming now to point (2) it is our opinion that a correct conservation policy can only be achieved through a correct policy for land use (Ricou, 1987; Gaspar, 1987). The fact that the insect fauna, at least locally, has recently benefited from laws on the agricultural use of pesticides, road and town development planning, or an improved domestic stock breeding, is too apparent to be easily dismissed. The fact that in some Mediterranean countries (and perhaps only for some insect groups), the general ecological situation may be spontaneously improving, and that some of this recovery may be simply the consequence of locally changing economic situations, does not lessen the results obtained by land planning. Rather, the present should be taken as a rare opportunity to work along the same line and create at least some of those natural parks that only a few years ago would have been almost unthinkable (Balletto and Kudrna, 1985; Balletto, 1989).

Unfortunately, spontaneous recovery is certainly not occurring throughout the whole Mediterranean basin. Local economic underdevelopment and a purportedly "traditional" use of the land cannot be taken as an indication of a reasonably good state of nature conservation. That no universal pattern for resolving conflicts of land use can yet be proposed should be kept in mind, particularly for Mediterranean climate ecosystems (Havel, 1986).

The absurdly increasing numbers of sheep and goats that year after year are put out to graze the pastures of some eastern Mediterranean countries are reducing huge areas of previously almost unspoiled land to barren expanses. The extensive and insanely massive use of pesticides that some southern Mediterranean governments used in 1988 to tackle a purported spread of agricultural pests has reduced some mountains to an entomological desert, while proving of little avail, if any, for its original purposes. The unbelievably fast touristic development experienced in the last few years on the eastern Mediterranean shores is destroying at the same speed not only the habitats of many highly specialized insect species, but also the last remnants of the breeding areas of the endangered monk-seal and Mediterranean sea turtles.

An apparently possible solution would be to promote internationally the creation of a network of natural parks and reserves, strategically distributed in the Mediterranean region so as to cover at least all major types of ecosystems. It is to be emphasized, however, that the practical difficulties generated by the high maintenance and surveillance costs, suggest that natural parks and reserves should not be extended over the territory and that all possible alternative or accompanying solutions should be sought. There are other drawbacks, to be carefully considered right from the start:

1. It is clear that not all European nations speak the same language where parks are involved. Requirements of these natural reserves should be made clear to individual governments, particularly in that no touristic development should be allowed to take place within their boundaries or immediate surroundings. Many cases could be quoted where the creation of a national park has resulted in extremely damaging consequences for the natural environment, even in terms of huge government-sponsored touristic developments. The size of individual caenoses (or plant and animal communities) comprised in every reserve should be big enough to be suitable for species survival. This not only requires that for every single species involved effective population numbers (Kimura and Crow, 1963) will never fall below theoretical values, but also that their numbers will never decrease below levels calculated for "minimum viable population sizes" (see Ewens *et al.*, 1987).
2. As a corollary, it is very important that reserves do not stand out, with respect to the natural environment, as oases in the desert.
3. Another corollary is that creating a natural park just because a region looks particularly attractive from an aesthetic or panoramic point of view may be important for a number of reasons, but it will be of little avail for animal and plant conservation, unless by mere chance. Among other things, this kind of action naturally involves those environments that for one or another reason remain of little interest for human development, while it has a tendency to skirt around those biotopes that really are threatened. For example, in Italy natural parks are now relatively easily created on depopulated montane areas or wooded regions. In constrast, it is only rarely possible to establish a reserve for the highly threatened residual minor wetlands (ponds, peatbogs, moors, fens, etc.) whose economic interest increased considerably in recent times, in view of their urbanization or transformation into ricefields. Coast areas, islands and some caves among the richest in endemics, are threatened by a psychotic kind of touristic development, often characterized by a very poor general quality. It is to be recalled that Italy, France and Spain have seen two-thirds of their Mediterranean coastal areas all but disappear under concrete and tarmac, in the small span of the last three decades. As the same process is swiftly accelerating in the eastern Mediterranean countries also (particularly in Greece and Turkey), the total destruction of coastal biotopes could become just one more sad reality within a very short time.

V. THREATENED ENVIRONMENTS OF THE MEDITERRANEAN AREA

Bearing all this in mind, together with the possibility that a synergism of correctly species-oriented and biotope-oriented approaches may play a beneficial role in conservation, we wish to close this chapter with a slightly provocative suggestion. At a time when conservation is broadly dominated by a continuous production of red data lists, each containing scores of animal species supposedly threatened with extinction, either at supranational, national or regional levels, we will endeavour to propose here a list of environments which we consider the most menaced of the Mediterranean area.

Environments, taken by very broad ecological categories, will be rated as "vulnerable", "threatened" or "endangered" on a scale of their progressive risk of disappearing over a few years in the Mediterranean area. For each, some information on the potentially most dangerous threats will be given, together with some suggestions for future action and information on the most directly threatened insect species.

A. Vulnerable Environments

This category comprises some environments which are still well represented in the Mediterranean area, but whose extent is declining. In this case, a conservative estimate would put the proportion of biotopes destroyed between 5 and 10% (but up to 50% on a local scale), compared with those available in the 1940s. Whenever possible, the most advisable action to be undertaken is promoting correct management, while the creation of natural reserves should be restricted to those cases where alternative options are impossible.

1. Caenoses of the High Altitudes of the Mountains of the Mediterranean Region

a. Pastures of the Alpine belt (particularly of the Alps and Pyrenees, but locally in southern Spain, Greece and Turkey); subalpine caenoses with Rhododendron, Vaccinium *and/or dwarf pine trees*. These regions, already scantily populated, are currently undergoing a progressive trend towards the withdrawal of many previously important human activities and particularly of domestic stock grazing. Here overgrazing has always been less important than in other areas, as the summer rains avoid excessive soil transpiration.

Threats. Threats are mostly represented by tourist development, particularly skiing facilities and related infrastructures. The subalpine belt is particularly affected.

Suggested action. Passing laws to provide limits to the opening of new chair lifts, ski lanes and ski stations, and preventing the cutting of shrubs of the subalpine belt; also creation of natural reserves and extension of those already existing to enclose some critical biotopes.

Affected insect fauna. Practically all mountain-dwelling (orophilous) species of the Mediterranean area (see later), many steno-endemics in all insect orders and a number of over-collected species particularly among the Coleoptera and Lepidoptera.

b. Pseudo-alpine pastures of the Italian peninsula; pseudo-pastures with Astragalus *of the mountains of north Africa, the major Mediterranean islands and of the Balkan, Iberian and Anatolian peninsulas.*

Threats. In most areas the major threats are overgrazing, particularly by sheep and goats, fires started with the purpose of extending pastures and soil erosion. All this has already resulted in many cases in the total destruction of the natural vegetation of these areas, and locally in near-desertification due to excessive soil transpiration.

Suggested action. Reducing sheep and goat populations, and also progressively withdrawing government-sponsored incentives, at least in selected areas.

Affected insects. Many mesoxeric and comparatively recent northern immigrants, a typical example of which, among the butterflies, is *Parnassius apollo* (some populations); many species of the high mountains of the Mediterranean region (see section II.B) including a very high number of neo-endemic species, such as some blue butterflies (*Polyommatus*) of the subgenus *Agrodiaetus*, etc.

2. Caves

Azonally distributed throughout the area, only more strictly (although not exclusively) linked to the presence of potentially water-soluble geological substrates.

a. Threats. Tourist use of the most spectacular caves, which are all too often also the largest and the richest in endemics, use as garbage dumps and pollution of subterranean waters are all potential threats.

b. Suggested action. Passing laws preventing misuse and the creation of subterranean reserves.

c. Affected species. A high number of insect species, principally belonging to the Diplura, Orthoptera, Coleoptera and other Arthropoda (Arachnida, Crustacea, Diplopoda and Chilopoda) of a particularly high scientific interest, often characterized by a very high level of adaptation to the subterranean environment. Many steno-endemics are also subject to excessive collection.

B. Threatened Environments

Environments in this category are characterized by intensive anthropic pressure: many of them (roughly between 30 and 50% of the original, but locally up to 80%) have already been altered beyond any possibility of recovery. The exploitation of the remainder is only just beginning. The general policy should be to try to revert the trend in cases where this is still possible, by promoting a type of management compatible with the survival of viable insect populations, as previously defined (IV.B). In some cases, however, the creation of natural reserves should be proposed.

1. Biotopes of the Pre-alpine Belts

These biotopes border the most important mountain chains (Atlas Mountains, Pyrenees, Alps, Apennines, Balkans, Pontic Alps, Taurus mountains), and other lower mountains of the Mediterranean area.

a. Threats. Tourist developments and urbanization as a consequence of the trend for second holiday homes; overgrazing; fires (see above); soil erosion; construction of infrastructures having a high environmental impact (big roadsystems, dams); afforestation with alien, generally fast-growing, species of trees.

b. Suggested action. Urban planning; passing better forestry laws; setting and enforcing hydrogeological limitations.

c. Affected species. Many endemic species; among the diurnal Lepidoptera, members of the genera (or subgenera) *Melitaea, Erebia, Agrodiaetus*, etc.

C. Endangered Environments

Natural environments in this category have disappeared from many broad areas, particularly of the more industrialized countries of the Mediterranean region. More than 90% of environments in this category have already been destroyed. Historically, their exploitation has always spread from one region to the next, from one country to its neighbour, and in the Mediterranean basin it has now reached almost everywhere. In most cases, the only possible solution will be to propose the creation of a natural reserve for the few still unspoiled biotopes.

*1. Wetlands**

Non-Ramsar, minor wetlands (retrodunal lakes, fens with *Molinia caerulea*, ponds, moors) certainly rank among the Mediterranean environments most directly threatened by extinction, as demonstrated by the fact that most have already disappeared from the area. Although these environments are protected by law in some European countries (Germany, Switzerland, etc.) this is generally not yet the case for the Mediterranean region.

a. Threats. Land drainage for agricultural use; direct transformation into rice fields; transformation into rubbish pits; water pollution; eutrophication; urbanization; sinking of the water table.

b. Suggested action. Creation of very strict natural reserves even in very small areas; prevention of all forms of water pollution.

c. Affected Insects. Endemic or relic species characterized by at least an aquatic or simply hygrophilous stage in their development. Examples are available from almost all insect orders: Ephemeroptera, Odonata, Plecoptera, Trichoptera, Lepidoptera, Coleoptera, etc., although this

---

* Under the Convention on Wetlands of International Importance (Ramsar Convention), wetlands of particular conservation value (often known as Ramsar sites) are nominated for special protection.

category is particularly abundant in the latter order. Among the Carabidae, not less than 500 Italian species, or about one-third of the total are wetland dwellers. Many of these are central European species that crossed the major mountain chains of southern Europe (Alps, Pyrenees) during the last glaciation and left some marginal isolates in the northern Mediterranean plains. These isolates are now among the very few last remnants of their species, at least in western Europe. This represents one of the most striking examples where the old "bird-oriented" approach to conservation has totally failed to promote even bare survival for other animal species. As a consequence, many of these have become extinct, while others still persist in a few peripheral colonies. Among the butterflies *Coenonympha oedippus, Lycaena dispar* and the hygrophilous species of *Maculinea* are worth recalling here. Many of these species were added to the appendices of the Bern Convention in 1987.

2. Hygrophilous and Mesophilous Forests

The original forests of the Padano-Venetian plains (northern Italy) were probably the biggest of this type within the Mediterranean region, although a similar situation also existed in northern Spain. As everywhere else in the basin, these forests have almost completely disappeared, due to a long history of deforestation, land drainage, crop cultivation and urban expansion, started well before Roman times. Only very few and normally small remnants are still in existence.

a. Threats. Deforestation consequent to hardwood price decrease; "natural" transformation of the forest consequent to sinking of the water table; transformation of riparian and other woods into groves of poplar trees; introduction of alien species of trees; cutting down the oldest trees.

b. Suggested action. Reversing the trend towards deforestation; preventing excessive sinking of the water table particularly during the summer, either by prohibiting excessive water utilization or by actively pumping water into the (normally) already existing canal systems; creation of natural reserves.

c. Affected species. As in the previous case, many central European species which managed to cross the major mountain chains of southern Europe (Alps, Pyrenees) during the last glaciation and left some marginal isolates in the northern Mediterranean plains. In some cases these isolates are now among the last few remnants of those species, at least in western Europe. Examples can be drawn from all orders of insects (Coleoptera,

Lepidoptera, Diptera, Hymenoptera). Many phytophagous species of the forests (Coleoptera: Lucanidae, Cerambycidae) and many steno-endemic species of the forest litter are also involved (Collembola, Diplura, Protura, Diptera, Coleoptera).

3. The Coastal areas of the Mediterranean Basin

Coastal environments were among the first to be exploited by humans. In many cases, and since Phoenicean and ancient Greek times, settlements in the Mediterranean region advanced inland only after the coasts were colonized. Initially, however, destruction did not proceed very far from the river mouths and the flattest regions of the coast that could be more easily colonized. Communication was often possible only by sea, at least up to the late Roman era, but, depending on the regions, also until well after Medieval times. It was only much later that the formerly well-spaced towns started to merge with each other along the shores. Now, the coasts of Southern Spain, southern France and much of Italy are covered in an almost continuous ribbon of concrete, and as the same pattern is spreading to many other Mediterranean countries (tourist resorts, second homes), the coasts have become an endangered environment.

a. Threats. Ever-increasing tourism and related facilities; urbanization, fires, water pollution.

b. Suggested action. Urban and extra-urban planning; creation of natural reserves.

c. Affected species. Numerous psammophilous and halophilous insects belonging to many groups; species of the maquis. Among the Coleoptera there are many endemic Mediterranean Cicindelidae, Carabidae, Scarabaeidae, Tenebrionidae. Also Hymenoptera (Sphecidae, Vespidae, Mutillidae, etc.); Diptera.

4. The Mediterranean Forests

Sclerophyllous forests with *Quercus ilex* (the holm oak), *Q. suber* etc. of the Mediterranean region. Although real mediterranean forests, with their big, century-old holm oaks are now very rare in the Mediterranean countries, some still survive, but in many cases they are extremely poorly managed.

Fig. 2. Some Mediterranean coastal biotopes: (upper) unspoiled (north-east Sardinia); (lower) after urbanization.

a. Threats. Overgrazing by domestic pigs. These animals unfortunately have the very damaging habit of swallowing the oak acorns together with much of the surrounding soil. Where pigs are overabundant, they can therefore easily destroy the soil, effectively preventing natural renewal of the forest. The obvious result is that the few remaining forests of this type are hopelessly ageing and, consequently, in many places they are now being logged.

b. Suggested action. Reversing the trend towards deforestation; prohibiting pig foraging at least in some selected areas, so as to allow the natural regeneration of the woodland; enforcing forest laws already in existence; creation of natural reserves.

c. Affected species. Many forest-dwelling, often palaeo-endemic insect species, such as some Coleoptera of the tribes Anillini and Scaritini, but also many Pselaphidae and Scydmaenidae (soil), several Lepidoptera (*Maniola nurag, Argynnis pandora, Charaxes jasius*, etc.), many parasitoid Hymenoptera.

5. The Mediterranean Islands

Same threats and suggested actions as for points V.B.1 and 5.C.3, only reinforced by the need to preserve situations made more vulnerable by the relatively small areas involved.

Affected species are all Mediterranean insular and micro-insular endemics; among the most well-known butterflies, *Papilio hospiton, Argynnis elisa* (species listed in the Bern convention), many *Hipparchia* etc.

Acknowledgements

It is a pleasure to thank Prof. G. Salamanna (Genova) who provided some very useful information about the Diptera and Dr W. Cameron-Curry (Torino) who carefully read and discussed the first draft of this paper.

This work was financially supported by the Ministry for Public Education (Italy) through its "40%" funding programme, research group on the biology of the Mediterranean Countries.

REFERENCES

Alvarez, W., Cocozza, T. and Wazel, F. C. (1974). Fragmentation of the Alpine orogenetic belt by microplate dispersal. *Nature (Lond.)* **248**, 309–314.

Balletto, E. (1990). Butterflies in Italy: their status, problems and prospects. "Proceedings of the International Congress on Future of Butterflies in Europe: Strategies for Survival". Wageningen, in press.

Balletto, E. and Kudrna, O. (1985). Some aspects of the conservation of butterflies in Italy, with recommendations for a future strategy (Lepidoptera Hesperiidae and Papilionoidea). *Boll. Soc. Ent. Ital.* **117**(1–3), 39–59.

Beng, H.-J. (1967). On the forest history of the Dalmatian coast. *Rev. Palaeobot. Palynol.* **2**(2), 271–279.

Caddy, J. F., Hemley, J., Morris, M., Malausa, J. C., Nagano, C. and Sawyer, R. (1987). Captive-breeding and farming of invertebrates. *Bull. IUCN* **18**(6–9), 17–18.

Carraro, F., Medioli, F. and Petrucci, F. (1975). Geomorphological study of the moraenic amphitheater of Ivrea, North West Italy. Quaternary Studies, Papers of 9th INQUA Congress, Christchurch, New Zealand, 1973, pp. 89–93.

Casale, A. (1973). Popolazione di *Antheraea Yamamai* (Guór. Men.) Spontaneamente ambientata in peculiare autonomo biotopo Italiano. *Annali Fac. Sci. agr. Univ. Studi Torino* **8**, 309–324.

Castri, F. (di) (1973). Mediterranean-type shrublands of the world. *In* "Mediterranean-type Ecosystems: Origin and Structure" (F. di Castri and H. A. Mooney, eds). Springer Verlag, Berlin.

Collins, N. M. (1987). The protection of insects by law in the countries of the European Community. *Cah. Liaison O.P.I.E.* **21**(4), 67, 17–30.

Collins, N. M. and Wells, S. M. (1987). Invertebrates in need of special protection in Europe. Council of Europe, Nature Environment series No. 35, 162 pp.

Darlington, P. J. Jr. (1957). "Zoogeography: the Geographical Distribution of Animals". John Wiley and Sons, New York.

Dell, B., Hopkins, A. J. M. and Lamont, B. B. (1986). Introduction. *In* "Resilience in Mediterranean-type Ecosystems". W. Junk, Dordrecht.

Erwin, T. L. (1982). Tropical Forests: Their richness in Coleoptera and other arthropod species. *Coleop. Bull.* **36**, 74–75.

Ewens, W. J., Brockwell, P. J., Gani, J. M. and Resnick, S. I. (1987). Minimum viable population size in the presence of catastrophes. *In* "Viable Populations for Conservation" (M. E. Soulé, ed.). Cambridge University Press, Cambridge, pp. 59–68.

Focarile, A. (1974). Aspetti zoogeografici del popolamento di Coleotteri (Insecta) nella Valle d'Aosta. *Bull. Soc. Flore Valdôt.* **28**, 5–53.

Gaspar, C. (1987). Protection ou gestion des Invertébrés. *Cah. Liaison O.P.I.E.* **21**(4), 67, 7–16.

Guilbot, R. (1987). Sensibilisation du public á la connaissance des insectes, des action pour mieu les protéger: l'example de l' O.P.I.E. *Cah. Liaison O.P.I.E.* **21**(4), 67, 81–84.

Havel, J. J. (1986). Land-use conflicts. *In* "Resilience in Mediterranean-type Ecosystems". W. Junk, Dordrecht.

Holdaus, K. (1954). Die spuren der Eiszeit in der Tierwelt. *Zool. Biol. Ges. Wien* **18**, 1–493.

Holdhaus, K. and Lindroth, C. H. (1939). Die europaischen Coleopteren mit boreo-alpiner Verbreitung. *Ann. Naturhist. Mus. Wien* **50**, 213–293.

Hsü, H. J., Montadert, L., Bernouilli, D., Cita, M. B., Erickson, A., Garrison, R. E., Kidd, R. B., Melierese, F., Müller, C. and Wright, R. (1977). History of the Mediterranean salinity-crisis. *Nature (Lond.)* **267**, 399–403.

Jeannel, R. (1942). "La genèse des faunes terrestres. Eléments de Biogéographie". Presses Universitaires de France, Paris.

Kimura, M. and Crow, J. F. (1963). The measurement of effective population numbers. *Evolution* **17**, 279–288.

Kral, F. (1972). "Grundlagen zur Entstehung der Waldgesellschaften in Ostalpenraum". G. Fischer Verlag, Stuttgart.

La Greca, M. (1964). Le categorie corologiche degli elementi faunistici italiani. *Atti Acc. naz. ital. Entomol. Rend.* **11**, 231–253.

Larsson, S. G. (1978). "Baltic Amber. A Palaeobiological Study", Entomograph, 1. Scandinavian Science Press Ltd, 192 pp.

Leduc, J. P. (1987). La protection des insectes en France. *Cah. Liaison O.P.I.E.* **21**(4), 67, 31–46.

Malausa, J. C. (1987). Les insectes dans les zones protegées de France. *Cah. Liaison O.P.I.E.* **21**(4), 67, 57–60.

Malausa, J. C. and Drescher, J. (1987). Elevage et acclimatation des Carabes; un example d'action pour la protection d'un groupe très collectionné. *Cah. Liaison O.P.I.E.* **21**(4), 67, 67–79.

May, R. M. (1988). How many species are there on Earth? *Science* **241**, 1441–1449.

Meregalli, M. and Osella, G. (1978). I coleotteri Curculionidae e Attelabidae delle oasi xerotermiche della Val di Susa. Atti XI Congr. Naz. Ital. Entomol., Portici-Sorrento, 1976, pp. 193–195.

Montoya, R. (1987). Protection de la fauna entomologica en España. *Cah. Liaison O.P.I.E.* **21**(4), 67, 61–65.

Pavan, M. (1966). "Une résolution culturelle européenne: la 'Charte des Invertebrées' du Conseil de l'Europe". Council of Europe, Strasbourg, 51 pp.

Pyle, R., Bentzien, M. and Opler, P. (1981). Insect conservation. *Annu. Rev. Entomol.* **26**, 233–258.

Quézel, P. (1957). Première étude palynologique de quelques paléosols sahariens. *Trav. Rech. Sahar.* **16**, 15–40.

Ricou, G. (1987). Effect de la pression humaine sur l'Entomofaune. *Cah. Liaison O.P.I.E.* **21**(4), 67, 3–5.

Rögl, F. and Steininger, F. F. (1983). Vom Zerfall der Tethys zu Mediterran und Paratethys. Die Neogene Paläogeographie und Palinspastik des zirkum-mediterranen Raumes. *Ann. Naturhist. Mus. Wien* **85(A)**, 135–163.

Rougeot, C. (1971). "Les Bombicoides de l'Europe et du Bassin Méditerranéen. 1. Faune de l'Europe et du Bassin Méditerranéen", Vol. 5. Masson, Paris, 156 pp.

Tassi, F. (1969). Problemi di conservazione nel campo dell'entomologia. *Mem. Soc. Ent. Ital.* **48**(3), 609–626.

7

Fragmentation, Isolation and Mobility of Insect Populations

J. P. DEMPSTER

I.	Introduction	143
II.	Isolation and Population Dynamics	144
III.	Examples of Insects responding to Increased Isolation	146
	A. *Papilio machaon*	146
	B. *Maculinea arion*	148
IV.	Conclusions	151
	References	153

I. INTRODUCTION

There can be little doubt that the largest single cause of changes in the distribution and abundance of insects in Britain over the past 50–100 years is loss of habitat resulting from changing land use. Of course, there have been big changes in the more distant past, with much of the original vegetation of Britain being destroyed, or greatly modified, before 1700, but the last 50 years has seen all but the most difficult terrain brought under agriculture or forestry. Ratcliffe (1984) records the following enormous losses in semi-natural vegetation in the recent past, as a result of man's activities. Of the permanent lowland grassland, only about 5% is now agriculturally unimproved, and only about 20% of pre-1940 calcareous grassland remains. Seventy-eight per cent of lowland heaths have been lost since 1830, and 40% since 1950. Lowland fens now exist as only a few scattered remnants, the largest expanse in East Anglia being reduced from over 3000 km^2 in 1930 to about 10 km^2 in 1984 (99.7% loss). Forest cover had already been reduced to about 5% of the land surface by 1924, and although it has since been almost doubled, this is largely the result of planting exotic conifers. Since the 1930s almost half of the remaining ancient semi-natural broadleaved woodland has been lost, mainly by conversion to conifer plantations.

Losses of this magnitude will have had an inevitable impact on the abundance of many species. If the habitat goes, so do all those species which depend on it. There will also, however, be indirect effects of such habitat loss on surviving populations, due to their increased fragmentation and isolation, and it is these indirect effects that are discussed in this paper.

Changes in land use have tended to be piecemeal, as land owners have responded to changing economic pressures, and this has led to an increased fragmentation of surviving habitats. One of the best documented examples of this is the Dorset heaths (Moore, 1962; Webb and Haskins, 1980). The maps of Isaac Taylor show that in 1759 heathland in the Poole Basin covered just under 40 000 ha in seven or eight large blocks. By 1978, under 6000 ha remained and this was divided between 768 small patches, 476 of which were less than 1 ha in area.

II. ISOLATION AND POPULATION DYNAMICS

Synecologists have long been interested in the effects of area and isolation of habitats on the numbers of species they contain. MacArthur and Wilson's (1963, 1967) classic study of island biogeography has formed the basis of such studies. This attempted to relate the number of species present on islands to the size of the island and to its distance from the source of colonists. It proposed that, for each island, the opposing forces of immigration and of extinction would balance, to give an equilibrium number of species. Autecologists have had little interest in this approach, in spite of the fact that the population dynamics of individual species must determine both extinction and immigration rates. I think that there are two reasons for this. Firstly, it is extremely difficult to quantify the impacts of movements on populations, and secondly, and perhaps more importantly, many ecologists believe that extinction is a rare event, because populations are buffered against it by density-dependent regulation.

Most attempts at quantifying dispersal involve the use of mark and recapture techniques, and these can give a very biased impression of lack of movement. Once a marked insect has left the place where it was marked, the chances of its recapture become less the further it moves, as larger and larger areas have to be searched to find it. Immigration is even more difficult to quantify, other than by studying the rates at which new habitats are colonized. In existing populations it is usually impossible to distinguish immigrant individuals from those bred on site. For these reasons, the majority of population studies ignore the effects of movement, by either assuming that the population is closed (that is, that movement does not occur), or that immigration and emigration balance one another, and so can be ignored. In my opinion, this is a major weakness to many studies, as it means that one of the determinants of insect abundance is ignored.

Results from the few population studies which have attempted to quantify movement suggest that it plays an extremely important role for many, if not most, insects. The scale of movement varies considerably between species, but it frequently appears to be a key factor affecting population fluctuations (cf. Dempster, 1989). Having said this, knowledge of the role of movement is very limited.

This lack of knowledge is carried across into much population theory. Most population models assume that the population is closed. They are concerned more with possible internal regulatory mechanisms to explain persistence than with the contribution of dispersal. Such models are based upon the concept of a population equilibrium, about which abundance is regulated by density-dependent increase or decrease. This is not the place to discuss the merits of different population theories, but many ecologists consider this concept to be ill-conceived and unrealistic (e.g. Milne, 1957; Pollard, 1981; Dempster and Pollard, 1981). Added to this, such regulatory mechanisms have proved extremely difficult to detect in field populations (*cf*. Dempster, 1983; den Boer, 1986, 1987, 1988). In my opinion, the ability of density-dependent factors to prevent population extinction has never been demonstrated adequately by theoretical, experimental or observational studies, yet it has become accepted as part of our ecological dogma.

Unfortunately, the frequency of natural extinctions is not well documented. For a start, it is difficult to be certain when a population has actually become extinct, and some species are so mobile that recolonization is rapid. Added to this, most population studies have been on large populations of common species, which have a lower probability of extinction. Nevertheless, many populations which have been studied have become extinct during the course of the study (see Dempster (1983) for examples in the Lepidoptera). Even so, there have been far too few attempts to estimate the rates of extinction of insect populations. One exception to this concerns the Carabidae, and den Boer (1981, 1985) has shown that local extinctions are a commmon feature of populations of these beetles. He estimated that population survival times commonly range from 10 to 40 years, rarely more than 100 years, and he considers that these relatively short survival times owe more to the instability of their populations (i.e. high levels of fluctuation) than to instability of their habitats. He also makes the point that ecologists get a false impression of the persistence of populations because autoecological studies rarely continue for more than 10–15 years.

I believe that there is now a growing body of evidence to suggest that insect populations are far more dynamic, both spatially and temporally, than current theoretical models would suggest, and that persistence owes more to repeated recolonization after extinction than to any internal population regulation. If this view is correct, it has considerable implications on how we should go about trying to conserve insects. This will be discussed later, after I have considered how insects are likely to respond to habitat fragmentation and isolation.

III. EXAMPLES OF INSECTS RESPONDING TO INCREASED ISOLATION

A. *Papilio machaon*

If, as seems likely, mobility is genetically determined, one might expect isolation of populations to increase selection against mobility, as emigrants fail to find suitable habitats and immigration of mobile individuals into the population stops. There is no direct way of testing this idea, but a study on the swallowtail butterfly (*Papilio machaon* L.) at Wicken Fen (Dempster *et al.*, 1976) suggests that the morphometrics of this insect may give some indication of its mobility.

Wicken Fen was once part of an extensive area of fenland, which was gradually drained and put under agriculture between the Middle Ages and the first half of the nineteenth century (Godwin, 1978). In the early 1800s, open fen habitats would still have been present around the surviving meres, such as Whittlesey, Trundle, Ugg and Ramsey. Almost certainly, swallowtail populations occurred in these habitats, and indeed, two specimens labelled Whittlesey 1819, 1821, are present in the collections of the Hope Museum, Oxford. Once these meres were drained (Whittlesey was the last to go in 1852), the Wicken population was completely isolated. The butterfly survived on the Norfolk Broads, but these are about 100 km from Wicken. The swallowtail eventually became extinct at Wicken in the early 1950s.

Morphometric studies on museum specimens showed that Wicken and Norfolk butterflies differed from one another in a number of measurements

(Dempster *et al.*, 1976). Differences in wing and thoracic measurements were particularly marked. Now differences in individual measurements may simply reflect differences in total body size, but changes in the ratio of measurements (i.e. shape) are more likely to reflect functional changes in the butterfly.

Figure 1 shows the means (±2 SE) of the ratio of thoracic width to length for male and female swallowtails from Wicken and Norfolk. Differences were particularly marked in the females between 1880 and 1920,

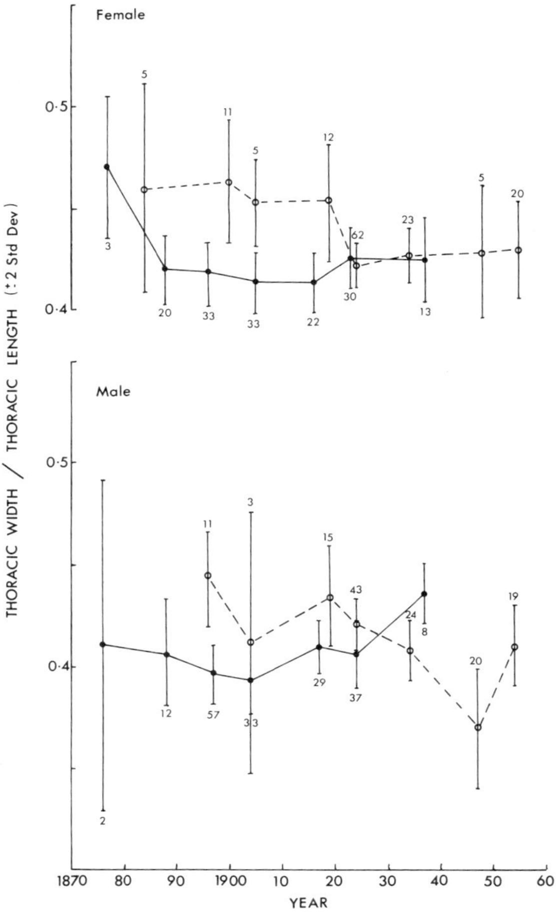

Fig. 1. Changes in the ratio of thoracic width to thoracic length in the swallowtail butterfly from Wicken Fen (—•—) and Norfolk (– –o– –). The small figures give the number measured.

with the ratio being significantly smaller in the Wicken specimens. Any explanation of these findings must be conjectural, but, since the thorax contains the wing musculature, it is tempting to think that a reduced thoracic size was linked to a reduced mobility, which in turn may have been the result of isolation of the Wicken site. The timing of the change in the Wicken butterfly would fit this hypothesis, as the fen became totally isolated around 1850. The ratio of thoracic width to length also declined in the Norfolk butterflies around 1920, so that there were no significant differences between the sites by the time the Wicken butterfly became extinct. This is less easily explained, but it seems probable that this coincided with a loss of suitable habitats in Norfolk, due to a reduction in the management of these fens during the 1914–1918 war. The hypothesis that these changes reflect differences in mobility is strengthened by an experiment carried out on Norfolk butterflies, which showed that individuals with low thoracic width to length ratios were weaker fliers (Dempster *et al.*, 1976).

B. *Maculinea arion*

I have recently had the opportunity to study similar morphometric data for the large blue butterfly (*Maculinea arion* L.), kindly supplied by Jeremy Thomas and Marney Hall. This butterfly became extinct in Britain in 1979. At one time the butterfly occurred on about 90 sites, stretching from Northamptonshire in the north, down through the Cotswolds and Quantocks, to Devon and Cornwall. About half of these sites were destroyed by ploughing or afforestation, and the others became unsuitable for the butterfly, and more particularly for its ant host *Myrmica sabuleti*, with the development of scrub, which resulted from lack of grazing, particularly by rabbits (Thomas, 1976, 1980; Heath *et al.*, 1984).

After its third instar, the larva of the large blue is predatory and lives within the nests of *Myrmica*, feeding on ant eggs and grubs. There is a large variation in the size of individual butterflies due to occasional overcrowding in the ant nests, and very small specimens can occur. Comparisons between ratios of measurements should, however, be less dependent upon larval food supply.

Many of the early specimens in collections have limited information concerning the locality where they were caught. Generally, locations are given which could apply to several populations. However, trends in their morphometrics, similar to those seen in the swallowtail, are clearly identifiable.

The best data set covers a total of 89 butterflies from Bude, in Cornwall. In this area, the butterfly was discovered in 1891 at Millook, just south of Bude (*cf.* Thomas, 1976). Very large numbers were collected each year from 1895 to 1914 from sites along this stretch of the coast, where the butterfly bred on rough grazings. By the turn of the century much of the flatter ground along the coast had been enclosed and ploughed, and the butterfly became restricted to the deep-sided coombs. Many sites became overgrown from lack of swaling (burning), but the butterfly remained locally abundant until about 1930, when it declined. Accounts of the habitat at the time suggest that many areas had become overgrown. The species hung on in low numbers in the area until the 1950s–1960s, the last specimen being seen at Crackington Haven in 1963 (Thomas, 1976). The butterfly was also discovered on a number of sites to the north of Bude in the late 1930s or early 1940s, but it is fairly certain that the data shown in Fig. 2 relate to sites to the south.

Figure 2 shows the changes that occurred in the ratio of thoracic width to thoracic length between 1900 and 1914, the period for which specimens are available. Data for males and females have been combined, as no significant differences were detected between the two sexes. A significant negative correlation with time ($r = 0.34$, $p < 0.01$) is seen, although there are reasonable data for only six individual years. Although they show a consistent trend, a *t*-test showed that the means for successive years were

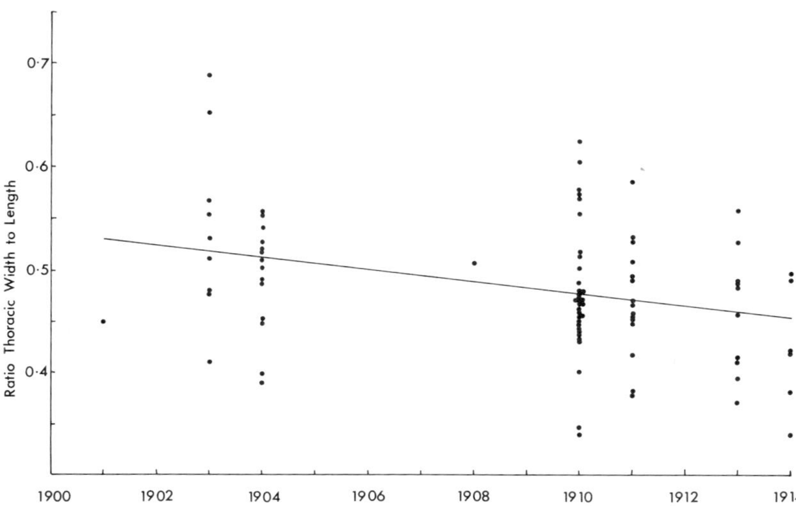

Fig. 2. Changes in the ratio of thoracic width to thoracic length in the large blue butterfly from Bude, Cornwall.

TABLE I. Ratios of thoracic width to thoracic length for the large blue butterfly from Bude, Cornwall.

Year	n	Mean ± 2 SD		
1903	9	0.541 + 0.173	0.512	
1904	14	0.493 + 0.105		$p < 0.05$
1910	33	0.479 + 0.128	0.477	
1911	15	0.473 + 0.110		$p < 0.08$
1913	10	0.461 + 0.121	0.448	
1914	6	0.426 + 0.123		

not significantly different, but the differences between pairs of years (1903 + 4, 1910 + 11, 1913 + 14) were close to significance (Table I). It is a pity that we do not have data after 1914, when the distribution of the butterfly became greatly restricted. However, the period covered by the data does coincide with the loss of habitat from the flatter high ground as this was put under the plough.

The large blue survived longest in south-east Dartmoor. Thomas (1976) considers that there were about six sites in this area, one near Ashburton, which became extinct around 1872, and five others, all within 1 km of each other. We have 40 specimens labelled Dartmoor covering the period 1932–1952, and a single specimen taken in 1973. Again, no significant differences were found between the sexes, and so the data were combined. The mean ratio of thoracic width to length for 1932–1935 was not significantly greater than that for 1950–1952, the two main periods for which there are data, but the ratio for the 1973 specimen was extremely low, and was outside the 95% fiducial limits of the mean for other years (0.28 compared with 0.52 ± 0.12).

Finally, it is perhaps interesting to look at data for the Cotswold sites, since these had an odd history of loss and resurgence of the butterfly. In all, 33 sites were recorded in this area, in the triangle between Gloucester, Stroud and Cheltenham (Thomas, 1976), but it is likely that some of these were offshoots from larger colonies, and probably no more than 12 sites existed at any one time. Soon after its discovery in 1850, the butterfly was extremely abundant in this area. However, it declined severely in the 1880s and was thought to be extinct for several years. It appears to have survived, however, and showed a remarkable recovery in the next 30 years to become locally common in the 1920s and 1930s. It declined again in the 1950s and the last certain record of it from this area was in 1960.

Figure 3 shows the data from this locality, grouped into 10-year periods. Again, there were no significant differences between the sexes. The ratio of thoracic width to length shows a trend which broadly fits what is known

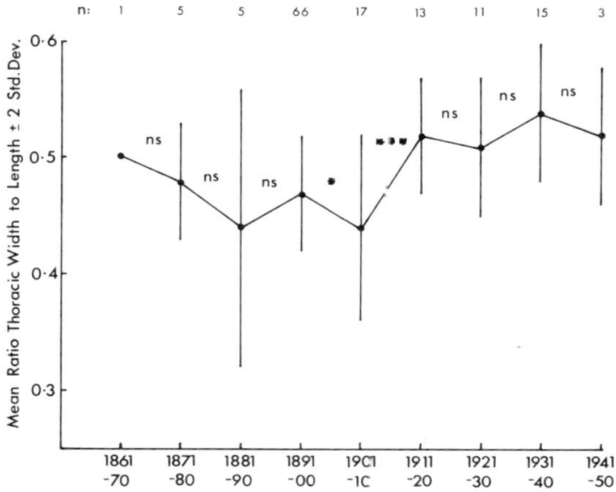

Fig. 3. Changes in the ratio of thoracic width to thoracic length (mean ± 2 SD) for the large blue butterfly from the Cotswolds (n = number of specimens measured; ns = non-significant difference; * $p < 0.05$; *** $p < 0.001$).

about the changing status of the butterfly in this area. The ratio declined after 1870, when the butterfly became rare, and increased again in 1910–1920, when the butterfly recovered and reoccupied sites from which it was previously lost.

IV. CONCLUSIONS

Any conclusions drawn from these data have to be tentative. There is no way of knowing whether morphometric changes of this sort have any connection with changes in behaviour. However, taking these data with those for the swallowtail, there does appear to be a consistent pattern of change which coincides with loss of habitat and isolation of populations, and which could well indicate a reduction in mobility of these butterflies.

Several authors have commented on the lack of movement of British butterfly species, and have suggested that, predominantly, they live in small closed populations (*cf.* Thomas, 1984, who suggests that 85% of UK butterflies live in closed populations). This may well reflect the current position in Britain, but one is left wondering whether this was always the case. If the morphometric changes reported here do reflect a change in mobility, the current lack of movement may simply be a symptom of the species being in trouble.

Where does all this lead us when planning a policy for conserving insects? In the long term, can we hope to maintain species on isolated areas of habitat in nature reserves? Evidence suggests that the answer to this question probably has to be "no". Again, if one looks at what has happened to the British butterflies, there is clear evidence of continuing losses of species from nature reserves. These losses are well documented for Monks Wood nature reserve (Steele and Welch, 1973) and for Castor Hanglands nature reserve (Collier, 1966), which have now lost most of the less mobile species which have very specific habitat requirements. There is no reason to believe that similar losses have not also occurred for other groups of insects. To conserve such species in the long term, we need to ensure that populations in reserves are kept as high as possible, so as to reduce the likelihood of local extinctions. This probably depends, more than anything else, upon the carrying capacity of the habitat. If the reserve contains sufficient habitat to meet the needs of a large population, the probability of its extinction will be reduced. I would stress that it is the carrying capacity that matters, not the absolute size of the reserve or of the habitat.

In the event of local extinction, the species can recolonize the site only if sources of colonists are sufficiently close to the site. Obviously, this will be dependent upon the mobility of the species concerned, but as habitats become more isolated, the ability of species to recolonize will be reduced. Added to this, it seems possible that isolation will cause selection of less mobile forms, and so accentuate the problem. For a small number of especially prized species, it is always possible to reintroduce them to reserves, but this cannot be the basis for conserving more than a handful of species. We have to create conditions which enable most species to look after themselves. To do this, we have to look outside our reserves to try to ensure that semi-natural habitats persist which can act as sources of migrants, or as "stepping stones" for migration. There is much to be said for trying to set up reserves within easy reach of one another. The vegetation on them should then be managed, out of phase with one another, so that one or more is always at any one stage of vegetational development. In this way, it may be possible to ensure the maximum availability of habitat in the right condition for different species, for as long as possible.

If we follow this strategy, one other management requirement is forced upon us. We need a far clearer set of management objectives for reserves than exist at present, and these must be co-ordinated with those of other reserves. What I am suggesting requires a far less insular approach to reserve management, with co-ordination between reserve owners to achieve maximum effect.

REFERENCES

Collier, R. V. (1966). Status of butterflies on Castor Hanglands NNR 1961–1965 inclusive. *J. Northampton Nat. Hist. Soc.* **35**, 451–456.

Dempster, J. P. (1983). The natural control of populations of butterflies and moths. *Biol. Rev.* **58**, 461–481.

Dempster, J. P. (1989). Insect introductions: natural dispersal and population persistence in insects. *Entomologist* **108**, 5–13.

Dempster, J. P. and Pollard, E. (1981). Fluctuations in resource availability and insect populations. *Oecologia* **50**, 412–416.

Dempster, J. P., King, M. L. and Lakhani, K. H. (1976). The status of the swallowtail butterfly in Britain. *Ecol. Entomol.* **1**, 71–84.

Den Boer, P. J. (1981). On the survival of populations in a heterogeneous and variable environment. *Oecologia* **50**, 39–53.

Den Boer, P. J. (1985). Fluctuations of density and survival of carabid populations. *Oecologia* **67**, 322–330.

Den Boer, P. J. (1986). Density dependence and the stabilisation of animal numbers. 1. The Winter Moth. *Oecologia* **69**, 507–512.

Den Boer, P. J. (1987). Density dependence and the stabilisation of animal numbers. 2. The pine looper. *Neth. J. Zool.* **37**, 220–237.

Den Boer, P. J. (1988). Density dependence and the stabilisation of animal numbers. 3. The winter moth reconsidered. *Oecologia* **75**, 161–168.

Godwin, Sir H. (1978). "Fenland: its Ancient Past and Uncertain Future". Cambridge University Press, Cambridge.

Heath, J., Pollard, E. and Thomas, J. A. (1984). "Atlas of Butterflies in Britain and Ireland". Viking, Harmondsworth.

MacArthur, R. H. and Wilson, E. O. (1963). An equilibrium theory of insular zoogeography. *Evolution* **17**, 373–387.

MacArthur, R. H. and Wilson, E. O. (1967). "The Theory of Island Biogeography". Princeton University Press, Princeton.

Milne, A. (1957). The natural control of insect populations. *Can. Entomol.* **89**, 193–213.

Moore, N. W. (1962). The heaths of Dorset and their conservation. *J. Ecol.* **50**, 369–391.

Pollard, E. (1981). Resource limited and equilibrium models of populations. *Oecologia* **49**, 377–378.

Ratcliffe, D. A. (1984). Post-medieval and recent changes in British vegetation: The culmination of human influence. *New Phytol.* **98**, 73–100.

Steele, R. C. and Welch, R. C. (1973). "Monks Wood: a Nature Reserve Record". The Nature Conservancy, UK.

Thomas, J. A. (1976). "The Ecology and Conservation of the Large Blue Butterfly". ITE, UK.

Thomas, J. A. (1980). Why did the large blue become extinct in Britain? *Oryx* **15**, 243–247.

Thomas, J. A. (1984). The conservation of butterflies in temperate countries: past efforts and lessons for the future. *In* "The Biology of Butterflies" (R. I. Vane-Wright, and P. R., Ackery, eds), pp. 333–353. Academic Press, London.

Webb, N. R. and Haskins, L. E. (1980). An ecological survey of heathlands in the Poole basin, Dorset, England in 1978. *Biol. Conserv.* **17**, 281–296.

8

Woodlands: Past, Present and Potential for Insects

M. S. WARREN AND R. S. KEY

I.	Introduction	155
II.	Historical Background	156
	A. The Post-glacial Wildwood	156
	B. The Age of Man — Prehistory to the Industrial Revolution	159
	C. The Last 200 Years	162
III.	Changes in the Status of Woodland Insects	167
	A. Insects of Old Forest Woodland Habitats	168
	B. Insects of Open Habitats within Woodland	171
IV.	Ecological Specialization among Insects of Mature Woodland and Open Spaces	176
	A. Niche Specialization among Mature Old Forest Insects	177
	B. Niche Breadth among Open Space Insects	184
	C. Mobility	187
V.	The Management of Woods for Insect Conservation	190
	A. The Determination of Priorities	190
	B. Habitat Continuity and the Maintenance of Traditional Regimes	193
	C. Simulation of Traditional Practices within Modern Wood Management	195
	D. Re-creatibility of Woodland Habitat — Opportunities and Limitations	198
VI.	The Need for Research and Education	200
	A. Research	200
	B. Education	201
VII.	Summary and Conclusions — the Future for Woodland Insects	201
	References	203

I. INTRODUCTION

Without the influence of man, forest is the natural climax vegetation for almost all of temperate Europe. Probably well over half of Europe's

150–200 thousand species of insect are dependent on this biotope. As a result of man's activities only a very small proportion of the original forest cover now remains, and virtually all of this has been heavily modified in some way by man. This has had far-reaching consequences for all woodland insects, but those habitats associated with the very earliest and latest successional stages have been altered most drastically. Problems in insect conservation are therefore particularly acute among the specialist faunas of the open spaces within woodland and with that of the most mature forest habitats, especially decaying wood on ancient trees. In contrast, habitats associated with the intermediate stages of the woodland succession — "mature" as opposed to "post-mature" woodland in modern forestry terminology (Kirby, 1984) — are widespread either under modern broadleaved high forest management or within neglected coppice woodland. The associated fauna, composed generally of less specialized species, has fared considerably better.

By focusing on the two contrasting ends of the woodland seral spectrum, we intend to highlight conservation problems that are relevant not only to other forest insects, but also to many other groups of plants and animals. Examples will be taken mainly from Great Britain, where the natural woodland has suffered considerably from human activity. Forest cover has been reduced to just 8% of the land area and, of this, only about one-quarter is ancient, semi-natural woodland (as defined by Peterken, 1981), while over half comprises plantations of non-native tree species. These changes are typical of a number of other countries in north-west Europe, for example The Netherlands and Ireland, where woodland cover has been reduced to 4% and 7%, respectively. Clearance of forest has been less complete in countries such as France, Germany, Italy, Spain and in East European countries where the forest cover remains at 20–40% of the land area and still mostly comprises native species (Peterken, 1981), but again the age structure of the remaining woodland has been heavily modified by man.

II. HISTORICAL BACKGROUND

A. The Post-glacial Wildwood

During the period shortly after the end of the last Ice Age, a more or less continuous cover of forest (the "wildwood") developed over the vast open spaces left after the glaciation (e.g. Godwin, 1975). The immediately post-glacial climate was somewhat warmer than at present and conditions outside of forest cover may well have been suitable for those species now restricted to warm microclimates in sheltered open space within woodland in Britain, but which are often associated with more open conditions in southern Europe (Dennis, 1977; Girling, 1982; Thomas, 1991). There is, however, some debate about the extent of open space within the primeval forest that subsequently developed. Natural openings occurred along river valleys, lakes, wetlands and cliffs and places where the underlying geology or drift deposits led to particularly thin or infertile soils (Bennett, 1988). Small gaps would also be created by the senescence and death of old trees, and larger gaps by natural catastrophic events such as fire and wind-throw during storms. Forest regeneration is likely to have been fairly rapid in most of these openings, but it seems likely that larger and more permanent openings would have been maintained by large herbivores including auroch, deer, bison and wild boar. Evidence compiled by Worsley (1978) and Coles and Orme (1983) also indicates that beaver cleared substantial areas of woodland about watercourses at this time. Beaver colonies usually move only a very short distance up or down stream, producing an open damp meadow surrounded by cleared woodland (Neff, 1957), in which herbivores may prevent the reinvasion of forest for several decades (Wilde *et al.*, 1950). This effectively provides a series of rotational clearances and semi-permanent glades which have similarities to the human practice of coppicing. Coles and Orme (1983) postulate that early man may have utilized these for his initial colonization of the European wildwood.

Together, these animals may have maintained substantial open spaces, possibly for long periods, long before man had his dramatic impact on the forest. This may help explain the existence of a large number of species of insect which are adapted to open biotopes in Europe, a continent usually thought to have been covered by almost closed forest in interglacial periods.

The age structure of the trees in the primeval forest was far more dominated by old and very old trees than in forest now influenced by man. This can still be observed in more or less natural remnants of forest in eastern Europe (e.g. the Bialowieza Forest national park on the Polish/USSR border (Falinski, 1978, 1986), certain natural forests remaining in Scandinavia (Heliovaara and Vaisanen, 1984) and in North America (Harmon *et al.*, 1986). Such an age structure leads inevitably to a very large accumulation of dead and decaying wood, both attached to living ancient trees, and also fallen to the forest floor (Table I). Harmon *et al.* (1986) estimate that as much as 17% of all organic material in "intact" woodland is in the form of dead wood, and that 6–25% of the soil surface normally has a piece of dead wood on it. The abundance, very large size and variety of conditions of decaying wood in natural forest

TABLE I. Fallen dead wood in various types of deciduous woodland in Europe (sources: Boddy and Swift, 1983; Christenson, 1977; Evans, 1984; Falinski, 1978; K. J. Kirby, unpublished data).

Woodland type	Quantity of fallen dead wood (tonnes ha^{-1})
Young plantation (lime/hornbeam)	1.4–3.4
Middle-aged plantation (beech)	4.8
Managed coppice (sweet chestnut + standards)	8.1
Abandoned coppice (various types)	0.2–15.8
Coppice singled to high forest (sweet chestnut)	15.4
"Natural" ancient forest (lime/hornbeam)	36–64

Where volume is given in m^3 a mean density of $0.685\,t\,m^{-3}$ is assumed, taken from Evans (1984).

provided a very diverse spectrum of niches for saproxylic (defined in Speight, 1989) invertebrates (Section IV.A) and many species of insect have become highly specialized to exploit this abundant resource.

Catastrophic events such as wind-throw, forest fire and disease epidemics would not only have produced gaps in the primeval woodland, but would also have led to a distinct patchiness in the distribution and abundance of fallen dead wood and damaged trees supporting saproxylic habitat (Cline *et al.*, 1980; Heliovaara and Vaisanen, 1984). This would have been intermixed with tree-fall gaps, beaver meadows and grazing areas to produce a large and small-scale mosaic of different forest successional stages in close juxtaposition. Probably as a direct result of this, many saproxylic species appear to have evolved to depend on both ends of the successional spectrum. Species whose larvae depend on dead wood frequently need open space as adults for mating and dispersal, and an abundant supply of flowers producing nectar and pollen as an energy and protein source for activity and egg laying. This dependence on different but adjacent successional tesserae within a mature woodland habitat mosaic, maintained by occasional catastrophic events, is not restricted to the invertebrates. The wider implications for other components of the flora and fauna are discussed by Remmert (1987).

B. The Age of Man — Prehistory to the Industrial Revolution

1. Prehistoric Clearance and Management

Human impact on the wildwood did not become severe until the Neolithic, 6500–4000 BP (see reviews by Peterken, 1981; Rackham, 1980). During this period, the extensive clearance of woodland to make way for agriculture produced possibly the single biggest effect on insects and other forest wildlife. Rackham (1986) estimates that about half of the British wildwood had been cleared by the Iron Age (2500 BP) and, at the time of the Domesday survey of 1086, woodland probably covered only about 15% of the land area. The destruction of natural forest has continued to the present day with ancient, semi-natural woodland now reduced to about 2%. As a result, a very large proportion of woodland insects must have declined substantially, and there is considerable subfossil evidence that many species became extinct (see Section III.A below) while many more now maintain a precarious existence in the small fragments of forest and woodland that remain.

At the same time as the forest was being cleared, early man began to exploit and utilize the remainder in an increasingly systematic way — continuous forest gave way to managed woodlands. The broad differences in biological characteristics between unexploited and exploited forests are outlined by Walter (1987). The impact on the woodland insect fauna would have been almost as great as that of forest clearance. A number of different woodland management regimes were developed.

2. Coppicing

The practice of coppicing started as long as 6000 years ago and, by the time of the Domesday survey of England in 1086, had become the dominant form of management in woods throughout lowland Britain (Rackham, 1980). Most coppiced woods were managed on a short rotation (e.g. Figure 2.4 in Peterken, 1981), often 5–10 years, while the larger standard trees were felled on a cycle of 80–150 years (Peterken, 1981). "Mature" (in modern forestry terminology) timber was highly valued for house and ship building, and few trees would have been allowed to reach the age where they provided good habitats for saproxylic insects. Even small branches and twigs were gathered and used as faggots for burning. Coppice therefore provided a large amount of open space and young growth habitat for insects, but little in the way of very ancient trees and decaying wood. Habitats for insects adapted to middle-aged woodland growth may also have been fairly restricted, but some species undoubtedly survived on the later stages of growth in the coppice cycle or on standard trees, which were sometimes quite numerous. Also, coppice rotation lengths were variable and, in some regions of the country, exceeded 30–40 years, providing a greater proportion of older growth and longer periods of shade which would have been more suited to this fauna. From the seventeenth century onwards, the amount of open space in many woods was increased by the creation and maintenance of rides, which were essentially a network of permanent, linear glades. These still exist in many woods, and are important access routes for forest management and timber extraction.

3. *Wood Pasture*

The other major system of forest management which became widespread in Britain was that of wood pasture. This has been defined by Harding and Rose (1986) as "areas of wooded land which have been used for the dual purpose of growing trees and grazing deer and livestock". Several different forms were developed, including wooded common, deer park, royal and baronial forest and chase, and winter-grazed woodland (Harding and Rose 1986). All forms contrasted greatly with coppiced woodland, providing a completely different habitat for insects. Grazing animals were excluded from most coppice, which was often fenced for several years after cutting to prevent browsing of the regrowth. The practice of pollarding was developed in wood pasture, where trees were cut 2–3 m above ground on a 5–20-year cycle to produce a coppice-style regrowth out of reach of livestock. The opportunities for tree regeneration in wood pasture were therefore much smaller and sporadic compared with coppice. This now has considerable implications for the continued survival of the habitat in the long term (see Section V.B).

The significance of pollarding to insects is that it considerably prolongs the life of a tree. Wind resistance is reduced by never allowing the development of a large crown of boughs and foliage, and the centre of gravity is lowered, thereby reducing the likelihood of wind-throw, catastrophic wind damage and lightning strike. The tree is also regularly rejuvenated by stimulating the production of vigorous young growth (Le Seur, quoted in Edlin, 1971) while, at the same time, exposed cut wood is continually created which can be colonized by saproxylic fungi and insects. The uneven structure of a pollard crown favours the retention of water, the development of rot holes and, eventually, heart rot and the bolling of the tree may contain significant quantities of decaying wood which continues to be replenished by new growth from the cambium. Pollard bollings may survive for 300–500 years (or 800 years for oaks), which is considerably longer than the "natural" life span of short-lived or shallow-rooted species such as beech (Edlin, 1971). Moreover, Cartwright and Findlay (1958) point out that trees in isolation, such as in parkland and in very open woodland, are more likely to lose branches through wind damage, and are more likely to become infected by fungi than are trees growing in closed canopy woodland. They are consequently likely to produce more microhabitats for saproxylic invertebrates. The wood pasture system therefore provided far more "post-mature" woodland habitats for insects than coppiced woodland, although in many coppiced woods the coppice panels were marked by boundary pollards, which also lived to a great age and may have served as refugia for some mature timber species.

4. Orchards

An extremely specialized form of plantation woodland management was also developed in the form of the orchard. Favoured fruit-bearing trees can be kept alive for much longer periods than their natural span by continual pruning to maintain fruit within picking range, much as in the system of pollarding. Some areas of Britain, notably Herefordshire and Worcestershire, have been traditional orchard-growing areas for centuries. This has led to some old orchards becoming important for saproxylic insects: at least one species in Britain, *Gnorimus nobilis* (Coleoptera; Scarabaeidae), is now almost dependent on old orchard fruit trees (Hyman, in preparation), although on mainland Europe it inhabits the dead wood of a variety of other trees species (Kelner-Pillaut, 1974).

5. Scottish and Irish Forests

While the lowland broadleaved forests of Scotland suffered a similar fate to that of their English counterparts, it is thought that the Caledonian pine forests of the Highland glens were largely unscathed until at least the beginning of the seventeenth century. There was then an increased demand for timber (Carlisle, 1977). Native pine forest in Ireland appears to have been completely cleared during this period, resulting in the total extinction of the "Caledonian" element of the pine feeding saproxylic fauna (Speight, 1985).

6. Hedgerows

Although the massive destruction of forest by man drastically reduced the total amount of woodland in Britain, and undoubtedly caused a number of extinctions and declines of woodland insects, the systems of woodland management that were developed maintained a considerable diversity of forest habitats. By the start of the nineteenth century, Britain still contained a rich mosaic of ancient semi-natural woodland, supporting a large number of invertebrate species associated with both ends of the woodland seral spectrum. While covering only about 4% of the land area, this lowland woodland was interconnected by an extensive network of hedges, some initially derived from the previous forest cover and thereby maintaining continuity of conditions. Most of these hedges also contained numerous mature trees, often including large pollards. Rackham (1986), quoting a survey of Thorndon in Suffolk in 1742, records a density of hedgerow pollards in normal farmland of 32 $acre^{-1}$ (79 ha^{-1}). This network of hedgerows was supplemented in some areas by the enclosures made between 1750 and 1850. By 1946, when the first systematic aerial photographic mapping of Britain took place, there were an estimated 13 miles of hedgerow per square mile of agricultural land in England and Wales (Pollard *et al.*, 1974). This extensive hedgerow network provided numerous important habitats for woodland edge invertebrates, and corridors between individual woods, compensating, at least to some extent, for the small proportion of forest cover.

C. The Last 200 Years

1. Loss of Coppice Woodland

The industrial revolution of the early nineteenth century produced another major change in British woodland. Initially there was an even greater demand for coppice products in the form of charcoal for iron smelting. However, charcoal and wood were soon displaced by coal as the main fuel for industry and domestic heating, causing traditional coppice and pollard woodland management to go into a rapid decline. By the time of the first national survey in 1905, coppice was practised in only 30% of woods in England and Wales (Peterken, 1981). Since then, it has declined still further to just 2% of woods, most of this being of non-native sweet chestnut in Kent and east Sussex (Forestry Commission, 1984). Over the last 50 years 82% of Britain's coppice has been neglected, but the coppicing of native species has declined by at least 94% (Table II). Throughout most of the country, coppicing and pollarding virtually ceased, and woodland became neglected or was promoted to high forest, either by clear felling and replanting or by "singling" of the coppice stools. Both changes have had radical effects on the insect fauna, as most modern high forest woodland is invariably far shadier than formerly, and the tree composition

TABLE II. Net changes in types of British woodland over the last *ca* 50–57 years (our calculations based on Countryside Commission, 1989; Forestry Commission, 1928, 1984; NCC, 1984; Peterken, 1981; Peterken and Allison, 1989).

Woodland type	*% change*
Total woodland	+76
Coniferous woodland (mainly non-native spp.)	+199
Deciduous woodland (mainly native spp.)	+3
Ancient semi-natural woodland (all native spp.)	−30–50
Managed coppice (all spp.)	−82
Managed coppice (native spp. only)	−94
Uncoppiced deciduous woodland (mostly "middle-aged")	+36
Uncoppiced ancient semi-natural woodland (mostly "middle-aged")	−2
(Hedgerows 1947–1985, England only)	(−36)

has often been completely changed, mostly involving the replacement of native with alien species, especially conifers. Species associated with early successional stages suffered a great loss of habitat, while species preferring intermediate successional stages of woodland benefited (see section III.B).

The extensive loss of open space through the decline of coppicing was partly ameliorated by the continued existence of the ride systems, but many of these also became heavily shaded as the surrounding trees grew taller under management as high forest. Nevertheless, rides have become increasingly important refuges for open space insects and as sources of nectar for mature wood species. However, ride vegetation is usually different from newly cleared woodland, and is not suitable for all early successional insects (see Section V and Warren and Fuller, 1990).

During the period that coppicing declined, the total amount of woodland in England almost doubled (Fig. 1). This, however, was largely the result of conifer afforestation of marginal land, notably of heathland and hill country. Very much larger areas were planted in the uplands of Wales and Scotland (Forestry Commission, 1984). Conifers have also been widely planted to replace native broadleaves in deciduous woodland, so that over the last 50 years there has been an almost four fold increase in the area of conifer woodland in Britain, while 30–50% of ancient semi-natural woodland has been lost (Table II). Nevertheless, because of the

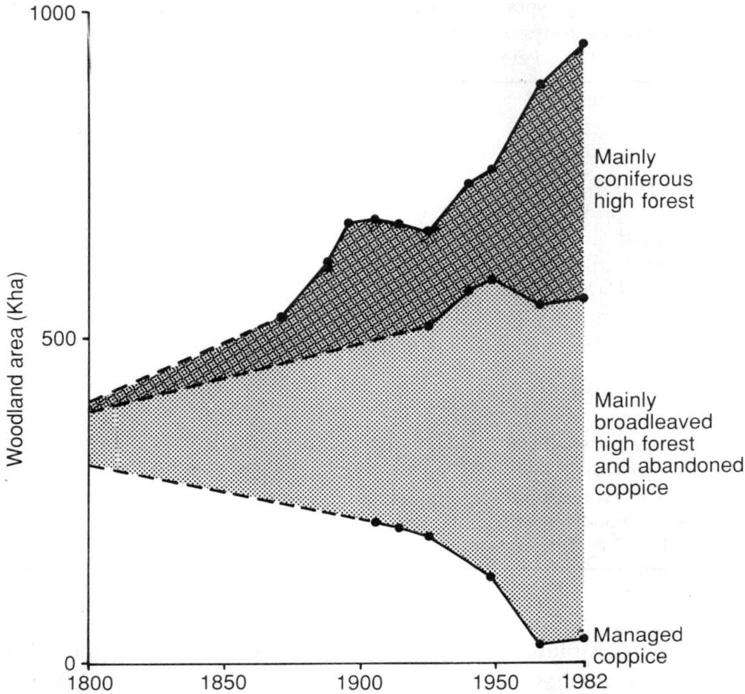

Fig. 1. Changes in the extent and composition of woodland in England since 1800 (sources: Forestry Commission, 1928, 1984; Peterken, 1981; Peterken Allison, 1989; Rackham, 1980). Dotted lines indicate estimated changes where census data do not exist.

abandonment of coppice, the area of uncoppiced deciduous woodland has increased by 36% over the last 50 years. Losses and changes in Britain's woodland and hedgerows are discussed by Peterken and Allison (1989).

2. Loss and Degeneration of Wood Pasture

Preceding and concurrent with the loss of coppice woodland was a similar decline in traditional wood pasture. As early as 1698 the process of pollarding was prohibited by Act of Parliament in the New Forest in Hampshire to promote the growth of long ship timbers (Tubbs, 1968). The process was reinstated to provide spring fodder for deer in the late eighteenth century, but ceased again in the mid-nineteenth century (Edlin, 1971). Pollarding was similarly outlawed by Act of Parliament in 1868 in the huge pollarded forests of Epping in Essex in order, so it was argued, to conserve the woodland (Ranson, 1978). Largely because it was a labour-intensive method of management, pollarding had almost ceased in British woods by the late nineteenth century (Edlin, 1971). Uncontrolled pollard regrowth following the last cutting has now resulted in large numbers of pollards becoming top heavy, and liable either to windthrow or the splitting of a pollard bolling unable to bear the strain of supporting the equivalent of a small grove of mature trees on its crown. Trees that were subjected to a smaller number of pollarding cycles, such as in the New Forest where they were pollarded perhaps no more than four times, seem now to fare worse than those trees that have had a longer history of pollarding. An example of the latter exists at Burnham Beeches, Buckinghamshire, where there were as many as 20 loppings per tree (Edlin, 1971). Large numbers of old pollards were, however, lost at both sites in the exceptional storms of 1987 and 1990. Another problem in neglected pollard woodland is the overtopping and shading out of the pollards by vigorous young maidens that have grown up through natural regeneration since the pollards were last lopped. Examples of this occur in parts of Epping and Hatfield Forests in Essex (Rackham, 1989), and in Windsor Forest in Berkshire.

During this century, large areas of wood pasture, ancient high forest remnants, old orchards, medieval deer parks, chase and royal forest have been cleared for arable agriculture, or felled and replanted, again most often with alien conifers. Of that which remains, the natural death of the ancient trees, coupled with a lack of regeneration as a result of past grazing, has led to the loss of the long-term ability of many sites to provide habitats for saproxylic invertebrates. This has often left increasingly unbridgeable age-class gaps in the structure of tree populations, spanning as much as a couple of hundred years. An example of this exists at Dinefor Deer Park in Carmarthenshire (Harding, 1981a).

3. Effects of Public Access to Sites with Ancient Trees

Ancient trees, especially pollards, are now viewed as potentially dangerous in areas with any form of public access (Edlin and Nimmo, 1956; Forestry Commission, 1987; Countryside Commission, 1988). The trees and fallen timber that are most suitable for saproxylic invertebrates are perceived either as dangerous, unsightly or as a potential source of infestation for pest species of fungi and invertebrates. This last concern is unnecessary given the almost complete lack of overlap of species between the saproxylic invertebrate fauna of ancient trees with the younger forest tree pest fauna (Heliovaara and Vaisanen, 1984) and the tendency of most pathogenic fungi to be excluded from trees already colonized by other saprophytic fungi (Swift, 1977).

The habitat for saproxylic invertebrates has suffered considerably from the tidying, felling, lopping and tree surgery of ancient trees, particularly since a significant proportion of the ancient parks and royal and publicly owned forests has been opened to the public. For example, the population of ancient oaks in Sherwood forest, which numbers fewer than 1000 trees and yet supports one of the richest old forest faunas in Britain, is threatened by felling and extensive tree surgery because the site has been developed as a popular country park, attracting over a million visitors a year into an area of only 181 ha (Squance and Harris, 1988). Harding (1976) even records ancient pollards in an old park that was converted to a safari park being killed by giraffes and baboons! The vandalism and unintentional damage perpetrated by some visitors to these sites has also become a problem. A serious example is the burning out of ancient hollow trees for use as sheltered sites for picnic fires, which has occurred at Sherwood Forest and at Dunham Massey Park in Greater Manchester. The continuing loss of saproxylic habitat through "tidying" at Windsor Forest and Great Park, despite its recognition as the most significant site for this fauna in Britain, is discussed by Chalmers-Hunt (1969) and by Owen (1988).

Conservation and landscape heritage organizations have also caused the loss of saproxylic habitats, though excessive attention to "tidiness" and by replanting with conifers. The Woodland Trust, a prominent British woodland conservation organization, continues to stress the necessity for tree surgery and felling of decaying trees, while Britain's largest conservation landowner, the National Trust, has felled ancient trees and cleared dead timber at significant sites for saproxylic invertebrates, such as at Dunham Massey Park and Calke Park in Derbyshire. The National Trust now has, however, a conservation team that has done much to change its policy in favour of conservation of this habitat.

Recent decades have seen a considerable increase in the use of wood-burning stoves for domestic heating. Concern has been expressed that saproxylic invertebrate habitat is at risk from official and unofficial collecting of standing and fallen dead wood from ancient forest sites (Harding, 1981b). A large firewood yard at Epping Forest in Essex, one of Britain's foremost sites for saproxylic invertebrates, regularly has excellent habitat on sale for burning, while firewood concessions to estate workers remain a problem in many ancient parklands. The wider implications to wildlife of Firwood collection in North America are discussed by Carey and Gill (1980).

4. Natural Fires

Natural fires create open space in woodland, adding to the amount of dead wood by killing or maiming trees. A distinctive assemblage of species specializes on wood exposed to fire or on those fungi which specifically grow on singed timber. The suppression of natural fires in forest to protect timber resources has therefore had a deleterious effect on this assemblage of forest invertebrates. Despite being highly mobile to pursue such an opportunistic life style, most of the species with this association are now scarce or threatened in Europe (Heliovaara and Vaisanen, 1984). Management involving the tolerance of natural fires is only practicable and desirable, however, in very large forests.

5. The Effects of Some Entomological Studies

The methods used to find saproxylic species by a small number of insect collectors may now pose some threat to their habitat. The use of invasive sampling methods can destroy the dead wood or leave it open to subsequent desiccation. This may have a minuscule effect in comparison with clearance, felling, lopping and the "tidying" of dead wood but, by targeting known key sites and the habitats of individual rare species, it is possible that the damage caused may jeopardize the long-term survival of particular species (Harding and Rose, 1986). Several European countries now list a significant number of saproxylic species as legally protected (Collins, 1987), but there are good reasons why this approach has not been followed in the UK (McLean, 1990). Voluntary codes suggesting ways of minimizing the dangers to the fauna while still studying it (e.g. by using emergence traps and recording adults at blossom) have been prepared (JCCBI, 1972; Key, in press).

6. Hedgerow Loss

The network of hedgerows that once linked woods has also been depleted, particularly in predominantly arable areas like East Anglia. Overall hedgerow loss in England between 1947 and 1985 is estimated at 36% (Countryside Commission, 1989b). Woodland insects have thus not only suffered from the destruction of large areas of native woodland and from radical change in management of the remainder, but also from the increasing isolation of the remaining fragments.

III. CHANGES IN THE STATUS OF WOODLAND INSECTS

Old records for most insects are sparse, the only exception being for "macro" Lepidoptera, especially butterflies. For all groups, few records extend back even as far as the beginning of the nineteenth century. The period between the mid-nineteenth century and the First World War saw a huge interest in insect collecting in Britain, and gave the first baseline of distribution information against which to assess subsequent change. The interest in many groups of invertebrates is now increasing very sharply, encouraged by new identification guides, biological recording schemes and numerous national and local societies.

A. Insects of Old Forest Woodland Habitats

Subfossil evidence of assemblages of saproxylic beetles dating back to the Bronze Age (3000–4000 years BP) indicates that a significant number of species have either become extinct or undergone major declines in Britain. Girling (1982) lists 20 species of saproxylic beetles represented in these subfossil assemblages which are now extinct here. Osborne (1964), Hammond (1974) and Speight (1989) summarize the decline in the British saproxylic beetle fauna, and a review of the ecology and past and present distribution of 45 species of saproxylic beetles found in Bronze Age peat deposits is given by Buckland (1979). Twelve of these species no longer occur in Britain, and most of the remainder are now extremely rare (Hyman, in preparation). Knowledge of the prehistoric distribution and status of most other saproxylic invertebrates, notably flies, is almost non-existent, both because they lack highly sclerotized parts and because they tend not to remain within the dead wood as adults. This has led to an almost complete absence of a fossil record. It is reasonable to assume, however, that saproxylic species in other groups fared no better than the beetles.

It seems therefore, that the major declines of saproxylic invertebrates occurred during the early woodland clearances, long before insect recording began. Many of the formerly widespread species described by Buckland (1979) were already very uncommon in Britain by the time the first catalogues of the British fauna were being compiled (e.g. Stephens, 1839; Fowler, 1887). Speight (1989) also points out that little credence was given in nineteenth and twentieth century works, to very early records of conspicuous saproxylic species which are no longer recorded in Britain. This has obscured the extinctions that occurred in the late eighteenth and early nineteenth centuries. Recent changes in status in the saproxylic species of beetle and fly are perhaps less easy to interpret as the level and geographical extent of current recording considerably exceeds that of earlier times. Modern records appear to indicate a spread of species that may, in actual fact, be declining considerably, such as

Hammerschmidtea ferruginea (Diptera; Syrphidae) (Falk, 1991). Many of the most renowned localities for saproxylic species of old have been totally destroyed (e.g. Wheatley Wood in Yorkshire), or rendered unsuitable for these species by inappropriate management (e.g. Darenth Wood in Kent). Hammond (1974) lists eight species of Coleoptera associated with mature forest that have become extinct in Britain in post-Linnaean times. Shirt (1987) includes 14 species of saproxylic insects as extinct in Britain, defined as those not seen since 1900, and a further 13 species that have not been recorded since 1950. Although a number of species have been rediscovered after several decades without records (Hammond, 1974), the large number of species not recorded for such long periods indicates that are saproxylic insects continuing to become extinct.

A large number of saproxylic invertebrates are also included in other European Red Data Books, including many species in common between the various countries. Certain species are much more threatened in some parts of their range than in others. For example, a number of the threatened saproxylic invertebrates of Britain are widespread in central Europe, and some that are threatened in Scandinavia (e.g. Andersson *et al.*, 1987 for Sweden) are less uncommon in Britain. Many saproxylic species that are not rare enough to be included on Red Data Book lists are nevertheless still markedly uncommon. These species are described as "Nationally Scarce" in Britain (NCC, 1989). Although they may still be quite widely distributed, it is now rare to find them together in rich assemblages. Where this occurs it is usually at sites that also support greater rarities. The number of sites with such rich assemblages is now very limited. Chalmers-Hunt (1969), Harding (1976, 1978a,b,c,d) and Alexander (in preparation) list the best national examples, and Garland (1983) has carried out a similar analysis for central northern England. Significant sites for scarce invertebrates, including those for old forest species, are documented for all counties of Great Britain in the Nature Conservancy Council's Invertebrate Site Register (e.g. Key, 1987).

The situation in Scottish woodlands is far less clear. The fauna of the best remnants of native pine forest is quite well documented (e.g. Owen, 1989), but that of other pine woodlands and of lowland broadleaved woodlands, is with a few exceptions, very poorly known (e.g. Crowson, 1962, 1964). The saproxylic fauna of 23 ancient Scottish birch woods was surveyed by White (1982).

Hammond (1974) has suggested that climatic change may be at least partly responsible for the decline and extinction of saproxylic species. While this cannot be discounted, the remarkably disjunct distribution of a number of relict populations of saproxylic invertebrates (e.g. *Schizotus pectinicornis* (Coleoptera; Pyrochroidae) only in the Welsh Marches and Scottish Highlands and *Ischnomera cinerascens* (Coleoptera; Oedemeridae) in single sites in Herefordshire, Buckinghamshire and North Yorkshire (Hyman, in prep.) indicates that the survival of these species in isolated ancient woodland has largely been a matter of chance. Scarce saproxylic insects remain only in sites with continuity of suitable conditions; there is no evidence of a retreat along climatic gradients toward centres of distribution remaining favourable for these species. This is further borne out by the number of saproxylic invertebrates that were formerly widespread, either in subfossil deposits or in historic times, but which are now restricted to single locations where there has been the greatest continuity of dead wood. A good example is of *Dryophthorus corticalis* (Coleoptera; Curculionidae) at Windsor Forest, the only site for a number of such species.

The loss of habitat for saproxylic invertebrates from all the causes discussed above has been so great that a ten-point recommendation for the conservation of their habitat has been prepared by the Council of Europe Committee of Ministers (Council of Europe, 1988).

Some saproxylic species have increased in Britain, both in abundance and in geographical range, during the period covered by historic recording, but most of these are specifically associated with decaying softwood timber and have benefited from the extensive planting of non-native conifers. Only a very small number of these are derived from the native Caledonian pine fauna (Hunter, 1977; Welch, 1986), for example *Xylota caeruleiventris* (Diptera; Syrphidae) (S. Falk, personal communication) and *Ampedus nigrinus* (Coleoptera; Elateridae) (Hyman, in preparation); the majority are clearly recent colonists from Europe (Hammond, 1974). Almost all of these species are eurytopic with respect to the conifer timber they can utilize. They often breed in young, stressed trees, and hence some have become forest pests. None is associated with very large, overmature trees. Speight (1985) pointed out that the saproxylic fauna of a Scots pine plantation differs markedly from that of a native pine forest, even in areas with natural pine forest. The ranges of a few of the scarcer species have also expanded as a result of the death of elms following the epidemic of the fungal disease *Ceratocystis ulmi*. Examples include *Aulonium trisulcum* (Coleoptera; Colydiidae) (Marshall, 1978) and *Siagonium quadricorne* (Coleoptera; Staphylinidae) (Girling, 1982). This, however, is likely to be a temporary phenomenon as dead elm timber disappears. One or two of other scarce species normally associated with "high quality" ancient forest sites also appear inexplicably to have expanded in range in recent years, for example *Agrilus pannonicus* (Coleoptera; Bupestridae).

B. Insects of Open Habitats within Woodland

Within woodland, most butterflies depend on open areas such as rides and glades while several species that are woodland specialists in lowland Britain usually breed in new clearings (Table III). Many butterflies associated with open areas within woodland were abundant and widespread

TABLE III. Main habitat preferences of butterflies breeding in British woods (after Warren and Fuller, in press. Shade figures follow definitions of Warren, 1985).

Very open, sunny rides or glades (e.g. less than 20% shade)	Lightly shaded rides or glades (e.g. 20–40% shade)
Carterocephalus palaemon[1]	*Aphantopus hyperantus*
Thymelicus sylvestris	*Leptidea sinapis*[1]
Thymelicus lineola	
Ochlodes venata	Fairly shaded rides or glades (e.g. 40–90% shade)
Erynnis tages	
Pyrgus malvae[2]	
Gonepteryx rhamni[3]	*Pararge aegeria*
Anthocaris cardamines	*Pieris napi*
Callophrys rubi[3]	
Thecla betulae[1,3]	Newly cut woodland (including coppice and ride margins)
Lycaena phlaeas	
Aricia agestis	
Polyommatus icarus	*Boloria euphrosyne*[1]
Celastrina argiolus[3]	*Argynnis adippe*[1]
Hamearis lucina[1,2]	*Mellicta athalia*[1]
Cynthia cardui	
Vanessa atalanta	Dappled shade within woodland or at wood or ride edges
Aglais urticae	
Inachis io	
Polygonia c-album[1]	*Ladoga camilla*
Eurodryas aurinia	*Argynnis paphia*
Boloria selene[1,2]	
Argynnis aglaja[2]	Tree-feeders, mostly confined to the canopy
Lasiommata megera	
Erebia aethiops	
Melanargia galathea	*Quercusia quercus*
Pyronia tithonus	*Strymonidia w-album*
Maniola jurtina	*Strymonidia pruni*
Coenonympha pamphilus	*Apatura iris*
	Nymphalis polychloros

[1] Species confined to early successional woodland or wood edges over a large part of their range.
[2] Also breed in newly cut woodland.
[3] Certain shrubs needed for breeding.

50–150 years ago (Heath *et al.*, 1984), and those most confined to this habitat have declined drastically in recent years. The most severe declines have been for the three species of fritillary (Lepidoptera; Nymphalidae) most closely associated with newly cleared woodland, *Argynnis adippe*, *Boloria euphrosyne* and *Mellicta athalia* (Fig. 1). In the past, such species thrived under traditional coppice management, but have since disappeared from large parts of the country now that this has ceased (e.g. Thomas, 1991; Warren, 1987c). *Mellicta athalia* and *Argynnis adippe* are now two of Britain's most threatened butterflies (Shirt, 1987).

Changes in forest management are also thought to be responsible for the decline of several other butterflies which are usually associated with woodland grassland (e.g. in rides or glades). These include *Carterocephalus palaemon* (Hesperiidae) (Collier, 1986), *Leptidea sinapis* (Pieridae) (Warren, 1984), *Hamearis lucina* (Lycaenidae) (Oates, in preparation) and *Boloria selene* (Nymphalidae) (Thomas, 1991). In contrast, the species associated with shady conditions in woodland, and those breeding in the tree canopy (Table III), have generally suffered rather less. Some, such as *Ladoga camilla* and *Pararge aegeria* (Nymphalidae), have even spread during this century (Emmet and Heath, 1989; Pollard, 1979). Of the five British canopy species, only *Apatura iris* and *Nymphalis polychloros* (Nymphalidae) have experienced major contractions in their range (Heath *et al.*, 1984). The former appears to need large blocks of woodland, and has probably suffered from forest clearance and fragmentation. The reasons for the decline of the latter species are unknown, but its past records in Britain have always been sporadic, and it may only form temporary populations following immigration from mainland Europe.

With the end of coppicing in most woodland, butterflies and other insects of woodland glades and clearings have increasingly relied on alternative types of habitat. Many butterflies have persisted in conifer plantations on ancient woodland sites, particularly in the large number of plantings that occurred during the 1950s and 1960s (Table 3 of Forestry Commission, 1984). These resemble young coppice in the early stages after planting. In central southern England, 50% of *Boloria euphrosyne* and *Argynnis adippe* colonies and 35% of *Boloria selene* colonies now breed in young conifer plantations which inherited rich ground floras from the deciduous woodland they replaced (Warren, in preparation). Unlike populations in traditionally worked coppice, however, these colonies have little future. The long rotation between planting and felling (*ca.* 60–80 years) means that alternative patches will not be created until long after the present colonies of butterlies have died out. Moreover, the composition of the ground flora will, by then, have changed so greatly due to the long period of intense shade and the change in soil chemistry

(Mitchell and Kirby, 1989) that the open space created in the next cycle of planting will be unsuitable for breeding. In Britain, 30% of colonies of *Mellicta athalia* occurred in young conifer plantations in 1980, but nearly all have since become extinct leading to serious overall decline of the species in the last 10 years (Warren *et al.*, 1984; Warren, 1989). Further declines can be predicted for all these woodland fritillaries as plantations mature, although these may be partly offset if there is a revival in coppicing or similar management aimed at conserving early successional species.

Another disadvantage of commercial plantations is that clear-fell areas tend to be very large and more widely separated by blocks of unsuitable habitat than when panels were cut under traditional coppice management (e.g. Mitchell and Kirby, 1989). This leads to problems of isolation, especially among sedentary species (see Section IV.D).

Several species of butterfly which traditionally bred in woodland grassland have also flourished, but again often only temporarily, in young conifer plantations. In the late 1970s, over 50% of British *Leptidea sinapis* colonies were breeding in commercial forestry plantations (Warren, 1984). As predicted from models of the butterfly's response to shading (Warren, 1985), many of these have now become extinct as the plantations have matured. For example, the status of this species on the Surrey/Sussex border, one of its recent strongholds, declined from at least 12 large colonies in the 1970s to just one small one in 1988 (Willmott, 1988).

A number of fritillary species that are associated with woodland in southern and eastern England regularly breed in more open habitats in western and northern Britain; utilizing rough grassland and scrub, as do *Mellicta athalia* (Warren *et al.*, 1984) and *Argynnis adippe* (M. Oates, personal communication) or moorland flushes as does *Boloria selene* (Thomson, 1980; Sutton and Beaumont, 1989). Some of these species have survived considerably better in these latter regions, whereas they have virtually disappeared from the whole of south-eastern and eastern Britain, where they depend totally on woodland clearings (Fig. 2).

Unlike woodland butterflies, which mostly breed on the woodland ground flora herbs, a greater proportion of woodland moths breed on trees and shrubs. This is estimated at 60% of woodland macro-moths (Waring, 1989). Many of these appear to prefer the young growth of woody species, and are associated with early successional, regenerating woodland. The habitat requirements of moths are relatively poorly known, but of the 125 "Nationally Notable" woodland macro-moths included in Hadley (1984), at least 60% seem to be associated with transitional woodland habitats such as open woodland, rides, clearings and scrub (Waring 1989). Many of these species were once more widespread in

(a)

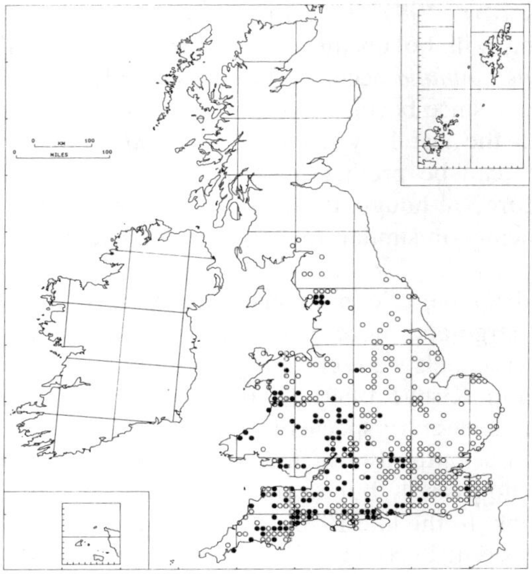

(b)

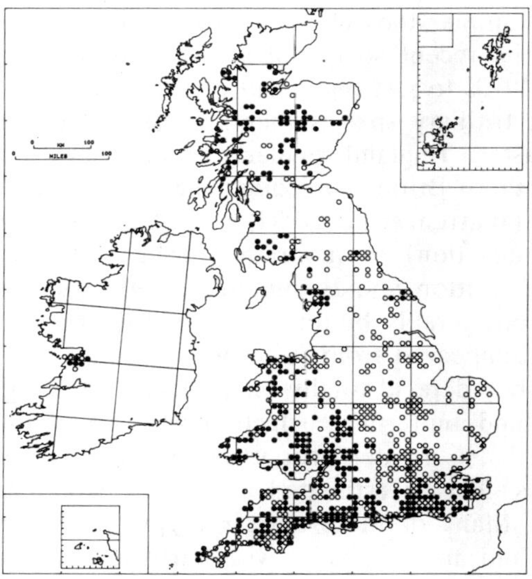

Fig. 2. Examples of insects thought to be declining following the cessation of coppicing: (a) *Argynnis adippe*; (b) *Boloria euphrosyne* (Lepidoptera; Nymphalidae) Distributions are plotted by 10 km grid square, with open and closed circles representing pre- and post-1970 records (reproduced courtesy of the Biological Records Centre, Institute of Terrestrial Ecology).

(c)

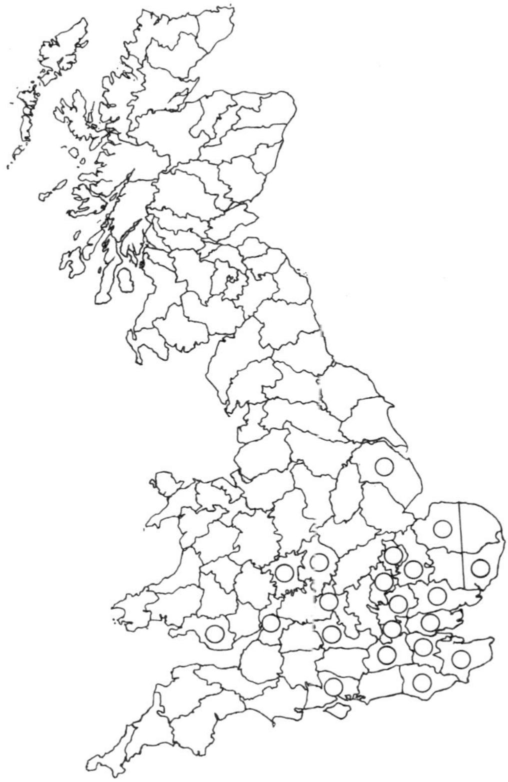

Fig. 2. Examples of insects thought to be declining following the cessation of coppicing (cont'd): (c) *Chrysomela tremula* (Coleptera: Chrysomelidae). This distribution is plotted by Watsonian vice-county, with all circles representing pre-1950 records (from Invertebrate Site Register data, Nature Conservancy Council).

Britain, and have declined in recent years due to changing woodland management.

Data on other orders is patchy, but it seems likely that a significant proportion of woodland insects are associated with early successional habitats. Many of the foliage-feeding species prefer young growth or warm sheltered conditions which are associated with regenerating coppice (e.g. Massee, 1965; Roskam *et al.*, 1991) In Germany, Heydemann and Muller-Karch (1980) found that open glades of less than 2–3 ha were by far the richest terrestrial ecosystems for animal species in general. Other studies on invertebrates in coppice are reviewed by Fuller and Warren (1990, 1991). Good data on the changes in status of early successionally stage woodland insects, other than Lepidoptera, are equally sparse. At least six scarce species of bug (Heteroptera) have declined in Britain, probably because of the loss of open woodland habitats (Kirby, 1991) and at least one species of beetle, *Chrysomela tremula* (Coleoptera; Chrysomelidae) may have become extinct due to the cessation of coppicing. Two other woodland shrub-feeding chrysomelid beetles, *Cryptocephalus coryli* and *C. nitidus*, were both quite widespread once, but have become very rare in the last 50 years. Their decline strongly suggests a direct association with coppice regrowth. The many other insects that have declined in woodland with the loss of active coppice include *Formica rufa* (Hymenoptera; Formicidae) (Welch, 1978), *Cheilosia semifasciata* (Diptera; Syrphidae) breeding on *Sedum telephium* in open woodland, and *Osmia pilicornis* (Hymenoptera; Megachilidae), nesting in dead wood in open conditions and nectaring preferentially on *Ajuga reptans* and *Viola* spp. (S. Falk, personal communication).

In general, ground-dwelling insects, especially Carabidae, have been found to be in greater abundance and diversity in recently cut coppice, whereas foliage-dwelling insects often increase in the later stages of regrowth, due to the increase in abundance and structural diversity of the foliage (Welch, 1969, 1978; Steel and Mills, 1988; Sterling and Hambler, 1988; Waring, 1988). Most of these studies, however, were restricted to particular taxonomic groups, or to species inhabiting a narrow range of structural components of the ecosystem, for example leaf miners and web-spinning spiders (Sterling and Hambler, 1988). Little account was taken of the scarcity of the individual species involved, and, in some instances, the studies were carried out in woodlands where the continuity of coppice management had been broken for a considerable period before recoppicing was initiated, with the likely elimination of stenotypic early successional species. While the numbers of individuals, biomass and, possibly, diversity of more widespread generalist species may increase with the ageing of the coppice, the more vulnerable species associated with the earliest stages are rapidly lost. It is therefore dangerous to extrapolate from these studies to the value of coppicing in general for invertebrate conservation, other than in the production of invertebrate biomass as a food source for other woodland animals.

IV. ECOLOGICAL SPECIALIZATION AMONG INSECTS OF MATURE WOODLAND AND OPEN SPACES

Threatened and locally distributed forest insects are characterized by extreme ecological specialization. Less specialized insects are often more widespread, and have usually declined far less seriously with the changing pattern of forestry. The extent of the specialization of early and late successional insects is considered below; the implications to conservation management are discussed in Section V.

A. Niche Specialization among Old Forest Insects

Although data on niche specificity among old forest insects is patchy, it is clear that the microhabitats used are extremely diverse, and that the species using them are often highly specific to them. Lists of saproxylic niches are given by Elton (1966), Chalmers-Hunt (1969), Stubbs (1972), Morris (1974), Dajoz (1974), Hunter (1977), Harding and Rose (1986), Harmon *et al*. (1986) and Speight (1989). As many as 20 broad categories of saproxylic habitat are described in these works.

A number of factors influence the abundance and distribution in space and time of these saproxylic niches, and their suitability for particular species. These factors may affect the invertebrates directly, such as the influence of temperature or humidity on insect larvae in dead wood, or indirectly by affecting the saproxylic fungi. When dead wood decays, as much as 35% of its dry weight may be converted to fungal biomass (Swift, 1973), upon which most saproxylic invertebrates depend directly or indirectly.

1. Size, Age and Number of Trees, and Area of Habitat

The number, distribution and diversity of such niches largely depends on the number, age and size of the forest trees on which they occur. Stubbs (1972) estimated that for the very specialized species, no more than 1% of ancient trees is likely to contain suitable habitat at any one time. This is borne out by species such as *Hypebaeus flavipes* (Coleoptera; Melyridae) and *Limoniscus violaceus*, which in Britain are known from only two or three individual ancient trees (Shirt, 1987; Hyman, in preparation). The largest and oldest trees are invariably those that support the most diverse faunas of saproxylic species, particularly of the more specialized, usually rarer species. This results from the increased complexity of external and internal architecture of the older trees, with features such as rot holes, heart rot, hollows, wood mould in different stages of development, stag-headed branches, and thick or loose bark. Sites that still have large numbers of very old trees almost invariably support higher numbers of saproxylic species, particularly of the more specialized and scarcer species, than those with fewer trees. This results from a much greater likelihood of continuity of particular niche conditions in past centuries. It is no coincidence that the foremost British sites for saproxylic invertebrates are not just those that have the longest history of old forest conditions, but are also those with the largest area and the largest numbers of very old trees. The three old forest sites in Great Britain found to support the richest assemblages of saproxylic beetles by Alexander (in preparation) are all larger than 700 ha and, of the 20 richest sites, seven are larger than 500 ha and 13 larger than 100 ha. All but one exceed 50 ha in extent. Of these sites, Windsor, Epping and the New Forest each contain thousands of very old trees, but elsewhere there are only hundreds, or even tens of ancient trees.

2. Tree Species

Specific associations with individual tree species among saproxylic invertebrates tend to decrease with the progression of the decay of the wood (Elton, 1966; Harmon *et al.*, 1986). Many "primary saproxylics" (i.e. initial colonizers into sound wood or bark of recently dead or stressed living trees) are quite specific to individual species or genera of trees; examples include many of the Scolytidae (Coleoptera). In contrast, the niches of saproxylic species inhabiting later decay stages are usually determined by microhabitat features within the decayed wood which are frequently not determined by the tree species. Some microhabitats on decaying beech, oak and lime support species of invertebrate specific to the tree species, but more usually microhabitats common to many broad-leaved tree species share a similar fauna which is largely distinct from that of apparently similar microhabitats on decaying conifer wood. Similarly, saproxylic species inhabiting the fruiting bodies of a wide range of wood-decaying fungi tend to occur in fungi with similar textures, porosity, moisture content, state of decay etc. Myxomycetes on dead wood do, however, tend to have a distinct fauna, this being dominated by beetles of the Sphindidae, Lathridiidae and Leiodidae, Anisotominae (Russell, 1979).

3. Microclimate: Humidity

The microclimate within dead wood is critical in determining its suitability for saproxylic invertebrates. The most important factors are the stability of temperature and the moisture content. These are largely determined by the size of a particular piece of dead wood and its position in relation to the surrounding vegetation and topography.

Microclimate particularly influences the saproxylic fungi which perform the major chemical decomposition of wood. Scheffer and Cowling (1966) noted that a wood moisture content of at least 26% of dry fibre weight is necessary before wood-decaying fungi can function, and that free water is essential before decay can take place. Saproxylic species are described as unusual among fungi in requiring such a high substrate water content. Kaarik (1974) noted that basidiomycete fungi, such as those causing red heart rot, function most effectively at moisture contents around 160% of the dry weight, and the wet white-rotting "fungi imperfecti", upon which some saproxylic Diptera such as *Ctenophora ornata* (Tipulidae) are dependent, function most effectively with a wood moisture content of 240% dry weight. The moisture content of dead wood is determined by several factors, notably the degree of contact with soil, insolation, the presence or absence of bark and overall size.

The moisture content of dead wood has been found to have a direct influence on saproxylic invertebrates, notably on larvae. Kelner-Pillaut (1974) found a high mortality among wood mould- feeding and predatory saproxylic beetle larvae when there were rapid changes in moisture content. These changes occur particularly in small pieces of wood which may alternate between complete saturation and being very dry within 24 hours (Boddy, 1983). Kelner-Pillaut (1974) also found that larvae of *Pseudocistela ceramboides* (Coleoptera; Tenebrionidae) and *Ludius ferrugineus* (Coleoptera; Elateridae) were unable to pupate if the moisture content of their wood mould habitat fell below 40–50%, while *Gnorimus nobilis* and *Prostomis mandibularis* (Coleoptera; Cucujidae) could not pupate with a moisture content below 60%. Dajoz (1974) found that certain species of cerambycid beetles were able to survive dry conditions in the dead wood that they inhabited, but this lengthened their life cycles from 2–3 years to 6–7 years. The moisture content of saproxylic fungi and slime moulds has also been found to affect the species composition of the fauna. Russell (1979) noted that beetles of the anisotomine Leiodidae usually inhabited moist myxomycetes under bark, while those of the Sphindidae and Lathridiidae fed on very dry myxomycetes. Maser and Trappe (1984) noted that the interiors of very large logs are buffered from fluctuations in moisture content and temperature. This happens especially when there is either a covering of bark or when the surface layers of the sap and heartwood of the log become sun dried and hardened into a moisture retentive "case" (Harmon *et al.*, 1986). Kelner-Pillaut (1974) also showed that the wood mould in cavities in very large willow and chestnut trees had a high moisture retention capacity, even when the hollow was exposed to the exterior. This buffering from fluctuations in substrate humidity may be one of the reasons why larger pieces of timber tend to support much richer faunas of saproxylic species than small branch wood.

Certain saproxylic species of insect only inhabit wood that has fallen into water. This must have been a more frequent habitat in the wildwood, before man began clearing wood from navigable waters and drainage streams. Dudley and Anderson (1982) record as much as $40\,kg\,m^{-2}$ of dead wood in streams flowing through natural old-growth conifer forest in North America. The European species *Cyanostolus aeneus* (Coleoptera, Rhizophagidae) lives largely under the thick bark of large pieces of timber in water or in very damp conditions (Hyman, in preparation), and Dudley and Anderson (1982) record a number of obligate aquatic wood-feeding species in North America among the larvae of Ephemeroptera, Trichoptera, Coleoptera and, in particular, the Diptera families Chironomidae, Stratiomyidae, Syrphidae, Tipulidae and Xylophagidae. Most species were found inhabiting the interface of the wood between air and water, although some, for example those of the dipteran family Axymyiidae, bore long galleries into the timber and have long respiratory siphons. *Nacerdes melanura* (Coleoptera; Oedemeridae) and *Pselactus spadix* (Coleoptera; Curculionidae) inhabit mainly salt water-soaked timber, while *Mesites tardii* (Coleoptera; Curculionidae), which inhabits firm dead wood mainly in coastal woodland, reputedly disperses in sea-borne driftwood (Read, 1982).

In contrast, other species of saproxylic invertebrates depend on dry conditions in dead wood. *Psilocephala melaleuca* (Diptera; Therevidae) is thought to develop in very dry, red-rotted wood (S. Falk, personal communication) as do the larvae of a number of beetles of the families Anobiidae and Dermestidae. *Ctesias serra* (Coleoptera; Dermestidae) lives only under very thick, dry bark where it feeds on spider webs and the husks of insects caught in them. Invertebrates associated with fire-scorched timber also seem mainly to be associated with very dry wood. Bugs of the family Aradidae, beetles of the families Carabidae, Histeridae, Leiodidae, Salpingidae and Cerambycidae, and flies of the Mycetophilidae and Platypezidae are typical of this habitat.

4. Microhabitat: temperature

Temperature has also been shown to have a profound effect on both the saproxylic fungi and their dependent invertebrates. Graham (1922, 1924, 1925) found that subcortical temperature in dead wood varied with a wide range of factors, notably air temperature, degree of shade, angle of incidence of sunshine, colour, thickness and rugosity of bark, position within the log, dampness of bark (leading to evaporative cooling) and the degree of ventilation of the wood created by the borings of insects. Kaarik (1974) noted that wood-decaying fungi are unable to function above 40°C. Savely (1939) found that the upper lethal temperature for saproxylic beetle larvae of the Pyrochroidae, which usually inhabit very moist and shaded dead wood, was 41°C, while that of the Bupestridae, which more frequently inhabit drier and sun-baked timber, was around 52°C. Craighead (1920) recorded temperatures of 60°C under the bark of logs exposed to direct sunshine; it has long been forestry practice to sterilize dead wood to prevent infestation by perceived pest species by exposing it to sunshine. Savely (1939) recorded mean daily temperatures of 43°C in logs exposed to direct sunshine in summer in North American temperate forest, with a maximum of 49°C. He also found that temperatures fluctuated much more highly between day and night in dead wood in the open than was the case with shaded dead wood. Dajoz (1974) found similar extreme fluctuations under the bark of standing trees in full sunshine, and stable temperatures under the bark of trees in the shade, while Kelner-Pillaut (1974) noted that certain beetle larvae in dead wood were unable to survive rapid fluctuations in temperature, and that a high water content in wood mould in the hollows of large old trees damped fluctuations in temperature by evaporative cooling. Holmquist (1931) found that the temperature inside oak logs was considerably buffered from the extreme cold of winter: it never fell below −1.5°C even when the air temperature was as low as −16°C. Kelner-Pillaut (1974) found that *Pseudocistela ceramboides* and *Ludius ferrugineus* normally continue their development throughout winter, yet are unable to grow at temperatures below 5°C.

5. *Nutrient Status of Dead Wood*

Certain saproxylic species are thought to be influenced by the nutrient status of decaying wood. Decay fungi are known to fix atmospheric nitrogen (e.g. Silvester *et al.*, 1982), and Kelner-Pillaut (1974) found a greater nitrogen content in the faeces of wood-feeding beetle larvae than in the source material. These faeces form the basis of the wood mould on which other species depend for food. He also found a consistently higher nitrogen content in wood mould in the hollows of willow compared with chestnut but this was because the leaves of both species accumulate in hollows, and those of willows contained more nitrogen. Externally derived nutrients are also known to be important in the nutrition of those invertebrates, including stenotypic species that inhabit water-filled hollows in trees (Kitching, 1969, 1971). The fauna of red-rotted and white-rotted wood appear to have few species in common. This is possibly because lignin remains in red-rotted wood but is broken down in white-rotted wood (Harmon *et al.*, 1986).

6. Position of Microhabitats

Insects have been found to subdivide niche availability by selecting microhabitats only in certain positions on the tree or fallen wood. Members of the genus *Criorhina* (Diptera; Syrphidae) mainly select oviposition sites in dead wood on exposed tree roots (S. Falk, personal communication) while another species of the same family, *Pocota personata*, selects rot holes high up on the tree trunk (Stubbs and Falk, 1983). Paviour-Smith (1964) found that *Tetratoma fungorum* (Coleoptera; Tetratomidae) inhabits bracket fungi that are between 1 and 5 m above the ground on tree trunks, and inferred that predation selected against individuals below this height, whereas there was an increased chance of desiccation above it. Graham (1925) noted that different species of insect selected the upper and lower halves of logs that were lying on the ground. This was related to differences in temperature and moisture content.

7. Location of Niches by Saproxylic Species

There appear to have been few studies of how saproxylic invertebrates search for their particular niches. Saproxylic aculeates, in common with those of other habitats, are thought to find their nesting sites visually, and may be unable to locate nest burrows if their surroundings are changed in any way. Gilbert and Norris (1968) found that bark beetles of the family Scolytidae dispersed randomly onto surrounding trees from their tree of emergence and were actively repelled by chemical substances specific to non-host species, rather than respond positively to chemicals from host species. Mycophagous species appear to be attracted by air-borne substances that are produced by fungoid wood or fungal fruiting bodies. Saproxylic beetles of the Nitidulidae, Staphylinidae and Leiodidae have been found to be rapidly attracted to aquatic extracts from fungoid wood in an old orchard (R. S. Key, unpublished).

8. Interdependence between Species

Many saproxylic species are known to be dependent on niches created by other saproxylic invertebrates. The solitary bees and wasps of dead wood most frequently nest in exit burrows of wood-feeding beetles, with different sized aculeates inhabiting the burrows of different beetle species. They also concentrate on dead wood in warm sunny conditions. Other insects have specific associations with saproxylics: *Epuraea angustula* (Coleoptera; Nitidulidae) occurs in the burrows of members of the genus *Xyloterus* (Coleoptera; Scolytidae) (Joy, 1932), and *Aulonium trisulcum* is predatory on *Scolytus scolytus* (Coleoptera; Scolytidae) (Marshall, 1978). *Limonniscus violaceus* may have some association with the debris that accumulates under bird nests made on the surface of red-rooted beech wood, although this is not an obligate association (Mendel and Owen, 1990). Donisthorpe (1924, 1926, 1927) recorded over 50 species of invertebrate, from various phyla, as occurring in the nest of the saproxylic ant *Lasius brunneus* (Hymenoptera; Formicidae); some had a stenotypic association with this species. The relationships between *Lasius brunneus* and its various inquilines and cleptoparasites are described by Donisthorpe (1927).

9. The Need for Nectar and Pollen

Nearly all saproxylic invertebrates depend on nectar as their prime source of energy, and on pollen to provide protein for egg laying. Many species favour the blossom of hawthorn, or various members of the Umbelliferae or Compositae. The adults of some insects whose larvae inhabit dead wood are predatory on other insects visiting such flowers. It appears that no study has been carried out on the dependency of saproxylic species on the supply of nectar, pollen or flowers to provide insect prey, nor on the distance that adult insects will travel to utilize the resource. It is only possible to say that large numbers of stenotypic saproxylic Coleoptera and Diptera may be observed at hawthorn blossom in late May and early June at those sites which have been found to support very rich faunas.

10. Non-saproxylic Old Forest Species

Relatively little is known of the association of non-saproxylic species with old forest conditions. A few rare phytophagous species, for example *Cryptocephalus querceti* (Coleoptera; Chrysomelidae) and *Cheilosia carbonaria* (Diptera; Syrphidae), and carnivorous species such as *Calosoma inquisitor* (Coleoptera; Carabidae) and *Didea fasciata* (Diptera; Syrphidae) appear to be very strongly associated with ancient forest sites in Britain (S. Falk, personal communication; Hyman, in preparation). Studies in Scandinavia also suggest that some ground beetles occur preferentially in ancient conifer forest conditions (Niemela *et al.*, 1988). An analysis of threatened Swedish old forest invertebrates of all ecological groups is given by Ehnstrom and Walden (1986). Middleton and Merriman (1985) found that the ground litter of virgin forest areas in Canada supported a higher diversity of species than that of managed forest, but a direct association of individual species with old forest conditions was not attempted. Szujecki (1979) noted that the soil faunas of mature forests recovered relatively slowly after clear felling, and that woodlands with a repeated history of felling and planting had an impoverished soil fauna. No one seems to have looked for an association between individual soil species and old forests. Associations of various mollusc species with old forests are included by Kerney and Stubbs (1980), Cameron *et al.* (1980) and Foster (1983), who indicate that old forest species are declining in Britain.

B. Niche Breadth among Open Space Insects

Early successional woodland insects fall into three broad categories, depending on whether they are associated primarily with newly created clearings in woodland, more permanent open grassland within woodland or with the young growth of woody species.

1. Ground Flora of Newly Created Woodland Clearings

This category includes species which breed on the woodland ground flora, but only where plants are exposed to sunlight as a result of the creation of large gaps in the canopy. Examples include the fritillary butterflies in Fig. 2, *Minoa murinata* (Lepidoptera; Geometridae) (P. Waring, personal communication), *Dicranocephalus medius* (Heteroptera; Stenocephalidae), *Oncochila simplex* (Heteroptera; Tingidae), *Eurydemia dominulus* (Heteroptera; Pentatomidae), *Sehirus biguttatus* (Heteroptera; Cydnidae) (Kirby, 1991; Massee, 1965) and *Cheilosia semifasciata* (Diptera; Syrphidae) (Falk, 1991). Other species, such as many solitary bees and wasps are associated with bare ground in new clearings.

2. Permanent Grassland in Woodland

Species associated with woodland grassland throughout a large part of their British range include the butterflies *Leptidea sinapis* (Warren, 1984) and *Carterocephalus palaemon* (Collier, 1986), and *Photedes fluxa*, *Apamea scolopacina* and *Apamea epomidion* (Lepidoptera; Noctuidae) (P. Waring, personal communication) and *Capsodes flavomarginatus* (Heteroptera; Miridae) (P. Kirby, personal communication). Many other insects that are associated with open habitats outside woodland, such as calcareous grassland or heathland, can also be abundant in woodland grassland (e.g. many of the butterflies listed in Table III), but the examples given above are more or less restricted to woodland throughout a large part of their British ranges. Species associated with newly cleared woodland and more permanent open spaces in woodland also include some that breed elsewhere in the wood but rely on open areas for their abundant supply of pollen and nectar. Examples are *Ladoga camilla* and *Argynnis paphia* (Lepidoptera; Nymphalidae) and many of the saproxylic species mentioned above.

3. Young Woody Growth on Shrubs and Trees

A large number of insects breed preferentially on the young growth of woody species. These include *Thecla betulae* (Lepidoptera; Lycaenidae) (Thomas, 1974), *Minucia lunaris* (Lepidoptera; Noctuidae) and *Furcula furcula* (Lepidoptera; Notodontidae) (Massee, 1965), *Byctiscus populi* and *B. betulae* (Coleoptera; Attelabidae), *Rutidosoma globulus* (Coleoptera; Curculionidae), *Cryptocephalus coryli*, *C. sexpunctatus*, *C. bipunctatus* (Coleoptera; Chrysomelidae) (Massee, 1965), *Agelastica alni* and *Chysomela tremula* (Coleoptera; Chrysomelidae) (Hyman, in preparation) and the gall-forming plant louse *Psyllopsis fraxini* (Homoptera; Psyllidae) (Roskam *et al.*, 1991). It is possible that these species select this younger growth because of differences in leaf chemistry between the foliage on younger shoots and that on more mature growth, or because of differences in plant architecture and position in relation to the surrounding vegetation and microclimate.

4. Position and Structure of Foodplant

Within each of the categories described above, species are often specialized to breed not only on one or a few plant species, but also on plants growing in a precise situation with regard to their age of growth, structure or relationship with the surrounding vegetation. *Mellicta athalia* selects plants of *Melampyrum pratense* growing in sunny situations (less than 70% average direct shade) and in areas of sparse vegetation, typically surrounded by 40–70% bare ground (Warren, 1987a). Such conditions are found only in the very early stages following woodland clearance, and persist for just 5 years after cutting in commercial coppices (Warren, 1987c). In contrast, the host plant is often abundant in older coppice and under deciduous high forest, but is never used. A similar specialization exists in Britain among several fritillaries which breed on *Viola* species growing in sunny situations in woods. Thomas (1991) found that each species occupied a slightly different niche in the succession following clearance. *Boloria euphrosyne* mainly selects small plants growing in bare ground or sparse vegetation during the first few years after clearance. *Boloria selene* selects larger plants in taller, grassy vegetation and *Argynnis aglaja* (Lepidoptera; Nymphalidae) selects large plants that are partially shaded or dominated by surrounding vegetation. *Argynnis paphia* selects plants growing in dappled shade beneath the woodland canopy, although adults still rely on open clearings to find mates and feed on flowers.

Insects that are confined to woodland grassland are also specialized with respect to the level of shade and vegetation structure. Some of the complexities involved are illustrated by *Leptidea sinapis*, which develops on a few species of Papilionaceae. In Britain, it selects taller plants that protrude above the general level of the vegetation and prefers partial shade (20–40% shade), usually along the edges of woodland rides (Warren, 1984, 1985a). Rides with less than 20% shade contain greater host plant densities but fewer eggs and adults, and rides with more than about 60% shade are too shaded to support any host plants. Warren *et al.* (1986) found that eggs transferred to the most open rides appear to survive equally well as those in the preferred conditions, they concluded that this selection is caused mainly by the indirect effect of shade on the growth form of its host plants, which influences the frequency of oviposition.

5. *Temperature and Insolation*

An important common feature of early successional woodland habitats appears to be that they provide a warm microclimate for thermophilous insects. This is particularly important at the northern edges of a species range, where climate may be a crucial limiting factor to the survival of populations. Many of the butterflies that are confined to fresh clearings or open grassland in British woodlands are typically found in non-woodland habitats further south in Europe (Warren, 1985b); for example elsewhere in Europe, *Leptidea sinapis* frequently occurs outside woodland, and its specialization in woodland grassland in Britain may be confined to the edges of its range. This aspect of niche breadth therefore varies throughout the range of each species, and different specializations are likely to have evolved in warmer climates. It is feasible that a species may seek out cooler, shadier habitats at its southern limit. This is supported by evidence among some grassland butterfly species, which are restricted to short, warm swards in Britain but occupy taller, cooler swards in southern Europe (Thomas, 1991).

A particularly important microclimatic effect of woodland clearance is the great increase in the amount of sunlight which reaches the ground. Data from coppiced woodland show that summer levels of insolation drop rapidly to only about 10% of that in the open 4–5 years after clearance, and 3% after 10 years (Ford and Newbould, 1977; Salisbury, 1924). Besides affecting light levels, coppicing affects the temperature, humidity and wind speed. Ash and Barkham (1976) have shown that the summer soil temperature in newly coppiced woodland is up to 10°C warmer than beneath a dense canopy, but that this declines rapidly to only 3°C warmer just 2–3 years after cutting.

As the canopy closes, the difference between day and night temperatures becomes less, providing a more stable microclimate. Geiger (1965) found that south-facing woodland edges and glades can be several degrees warmer than in the open or in the interior of a wood, during both the day and night. Similar values can be expected for sunny woodland rides, but no direct measurements have been published. Sheltered glades can, however, be cooler than the interior of woods during cloudy weather, or during the winter when insolation is low, when they can also act as frost hollows.

C. Mobility

As Dempster (Chapter 7) points out, the mobility of insects is extremely hard to measure and may have been underestimated in many population studies. There is also some evidence that dispersal rates may be highly variable between years, and that they may be considerably higher in warm weather. Nevertheless, the weight of evidence supports the generalization that many of Europe's threatened woodland insects have very limited powers of dispersal, and that this should be a major consideration when considering an overall conservation strategy (see Section V). Dempster also raised the concern that habitat fragmentation in recent times may have caused (and still be causing) the evolution of local isolated races with reduced powers of dispersal.

The niches occupied by early successional insects tend to be very short lived but widely dispersed under traditional forms of woodland management while those of late successional species appear to be very restricted in space but relatively long lived. Both situations require the species to move periodically to colonize new areas where its habitat has become available.

1. Early successional species

Colonies of species breeding in newly coppiced woodland may need to move from one habitat patch to another every 4–5 years. Given the transient nature of their habitat, it is reasonable to expect the species to have evolved a high degree of mobility. The reverse seems, however, to be the case, at least for many of the more localized and threatened species.

Few species have been studied in detail, but those that have seem to be extremely sedentary and form closed populations in restricted areas. For example, mark–recapture experiments on three adjacent colonies of *Mellicta athalia* within a large block of woodland have shown that the average adult range was only 118 m for males and 89 m for females (Warren, 1987b). The proportion of the total population estimated to have migrated between the colonies, an average linear distance of about 800 m, was only 1.7% for males and 1.3% for females. Detailed monitoring of numerous newly cut areas spread over 1000 ha of continuous woodland has shown that not all suitable habitats were colonized by the species. Habitats that were created within about 300 m of existing colonies were colonized almost immediately, but colonization was delayed and frequently did not occur at all in habitats over 600 m away (Warren, 1987c). On this basis, it can be predicted that sites more than 10 km from a source colony would effectively never be colonized, particularly if the intervening vegetation were open farmland.

A similar lack of mobility has been measured by Thomas (1991) for *Boloria selene*, but the information available for other species is largely anecdotal. Some evidence comes from the success, albeit often temporary, of man-made attempts at insect re-establishment. The outcome of 323 attempts for butterflies has been documented by Oates and Warren (1990), 38% establishing colonies which persisted for over 3 years. The success of so many re-establishments, which include several early successional species such as *Mellicta athalia*, suggests that many suitable habitats currently exist for woodland insects, but have not yet been colonized naturally. The spread of introduced colonies of sedentary woodland species is also revealing. For example, Thomas (1989) describes the spread of an introduced colony of the butterfly *Strymonidia pruni* (Lepidoptera; Lycaenidae) over more or less continuous habitat at a rate that averaged only 1 km each decade.

2. *Old Forest Species*

In contrast, remarkably little is known of the mobility and dispersal abilities of invertebrates associated with the mature timber habitat. One feature of many saproxylic species that may make them especially vulnerable to local extinctions could be poor powers of dispersal (Hammond, 1974), or the tendency of species with obvious powers of flight to oviposit close to the site of emergence. While there may be evolutionary advantages in this strategy, which increases the probability of a female finding appropriate habitat close to her larval development site, it is unlikely that niche availability was a limiting factor in primeval forest conditions with its abundance of dead wood. Perhaps strong powers of dispersal were not developed in many saproxylic species simply because of this high niche availability — there simply was no need to range widely to find appropriate habitat.

It appears that no research has been done on mobility of these species. The difficulties in carrying out such studies are particularly challenging, given the small size, inconspicuousness of and difficulty in sampling many saproxylic species. Certainly, some common saproxylic invertebrates have demonstrated abilities to disperse efficiently. *Euophryum confine* (Coleoptera; Curculionidae) has spread widely in Britain in woodland, hedgerow trees and in seasoned timber indoors since its introduction from New Zealand in the 1930s (Buck, 1948), and *Aulonium trisulcum*, which was formerly scarce, expanded rapidly to take advantage of the abundance of habitat created by the death of elm trees as a result of Dutch elm disease (Marshall, 1978). Even scarcer species turn up as isolated individuals far from appropriate habitats; for example, one individual of *Mallota cimbiciformis* (Diptera; Syrphidae) was found in a conifer plantation in a sand dune on the island of Anglesey in 1987 (S. Falk, personal communication). This species breeds in rot holes in large old deciduous trees and there would appear to be no suitable habitat for many miles. Many other species, however, occur only in a single site, often only in a small area within that site, or even on a single tree; examples include *Limoniscus violaceus* and *Hypebaeus flavipes* (see Section IV.A). There is very little evidence for the colonization anew of woodland sites by the more demanding, usually scarcer saproxylic species, although the differentiation between colonization and the discovery of an overlooked population of a species at a site can be remarkably difficult. Some park woodland created in the eighteenth century by Capability Brown, Repton and others, now superficially resembles some of the ancient dead wood invertebrate sites in terms of the abundance of dead wood and the size of trees, but remains remarkably impoverished in saproxylic species (K. N. A. Alexander, personal communication).

There is evidence that some saproxylic species are polymorphic in their wing length, and that the polymorphism is switched by changes in suitable habitat. Hamilton (1978) showed that after dead wood was colonized by winged adults of some species of ptiliid beetle, these were flighteless in subsequent generations and did not leave the wood, forming an inbreeding "multiplication phase" population. Winged adults were produced when saproxylic conditions deteriorated, and these "dispersal phase" individuals did not mate until after leaving the dead wood, presumably to increase their chances of exchanging genetic material. It is uncertain how widespread this life strategy is among saproxylic invertebrates, but it could have considerable implications in how fallen dead wood is dealt with on conservation sites. It is often suggested by conservation workers that timber which falls in the winter is left only until the following summer, when the generation of larvae contained therein is expected to have dispersed, or that fallen timber is left for a finite number of years before being disposed of. These recommendations may be potentially harmful in that the insects may be unable to leave the habitat and disperse if conditions have not changed to the extent that macroptery is induced. The removal or destruction of fallen timber therefore becomes a drain on the invertebrate populations.

3. Implications of Climatic Change

Hengeveld (1986) emphasizes the dynamic nature of insect geographical distribution throughout the post-glacial period, pointing out that species' survival potential decreases away from the centre of distribution, being lowest at the extreme edges of range. The fragmentation and isolation of the remaining sites where old forest faunas persist has considerable implications for the future survival of these species now that climatic change is predicted. The apparent lack of mobility of many woodland species, coupled with the extreme scarcity of suitable habitat, means that they are unlikely to migrate to follow changes in the distribution of major vegetation types, such as are predicted by the more extreme models of climatic change. Moreover, these species are unlikely to be able to take advantage of favourable climatic conditions beyond the edges of their current range, should these develop.

V. THE MANAGEMENT OF WOODS FOR INSECTS CONSERVATION

A. The Determination of Priorities

It is necessary to divert the planning of conservation management more towards these vulnerable invertebrates and away from well-intentioned, but potentially damaging, management operations in woodland. It is still common to see valuable open space in woodland nature reserves planted up with young trees, decaying trees felled and dead wood tidied and burnt, while some woodland management manuals continue to recommend operations that are highly destructive to the woodland fauna.

It is not the intention of this paper to form a comprehensive handbook of the various management options and techniques for conserving woodland invertebrates. Objectives for the conservation of other biological aspects in woodland have been described in detail elsewhere (e.g. Peterken, 1981) as have woodland management techniques (e.g. Brooks, 1980). Rather, the intention is to highlight the importance of historical factors when determining site selection and management priorities for woodland insect conservation.

1. Site Selection

An obvious priority for the conservation of woodland insects is the safeguarding of a full range of woodland types. In Britain, this is done through the protection of a representative series of Sites of Special Scientific Interest (SSSI) and National Nature Reserves by the statutory conservation body, the Nature Conservancy Council. Although the needs of invertebrate are now recognized in site selection and evaluation criteria (NCC, 1989), these sites were initially selected largely on the basis of higher plant communities and, consequently, the special needs of early and late successional woodland insects remain under-represented. For example, of the 45 woodlands in central southern Britain calculated to be of national or regional importance for butterflies (mostly early successional species), 19 (42%) were not scheduled as Sites of Special Scientific Interest and none was a National Nature Reserve (Warren, in preparation). Most of these are commercial forests with a large conifer element, where the main interest was in the rides, glades and young plantations. They were given a low priority because they are no longer semi-natural. Five of them had been partly scheduled, but the boundary had been determined to exclude conifer plantations even though these included the main breeding areas of threatened species. Moreover, only one National Nature Reserve has been declared in Britain as an example of a woodland with a unbroken history of coppicing. Very few examples of unbroken coppice exist, but relatively few of these have been notified as Sites of Special Scientific Interest, in contrast to the middle-aged, neglected coppice woodlands which make up the majority of woodland reserves and SSSIs. Only a single National Nature Reserve has been created for the conservation of old forest insects, although there is now better representation as SSSIs of the richest sites for this fauna. However, management for saproxylic invertebrate habitat on SSSIs remains a point of contention between the landowners and the Nature Conservancy Council on a majority of the richest sites for this fauna in Great Britain.

Ancient woodland sites are clearly a major priority when considering an overall conservation strategy for forest wildlife, and this is now widely recognized by British nature conservation bodies as well as in the Forestry Commission's recent broadleaves policy. The survival of most woodland insects that are associated with very early and very late successional stages is, however, dependent on the historical continuity and future implementation of specific types of management, most especially on coppicing and the various types of wood pasture management.

2. The Use of Indicator Species

Harding (1976, 1978a–d, 1981a,b) and Harding and Rose (1986) introduced the concept of using insect species as indicators of continuity of the mature timber habitat, in a similar way that botanists have identified plant assemblages that are indicative of ancient woodland. Invertebrates, particularly beetles, have been graded as to the degree of their exclusiveness in association with sites that have had an exceptional continuity of old forest conditions. Similar analyses have been carried out for hoverflies (Stubbs, 1982) and for molluscs (Kerney and Stubbs, 1980). The degree of specialization of the saproxylic invertebrates, coupled with their limited powers of mobility and fairly short life cycles, make them particularly vulnerable to even short periods of inappropriate management. Few specialized saproxylic species can survive the clear felling and replanting or regeneration of a whole woodland. This "habitat continuity indicator" approach allows some assessment to be made of a site's fauna in relation to its known history. In an analysis of all those British woodlands still found to support exceptionally diverse assemblages of saproxylic beetles, Alexander (in preparation) shows that the 20 richest sites date back to at least the seventeenth century. Most, if not all, are much older, originating either as medieval deer enclosures, ancient royal forests or other pasture woodland sites. Garland (1983) extended this analysis in a regional context for beetles in central northern England, demonstrating in particular that species near the edges of their range are more fastidious in their habitat specificity and consequently more useful indicators of continuity. In contrast, Speight (1986, 1989) has used this approach to identify the species most indicative of continuity of old forest conditions in Europe as a whole, setting out useful criteria for determining which species have greatest potential as bioindicators.

3. Management Priorities

Ancient woodlands which meet the criteria for selection as prime conservation sites do not automatically cater for the needs of these specialized invertebrates. The priorities in woodland insect conservation should therefore include:

1. The identification of those sites where there is the longest continuity of suitable management extending up to the present day.
2. The perpetuation or simulation of traditional management under which the assemblages of species have survived for centuries or millennia.
3. The managing of these sites to favour those aspects of their fauna (and flora) dependent upon this continuity of management. Attempts to diversify the habitat on these sites through management need to be guarded against. Diversification of habitat is often regarded as desirable by conservation organizations, even in relatively intact semi-natural sites; this can sometimes be detrimental to the intrinsic value of the site.

These are difficult tasks. Some of the problems and solutions are discussed in Section V.B. Because many traditional regimes are currently uneconomic, it is a regrettable likelihood that they will be perpetuated only on a few key sites. The selection of key sites for this treatment is therefore extremely important.

B. Habitat Continuity and the Maintenance of Traditional Regimes

Given the vulnerability of many invertebrates even to short breaks in habitat continuity, the major factor when considering the conservation of invertebrates of early and late successional woodland is the need for continuity over long periods, effectively in perpetuity. This is particularly important when considering small, isolated sites.

1. Coppice

Details for the rotational management of coppiced woodland are given by Fuller and Warren (1990). To a large extent these follow the traditional practices described in Section II.B, although on most conservation sites there is less room for variation or error because of their small size or the lack of resources to manage large areas. Briefly, new panels should be at least 0.5 ha in size and new ones initiated regularly, preferably every year, to provide a constant supply of new habitat. Moreover, successive panels need to be cut fairly close together, ideally in rotational sequence so that colonies of less mobile species can move easily from one to another. The presence of a network of interconnecting rides is advantageous because these can act as flight paths which facilitate movement and the rapid colonization of new coppice panels, particularly if successive cuts are widely separated. They may be especially important in modern forestry systems where canopy gaps are separated far more widely in both time and space. The management of the rides themselves, which ideally should also be rotational, is considered in Section V.C.

2. Wood Pasture

Apart from the continuing loss of habitat, the greatest problem confronting the conservation of ancient forest invertebrates is the existence of gaps in the trees' generation structure. These are apparent at many of the most important sites, especially in old parklands. While new parkland trees have been planted recently at many of these sites, the greater problem lies in ensuring that the older trees survive in sufficient numbers until this younger generation becomes old enough to start producing suitable habitat. For oak maiden trees, this may be in the region of 200 years, although it is usually less for other species. There is much scope for the "prematuration" of young and mature (as opposed to post-mature) trees, as discussed in Section V.C. Conservation planning for time scales measured in centuries is a difficult concept for conservation workers, especially when there is uncertainty over funding and the continuity of staff.

Traditional management has been revived recently on some of the old pollard wood pastures in southern England. Some experimental repollarding has been reinstated at Hatfield Forest (Rackham, 1989) and Hainhault and Epping Forests (Mitchell, 1989) in Essex, to determine how well trees will repollard after 100 or so years of crown growth (see Section II.C above). Mitchell (1980) noted that certain species of tree repollarded more successfully than others, notably willow, ash and hornbeam. Oak was variable in its response to repollarding, whereas beech was least successful under the conditions in the study. By allowing a single vertical branch to remain when pollarding beech, a far better success rate has been achieved at Burnham Beeches in Buckinghamshire (H. Read, personal communication). There remains, however, considerably more scope for experimentation in this field; meticulous records are being kept by the London Ecology Centre of the experimental repollarding plots at Hainault Forest to determine what factors most favour a successful response (D. Pape, personal communication). New pollards are also being created at this site and at Burnham Beeches, which are essential if continuity is to be maintained in the long term. Many new pollards have also been fortuitously created throughout south-east England by tree surgery following the severe storms that struck southern England in 1987 and 1990. There is, however, likely to be little documentation of the results of this pollarding, and it is doubtful whether these trees will continue to be pollarded in future. Mitchell (1989) serves as a preliminary manual for pollarding and repollarding, but much more experimental work is necessary in this field.

C. Simulation of Traditional Practices within Modern Woodland Management

Sections II and III contained an outline of how early and late successional habitats of the original wildwood were perpetuated under traditional management regimes. In this section we consider the potential for simulating the most crucial aspects of such regimes within the framework of modern forestry.

1. Open Space in Woodland

Many insects already rely on open habitats that have been created fortuitously within commercial forests, notably the rides and glades, and there is great potential for enhancing these specifically for conservation. Forest rides are maintained primarily for timber extraction or for game management, but they can provide a wide variety of early successional habitats including open grassland, scrub and young tree growth, as well as the transitional edge zone between forest and grassland, which is particularly rich in insects. Permanent forest glades provide similar habitats to rides, but are often larger and more open, thus providing opportunities for additional insects. Both can be incorporated and managed for conservation with little interference to the commercial management of the forest, and can be of additional benefit to game management, access, fire control and scenic enjoyment by visitors.

Because different species are adapted to distinct combinations of the level of shade and both the species composition and physical structure of the vegetation, the main aims of management are to maintain a variety of conditions along these internal edges. The easiest way to achieve variety is to manage the rides in parallel strips by cutting on rotations of different lengths. Further variation can be created by cutting bays or "scallops" into the woodland edge, thereby creating a longer and more sheltered edge. Practical details and recommended options are described by Warren and Fuller (1990). The amount of shade in woodland rides has a profound effect on insects, notably butterflies (Hall and Greatorex-Davies 1989;

Warren, 1985). Most of the latter breed only in open, sunny conditions, and minimum recommended ride widths for the most light-demanding species are 1–1.5 times the height of surrounding trees. In even-aged plantations, this means the maintenance of some rides 25–40 m wide in order to conserve such species. In addition, it may be essential to manage the ride edge scrub by cutting sections on a 7–15-year rotation to produce a linear coppice.

Recommended ride-cutting regimes have only been tested for short periods so their effectiveness is largely unknown, although immediate benefits have been recorded for some butterflies (e.g. Pollard, 1982; Pollard *et al.*, 1986). It seems unlikely, however, that ride management can provide suitable habitats for all early successional insects, as it is carried out on a much smaller scale than traditional coppice. Some species that were formerly abundant in coppiced woodland also do not appear to use ride edges, at least in their present form. An example is one of Britain's most endangered butterflies, *Argynnis adippe*. The reinstatement of coppice management may therefore be of greater significance for some rarities than others. There are signs that this may now be more economic (e.g. Kirby, 1988; Tabor, 1989). New uses are being sought for coppice products, and woodland management techniques are being developed which in some ways mimic the effects of coppice management, for example in the growing of edible wood-decaying fungi, notably shi-itake and oyster mushrooms in cut oak logs (Chang and Hayes, 1978; Campbell and Slee, 1985), and the use of smallwood as a biomass energy resource (de Groot, 1989).

2. *Old Forest Conditions*

The integration of habitats for old forest invertebrates into modern commercial woodland is a more challenging problem. It is necessary to allow very large numbers of trees to senesce and die naturally over time scales measured in centuries. These trees cannot realistically contribute to the commercial management of the forest for timber production, apart from the use of the thinnings in the early stages of the establishment of a cohort of trees, or unless a commercial market can be created for pollard products. They may add to the amenity value of the wood if there is public access, but even this may eventually lead to conflict over safety considerations, once the trees get into the ideal state to produce habitats for saproxylic invertebrates (see Section II.C, above). While it is possible to leave thinnings, loppings and the occasional felled tree for conservation purposes, even creating "habitat piles" as recommended by Brooks (1980), this is of relatively little value for the highly specialized, old forest species discussed above. Only those opportunistic, ubiquitous dead wood species with less demanding niches are likely to take advantage of such habitats. While it is good that an increase in these species may result from such efforts, such methods are no substitute for the retention and eventual replacement of ancient, living trees. It is sometimes suggested that the saproxylic invertebrates can be accommodated either by the felling of "weed" trees, thinnings etc., or even by the importation into the wood of timber felled elsewhere. The limitation of either of these approaches is that it does not provide decaying timber either of sufficient size or quality to satisfy the more demanding species. Felled and standing dead sound timber does not produce the diversity of niches necessary for these species. Decay starts from the exterior of the log and penetrates inward, heart rot and rot holes do not develop, and the duration of individual pieces of timber is much shorter than that of decaying wood on living trees which, in effect, is a renewable resource.

The long-term survival of isolated assemblages of saproxylic invertebrates on small sites could be greatly encouraged by planting surrounding land with appropriate trees and managing them in the long term to produce old growth, allowing them to age and die naturally. There are legal and financial problems in achieving this, however. In Britain, no mechanism exists whereby the statutory conservation body can grant-aid the planting or purchase of this surrounding land. A more realistic objective might be the replanting and appropriate management of blocks of former parkland, pollard or other form of wood pasture forest, that have recently either been converted to arable land or planted with conifers but which remain an integral part of a quality site. Even in these instances, however, such blocks have usually been excluded from the area afforded statutory protection and no mechanism exists by which such conversion can be encouraged.

In sites where the number of very old trees is dangerously small, but where there are sufficient mature trees which are not yet suitable for producing dead wood habitat, or where there are generation gap problems, it may be possible to "pre-senesce" younger trees so that the cycle of decay is initiated prematurely. This is again dependent upon achieving the co-operation of the landowning interest, who is likely to need some persuading that the maiming of sound trees "in their prime" is in the best interests of conservation.

Speight (1989) recommends the deliberate mutilation of trees to allow ingress of water, fungi and saproxylic invertebrates. Little experimentation in this direction appears to have been conducted in Europe but considerable research has been carried out in the USA on the management of "snag" trees (standing dead trees), and the creation of decay cavities in dead and living trees. While this has been done primarily for the provision of nesting sites for hole-nesting birds and den sites for various species of mammal, the methodology would seem to be adaptable for the "pre-senescence" of trees for old forest invertebrates. Methods include the use of chain saws to initiate rot hole formation (Sanderson, 1975; Carey and Sanderson, 1981; Carey and Gill, 1983), the inoculation of heart rot-inducing fungi to create hollow trees (Silverborg, 1959; Toole, 1965; Conner *et al.*, 1983), the use of ring barking ("girdling"), fire and herbicides to create standing dead trees (Styskel, 1983; Conner *et al.*, 1981), and even the use of explosives to produce standing dead trees, shattered boles and branch stumps to initiate fungal cavity formation (Bull *et al.*, 1981). Wood-decaying fungal inoculation techniques have also been developed for the production of edible fungi (Toole, 1966; Chang and Hayes, 1978; Campbell and Slee, 1985). All of these techniques could be applied towards the creation of dead wood habitat for insects in mature trees, given sufficient resources and willingness of woodland managers. Clearly there is much further research needed into the positive management of habitat to accommodate this specialized fauna.

D. Re-creatability of Woodland Habitats — Opportunities and Limitations

The re-creation of early successional habitats is comparatively quick and potentially quite easy, for example by restarting the coppice cycle in neglected woods. The ground flora may not, however, respond satisfactorily if there has been a long break in the cycle and there may be problems associated with poor or patchy tree regrowth (e.g. Fuller and Warren, 1990). Moreover, the response of the invertebrate fauna will depend on the historical continuity of open space woodland habitat within the immediate vicinity of the newly managed block. If continuity has been broken for more than 10–20 years, many stenotypic early successional stage species will no longer be present in that area, and the fauna that remains will consist only of more common, eurytopic species rather than the fauna for which the management is intended.

Several recent policy initiatives in Britain suggest a future increase in the rate of broadleaved planting in the lowlands. These include the Woodland Grant Scheme which gives an enhanced rate for broadleaved planting (Forestry Commission, 1988) and an ambitious plan to plant 12 new Community Forests on the edges of major urban areas, totalling 400 000 ha (Countryside Commission, 1989a). There therefore exists the potential for these new woodlands to be colonized by some woodland invertebrates. The continuity of habitats in ancient woodland means, however, that it is likely to remain a far richer habitat for woodland insects for many centuries. The value of new plantations will be greatest if species native to the area are planted and if a diversity of age structures is created. The benefits to early successional insects will accrue largely to species associated with young woody plant growth, but will be very limited for the large number of species which depend on plants confined to long-established woodland. In contrast, planting of woodland on marginal agricultural land, such as is frequently suggested, may result in a considerable net loss in insect conservation value. A fauna associated with unimproved grassland, hill pasture, heathland or ruderal plants in marginal arable land may be replaced by an impoverished woodland fauna containing only the most vagile species. Species associated with open grassland or heathland may survive if rides and glades are left within the planted area, but this is a poor substitute for the appropriate management of these biotopes in their original form. For the old forest invertebrates of ancient woodland, these new woodlands will be almost irrelevant unless they form part of a wider strategy to enlarge and link existing important sites for these species, and unless there are very long-term management plans that specifically set out to allow trees to reach post-maturity. The extremes of specialization and apparent lack of mobility of old forest invertebrates means that most isolated new woods simply will not be colonized by these species.

There is also the possibility of deliberately re-establishing populations of a few species that are favoured by the new management. Several successful re-establishments of threatened woodland butterflies have been conducted recently as part of an overall conservation strategy (e.g. Oates and Warren, 1990). Speight (1989) recommends the re-establishment of saproxylic species to their former sites, ideally as larvae in logs or trunks which are transferred between sites. Re-establishments are, however, only likely to be successful for species whose ecology is well known, and whose popularity enables the expenditure of considerable time and effort, and where habitat management can be maintained in the long term. Thus for most insect species this option is impracticable, and their conservation will depend on maintaining suitable management on existing sites.

A dependency on re-establishment as a strategy for the conservation of popular species may bring the danger of complacency. Moreover, populations of less popular and less well-studied species, which coexisted with the original lost populations of the species to be reintroduced, are rarely, if ever, translocated in the same exercise, but may be equally or more threatened.

VI. THE NEED FOR RESEARCH AND EDUCATION

A. Research

A strategy for conservation can only be effective if it is based on accurate information on species and habitats. This is currently lacking for many of the crucial issues concerning forest insects. Some headway has been made in Britain in the evaluation of habitat features in early successional woodland habitats and by monitoring certain species (e.g. Pollard, 1982), but little research has been carried out into monitoring the mature timber habitat resources, standardizing methods of assessing the quality of dead wood habitat, or in monitoring their species. Some techniques have been developed in North America and Europe for quantifying standing and fallen dead wood (Warren and Olsen, 1964; Van Wagner, 1968; Brown, 1974), for quantifying rot holes in forests (Carey, 1983) and for assessing the quality of individual standing or fallen dead trees (Dajoz, 1974; Lang and Forman, 1978; Cline *et al.*, 1980; Lambert *et al.*, 1980; Trisca and Cromack, 1980; MacMillan, 1981; Sollins, 1982) although almost none of these methods have been applied in Britain. There is consequently an urgent need for further practical research on a wide variety of topics. The list given below is not exhaustive, and no attempt has been made to allocate priority:

1. Site surveys and evaluation to identify priorities for special conservation measures. Woodlands with a long continuity of similar management should receive special attention.
2. Development of standardized methods of sampling forest insects, including the development of non-destructive methods for sampling saproxylic insects (e.g. emergence traps) to allow valid comparisons between different sites and long-term monitoring of species.
3. Development of methodology for monitoring of microhabitat quality to gauge the effectiveness of management of prime sites.
4. Comprehensive studies on the responses of forest insects to traditional management, especially coppicing and pollarding, and to modern forms of sylviculture. The former should be concentrated on sites with a long history of traditional management.
5. Ways of attempting to incorporate early and late successional habitats into modern forestry with minimum economic impact.
6. Methods of measuring the mobility of woodland insects, including the importance of barriers to dispersal and the use of habitat corridors (i.e. rides, hedgerows and edges). Saproxylic species should be given a high priority.

7. Possible role of re-establishments in conserving threatened species.
8. Autecological studies of early and late successional woodland insects, to give further insights into niche specialization and the effects of management. Long-term monitoring of individual populations to determine natural fluctuations and to differentiate the effects of management and habitat change.
9. Role of microclimate in habitat specificity in different parts of species' ranges.

B. Education

There is still some way to go in encouraging a positive attitude towards most insects and their distinctive habitat requirements. It is essential to communicate general principles of invertebrate conservation, as well as some of the results of research and the anecdotal observations of entomologists, to landowners, foresters and the general public. There are still few publications readily available which address these subjects; many of those quoted here are now out of print. The importance of woodland that contain post-mature habitats has recently been recognized by the Council for Europe (Council of Europe, 1988; Speight, 1989) but there are still huge problems in implementing the recommendations. Similar recognition of early successional woodland habitats is urgently needed, particularly as the problems experienced in Britain may well spread if coppicing declines in other European countries. The Nature Conservancy Council has recently attempted to increase awareness of the plight of many invertebrate species by a series of workshops on invertebrate conservation given to other conservation organizations and statutory and voluntary bodies involved in woodland management. Much more publicity is needed, however, if attitudes are to be changed.

VII. SUMMARY AND CONCLUSIONS — THE FUTURE FOR WOODLAND INSECTS

Forest insects have experienced considerable changes in both the extent and composition of their habitat since the last Ice Age. Human impact has been severe for thousands of years, but for many centuries traditional regimes provided habitats for a rich assemblage of species associated with early successional stages, and allowed a proportion of the old forest fauna also to survive. There have, however, been radical changes in British woodland over the last 200 years, with an almost total shift away from coppicing and wood pasture management to systematically cropped high forest systems, mainly comprised of non-native species. The changes over the last 50 years have been unprecedented in their rapidity, and it has become apparent that they are causing a massive decline in many insect species. If this situation continues, our insect fauna, which has already been impoverished due to extensive forest clearance and centuries of human management, will experience further substantial losses. In this paper we have attempted to highlight the particular plight of early and late successional forest insects, and suggest they should be given a high conservation priority in the planning of future forest strategy.

In recent years, there has been a greatly increased awareness among both conservationists and foresters of the value of open spaces and dead wood in forests, largely due to the popularity and needs of butterflies and insectivorous birds. However, the requirement by some insects for a continual supply of new clearings, as occurs with coppicing, still represents a major problem both on and off nature reserves. Even greater problems face saproxylic insects; their requirements are nowhere near so widely recognized.

It is vital that sites are managed appropriately where a coppice-wood or old forest fauna survives. We need, in particular, to educate forest and estate managers and tree surgeons to be sympathetic towards the importance and vulnerability of these faunas. Such sympathy is still far from widespread but, with it, there is much that could be done to enhance the value of these existing sites to ensure the survival of reservoir populations of the most vulnerable species.

There remain huge problems, both of attitudes and resources, facing the conservation of early successional and old forest insects in Britain, and in Europe. While many of Britain's early successional species remain relatively common in parts of mainland Europe, there are signs that countries like The Netherlands may have lost even more open space woodland butterflies than Britain (Geraedts, 1986), possibly also due to the cessation of coppicing. The saproxylic fauna associated with old forest is threatened throughout the continent. With the prospect of a changing climate there are potentially greater problems and opportunities for both these faunas in the future, but in the meantime the priority must be to ensure the survival of the key sites that have historical continuity of management, and to ensure that this management continues.

Acknowledgements

We thank Dr K. N. Alexander, Mr S. J. Falk, Dr K. J. Kirby, Dr P. Kirby, and Dr P. Waring for providing information, and Mr J. H. Bratton, Mrs R. J. D. Key and Dr I. F. G.

McLean for comments on earlier drafts. This work was commissioned by the Terrestrial Invertebrate Zoology Branch of the Chief Scientists' Directorate of the Nature Conservancy Council as part of its programme of research into invertebrate conservation.

REFERENCES

Andersson, H., Coulianos, C.-C., Hammarstedt, O., Imby, L., Janzon, L. A., Lindelow, A. and Walden, H. W. (1987). Hotade evertebrate i Sverge. [Threatened invertebrates in Sweden.] *Entomol. Tidskr.* **108**, 65–75.

Ash, J. E. and Barkham, J. P. (1976). Changes in the field layer of a coppiced woodland in Norfolk, England. *J. Ecol.* **22**, 697–712.

Bennett, K. D. (1988). A provisional map of the forest types for the British Isles 5000 years ago. *J. Quatern. Sci.* **4**, 141–144.

Boddy, L. (1983). Microclimate and moisture dynamics of wood decomposing in terrestrial ecosystems. *Soil Biol. Biochem.* **15**, 149–157.

Boddy, L. and Swift, M. J. (1983). Wood decomposition in an abandoned beech and oak coppiced woodland in S E England. I Patterns of wood litter fall. *Holarct. Ecol.* **6**, 320–332.

Brooks, A. (1980). "Woodlands. A Practical Conservation Handbook". British Trust for Conservation Volunteers, Reading.

Brown, J. K. (1974). "Handbook for Inventorying Downed Woody Material". USDA Forest Service General Technical Report INT-16. Interstate Forest and Range Experimental Station, Ogden, Utah.

Buck, F. D. (1948). *Pentarthrum huttoni* Woll. (Col., Curculionidae) and some imported Cossoninae. *Entomol. Monthly Mag.* **84**, 152–154.

Buckland, P. C. (1979). "Thorne Moors: a Palaeoecological Study of a Bronze Age Site". Occasional publication No. 8, Dept Geography, University of Birmingham.

Bull, E. L., Partridge, A. D. and Williams, W. G. (1981). "Creating Snags with Explosives", pp. 1–4. USDA Forest Service Research Note PNW-393. Pacific-Northwest Forest and Range Experimental Station, Portland, Oregon.

Cameron, R. A. D., Down, K. and Pannett, D. J. (1980). Historical and environmental influences on hedgerow snail faunas. *Biol. J. Linn. Soc.* **13**, 75–87.

Campbell, A. C. and Slee, R. W. (1985). Commercial cultivation of Shiitake in Taiwan and Japan. *Farmers Club J.* **1985**, 29–35.

Carey, A. B. (1983). Cavities in trees in hardwood forests. *In* "Snag Habitat Management – Proceedings of the Symposium" (J. W. Davis, G. A. Goodwin and R. A. Ockenfels, eds), pp. 167–184. USDA Forest Service General Report RM-99. Rocky Mountains Forest and Range Experimental Station, Fort Collins.

Carey, A. B. and Gill, J. D. (1980). "Firewood and Wildlife". USDA Forest Service Research Note NE-299. Northeastern Forest Experimental Station, Broomall, PA.

Carey, A. B. and Gill, J. D. (1983). Direct habitat improvements – some recent advances. *In* "Snag Habitat Management – Proceedings of the Symposium" (J. W. Davis, G. A. Goodwin and R. A. Ockenfels, eds), pp. 80–87. USDA Forest Service General Report RM-99. Rocky Mountains Forest and Range Experimental Station, Fort Collins.

Carey, A. B. and Sanderson, H. R. (1981). Routing to accelerate tree cavity formation. *Wildl. Soc. Bull.* **9**, 14–21.

Carlisle, A. (1977). The impact of man on the native pinewoods of Scotland. *In* "Native Pinewoods in Scotland" (R. G. H. Bunce, and J. N. R. Jeffers, eds), pp. 70–77. Proceedings of Aviemore Symposium, 1975. Institute of Terrestrial Ecology, Cambridge.

Cartwright, K. and Findlay, W. P. K. (1958). "The Decay of Timber and its Prevention", 2nd edn. Thames and Hudson, London.

Chalmers-Hunt, J. M. (1969). Insect conservation in mixed woodland and ancient parkland. *Entomol. Rec. J. Variat.* **81**, 156–163.

Chang, S. T. and Hayes, W. A. (1978). "The Biology and Cultivation of Edible Mushrooms". Academic Press, New York.

Christensen, O. (1977). Estimation of standing crop and turnover of dead wood in a Danish oak forest. *Oikos* **28**, 177–186.

Cline, S. P., Berg, A. B. and Wight, H. M. (1980). Snag characteristics and dynamics in Douglas fir forests of western Oregon. *J. Wildl. Manage.* **44**, 773–786.

Coles, J. M. and Orme, B. J. (1983). *Homo sapiens* or *Castor fiber? Antiquity* **58**, 95–102.

Collier, R. (1986). "The Conservation of the Chequered Skipper in Britain". Focus on Nature Conservation, No. 16. Nature Conservancy Council, Peterborough.

Collins, N. M. (1987). "Legislation to Conserve Insects in Europe". Amateur Entomologists' Society Pamphlet No. 13, AES, London.

Conner, R. N., Dickson, J. G. and Lock, B. A. (1981). Herbicide killed trees infected with fungi: potential cavity sites for woodpeckers. *Wildl. Soc. Bull.* **9**, 308–310.

Conner, R. N., Dickson J. G. and Williamson, J. H. (1983). Potential woodpecker nest trees through artificial inoculation of heart rots. *In* "Snag Habitat Management — Proceedings of the Symposium" (J. W. Davis, G. A. Goodwin and R. A. Ockenfels eds), pp. 68–72, USDA Forest Service General Report RM-99. Rocky Mountains Forest and Range Experimental Station, Fort Collins.

Council of Europe (1988). Recommendation No. R(88)10 of the Committee of Ministers to members states on the protection of saproxylic organisms and their biotopes. Council of Europe, Strasburg.

Countryside Commission (1988). "Task Force Trees". Countryside Commission, London.

Countryside Commission (1989a). Forest for a new age. *Countryside Commission News* **40**, 1.

Countryside Commission (1989b). "Notes on changes in hedgerow length in England and Wales". Unpublished report. Countryside Commission. Cheltenham.

Craighead, F. C. (1920). Direct sunlight as a factor in forest insect control. *Proc. Entomol. Soc. Wash.* **22**, 106–108.

Crowson, R. A. (1962). Observations on Coleoptera in Scottish oak woods. *Glasgow Nat.* **18**, 177–195.

Crowson, R. A. (1964). Additional records of Coleoptera in Scottish oak woods. *Glasgow Nat.* **18**, 371–375.

Dajoz, R. (1974). Les insectes xylophages et leur role sans la degradation du bois mort. [Wood feeding insects and their role in the decomposition of dead wood.] *In* "Ecologie Forestiere" P. Pesson, ed.) pp. 257–307. Ganthier-Villars, Paris.

De Groot, P. (1989). Plant power: fuel for the future. *New Sci.* **1695** 16 December, 30–33.

Dennis, R. L. H. (1977). "The British Butterflies. Their Origin and Establishment". E. W. Classey, Faringdon.

Donisthorpe, H. (1924). Notes on the myrmecophiles found with *Acanthomyops (Donisthorpea) brunneus*, Latr., in Britain. *Entomol. Rec. J. Variation* **36**, 121–122, 133–139.

Donisthorpe, H. (1926). A further study of the habits of *Acanthomyops (Donisthorpea) brunneus*, Latr., and the myrmecophiles inhabiting its nests. *Entomol. Rec. J. of Variation* **38**, 40–43, 52–55.

Donisthorpe, H. (1927). "Guests of British Ants". G. Routledge. London.

Dudley, T. and Anderson, N. H. (1982). A survey of invertebrates associated with wood debris in aquatic habitats. *Melanderia* **39**, 1–22.
Edlin, H. L. (1971). Woodland notebook: Goodbye to the pollards. *Q. J. For.* **65**, 157–165.
Edlin, H. L. and Nimmo, M. (1956). "Tree Injuries – Their Causes and Prevention". Thomson & Hudson, London.
Ehnstrom, B. and Walden, H. W. (1986). "Faunavard I Skogsbruket. Del 2 – Den lagre faunan". Skogsstyrelsen. Jonkoping (in Swedish).
Elton, C. S. (1966) "The Pattern of Animal Communities". Methuen, London.
Emmet, A. M. and Heath, J. (1989). "The Moths and Butterflies of Great Britain and Ireland, Vol. 7, Part 1, Hesperiidae – Nymphalidae. The Butterflies". Harley Books, Colchester.
Evans, J. (1984). Silviculture of broadleaved woodland. *For. Commission Bull.*, vol. 62. HMSO, London.
Falinski, J. B. (1978). Uprooted trees, their distribution and influence in the primeval forest biotope. *Vegetatio* **38**, 175–183.
Falinski, J. B. (1986). "Vegetation Dynamics in Temperate Lowland Primeval Forests". Junk, The Hague.
Falk, S. J. (1991). "A Review of the scarce and threatened flies of Great Britain (Part 1)." Research and survey in Nature conservation, Nature Conservancy Council, Peterborough, in press.
Ford, E. D. and Newbould, P. J. (1977). The biomass and production of ground vegetation and its relation to tree cover through a deciduous woodland cycle. *J. Ecol.* **65**, 201–212.
Forestry Commission (1928). "Report on Census of Woodlands and Census of Production of Home-grown Timber 1924". HMSO, London.
Forestry Commission (1984). "Census of Woodland Trees 1979–82: Great Britain". Forestry Commission, Edinburgh.
Forestry Commission (1987). "The Recognition of Hazardous Trees". HMSO, London.
Forestry Commission (1988). "Grants for Woodland Planting". HMSO, London.
Foster, A. P. (1983). National review of non-marine molluscs. Invertebrate Site Register report No. 14. Nature Conservancy Council. London (unpublished).
Fowler, W. W. (1887). "The Coleoptera of the British Islands", 6 vols. Reeve, London.
Fuller, R. J. and Warren, M. S. (1990). "Coppiced Woodlands: Their Management for Wildlife". Nature Conservancy Council, Peterborough.
Fuller, R. J. and Warren, M. S. (1991). "Conservation Management in Ancient and Modern Woodlands: Responses of Fauna to Edges and Rotations. Proceedings of 29th Conference of the British Ecological Society, Southampton 1989. Blackwell Scientific, Oxford, in press.
Garland, S. P. (1983). Beetles as primary woodland indicators. *Sorby Rec.* **21**, 3–38.
Geiger, R. (1965). "The climate near the Ground". Harvard University Press, Cambridge, Massachusetts.
Geraedts, W. H. J. M. (1986). "Voorlopige atlas van de Nederlandse dagvlinders – Rhopalocera". Stichting Vlinderonderzoek, Wageningen (in Dutch).
Gilbert, B. L. and Norris, D. M. (1968). A chemical basis for bark beetle (Scolytidae) distribution between host and non-host trees. *J. Insect Physiol.* **14**, 1063–1068.
Girling, M. A. (1982). Fossil insect faunas from forest sites. *In* "Archaeological Aspects of Woodland Ecology. Symposium for the Association for Environmental Archaeology" (M. Bell and S. Limbrey, eds), No. 2, pp. 129–146. B.A.R. International Series No. 146.
Godwin, H. (1975). "The History of the British Flora: a Factual Basis for Phytogeography", 2nd edn. Cambridge University Press, Cambridge.

Graham, S. A. (1922). Effect of physical factors in the ecology of certain insects in logs. *Minn. State Entomol. Rep.* **19**, 22–40.

Graham, S. A. (1924). Temperature as a limiting factor in the life of subcortical insects. *J. Econ. Entomol.* **17**, 377–383.

Graham, S. A. (1925). The felled tree trunk as an ecological unit. *Ecology* **6**, 397–411.

Hadley, M. (1984). A national review of British macrolepidoptera listing both national and regional rarities. Invertebrate Site Register Report Number 46, NCC, London (unpublished).

Hall, M. L. and Greatorex-Davies, J. N. (1989). "Management Guidelines for the Conservation of Invertebrates, especially Butterflies, in Plantation Woodland". Institute of Terrestrial Ecology, Monks Wood, Huntingdon, (unpublished).

Hamilton, W. O. (1978). Evolution and diversity under bark. (L. A. Mound and N. Waloff eds) "Diversity of Insect Faunas". Symposium of the Royal Entomological Society of London, **9**, 156–175.

Hammond, P. M. (1984). Changes in the British Coleopterous Fauna. *In* "The Changing Flora and Fauna of Britain" (D. L. Hawksworth, ed.) Proceedings of a Symposium of the Systematics Association at the University of Leicester, April 1973, pp. 323–369. Academic Press, London.

Harding, P. T. (1976). The invertebrate fauna of the mature timber habitat: an inventory and survey of areas of conservation value. Unpublished CST Report No. 65. Nature Conservancy Council, Banbury.

Harding, P. T. (1978a). The invertebrate fauna of the mature timber habitat: survey of areas — site reports: England. Unpublished CST Report No. 162. Nature Conservancy Council, Banbury.

Harding, P. T. (1978b). The invertebrate fauna of the mature timber habitat: survey of areas — site reports: Scotland. Unpublished CST Report No. 163. Nature Conservancy Council, Banbury.

Harding, P. T. (1978c). The invertebrate fauna of the mature timber habitat: survey of areas — site reports: Wales. Unpublished CST Report No. 164. Nature Conservancy Council, Banbury.

Harding, P. T. (1978d). An inventory of areas of conservation value for the invertebrate fauna of the mature timber habitat. Unpublished CST Report No. 160. Nature Conservancy Council, Banbury.

Harding, P. T. (1981a). The conservation of pasture-woodlands. *In* "Forest and Woodland Ecology — an Account of Research Being Done in ITE" (F. T. Last and A. S. Gardiner, eds), pp. 45–48. ITE Symposium No. 8. Institute of Terrestrial Ecology, Cambridge.

Harding, P. T. (1981b). Burn wisely — a plea for the conservation of old trees and dead wood. *Nature in Wales* **17**, 144–148.

Harding, P. T. and Rose, F. (1986). "Pasture-woodlands in lowland Britain. A Review of their Importance for Wildlife Conservation". Institute of Terrestrial Ecology, Huntingdon.

Harmon, M. E., Franklin, J. F., Swanson, F. J., Sollins, P., Gregory, S. V., Lattin, J. D., Anderson, N. H., Cline, S. P., Aumen, N. G., Sedell, J. R., Lienkaemper, G. W., Cromack, K. and Cummins, K. W. (1986). The ecology of coarse woody debris in temperate ecosystems. *Adv. Ecol. Res.* **15**, 133–302.

Heath, J., Pollard, E. and Thomas, J. (1984). "Atlas of Butterflies in Great Britain and Ireland". Viking, Harmondsworth

Heliovaara, K. and Vaisanen, R. (1984). Effects of modern forestry on northwestern european forest invertebrates: a synthesis. *Acta For. Fenn.* **189**, 1–32.

Hengeveld, R. (1986). "Insect Conservation and Biogeographical Processes. Proceedings of the 3rd European Congress on Entomology", pp. 449–452.

Heydemann, B. and Muller-Karch, J. (1980). "Biolegisher Atlas Schleswig-Holstein". Karl Wachnoltz, Neumunster.
Holmquist, A. M. (1931). Studies in arthropod hibernation. III. Temperatures in forest hibernacula. *Ecology* **12**, 387–400.
Hunter, F. A. (1977). Ecology of pinewood beetles. *In* "Native Pinewoods in Scotland. Proceedings of Aviemore Symposium, 1975". (R. G. H. Bunce and J. N. R. Jeffers, eds), pp. 42–55. Institute of Terrestrial Ecology, Cambridge.
JCCBI (1972). "A Code for Insect Collecting". Leaflet published by the Joint Committee for the Conservation of British Insects.
Joy N. H. (1932, reprinted 1976). "A Practical Handbook of British Beetles", 2 vols. E. W. Classey, Oxford.
Kaarik, A. A. (1974). Decomposition of wood. *In* "Biology of Plant Litter Decomposition", (C. H. Dickson and G. J. E. Pugh, eds), Vol I, pp. 129–174. Academic Press, London.
Kelner-Pillaut, S. (1974). Peuplement entomologique des terraux d'arbres creux. [Insect populations in wood-mould of hollow trees.] *Bull. Ecol.* **5**, 123–156.
Kerney, M. and Stubbs, A. E. (1980). "The Conservation of Snails, Slugs and Freshwater Mussels". Nature Conservancy Council, London.
Key, R. S. (1987). Review of the Invertebrate Sites in England: North Yorkshire. Unpublished Invertebrate Site Register Reports No. 77, 78 and 79. Nature Conservancy Council, Peterborough.
Key, R. S. (1991). Conservation and the coleopterist. *In* "Coleopterists' Handbook" (J. Cooter, ed.). Amateur Entomologists' Society, in press.
Kirby, K. J. (1984, reprinted 1987). "Forestry Operations and Broadleaf Woodland Conservation". Focus on Nature Conservation No. 8. Nature Conservancy Council, Peterborough.
Kirby, K. J. (1988). Conservation in British woodland — adapting traditional management to modern needs. *In* "The Cultural Landscape — Past, Present and Future" (H. H. Birks, H. J. B. Birks, P. E. Kaland and D. Moe, eds). Cambridge University Press, Cambridge.
Kirby, P. (1991). "A Review of the Scarce and Threatened Hemiptera of Great Britain." Research and Survey in Nature Conservation No. 35. Nature Conservancy Council, Peterborough.
Kitching, R. L. (1969). A preliminary note on the fauna of water filled tree holes. *Entomologist* **102**, 7–9.
Kitching, R. L. (1971). An ecological study of water-filled tree holes and their position in the woodland ecosystem. *J. Anim. Ecol.* **40**, 281–302.
Lambert, R. C., Lang, G. E. and Reiners, W. A. (1980). Loss of mass and chemical change in decaying boles of a subalpine balsam fir forest. *Ecology* **61**, 1460–1473.
Lang, G. E. and Forman, R. T. T. (1978). Detritus dynamics in a mature oak forest: Hutchinson Memorial Forest, New Jersey. *Ecology* **59**, 580–595.
MacMillan, P. C. (1981). Log decomposition in Donaldson's Woods, Spring Mill State Park, Indiana. *Am. Midl. Nat.* **106**, 335–344.
Marshall, J. E. (1978). The larva of *Aulonium trisulcum* (Fourcroy) (Coleoptera; Colydiidae) and its association with elm bark beetles (*Scolytus* spp.). *Entomol. Gaz.* **29**, 59–69.
Maser, C. and Trappe, J. M. (1984). The seen and unseen world of the fallen tree. USDA Forest Service General Technical Report PNW-164. Pacific Northwest Forest and Range Experimental Station (unpublished).
Massee, A. M. (1965). Some features of conservation interest arising from surveys. *In* "The Conservation of Invertebrates" (E. A. G. Duffey and M. G. Morris, eds). Monks Wood Symposium No. 1, Nature Conservancy, Abbots Ripton.
McLean, I. F. G. (1990). The role of legislation in the conservation of invertebrates. *In*

"European Invertebrate Survey Colloquium: The Bern Convention Invertebrates and Their Conservation. Conclusions and Summaries". pp. 36–38. Council of Europe, Strasbourg.

Mendel, H. and Owen, J. A. (1990). *Limoniscus violaceus* (Muller) (Col.: Elateridae), the violet click beetle, in Britain. *Entomologist* **109**, 43–46.

Middleton, J. and Merriman, G. (1985). The rationale for conservation: problems from virgin forest. *Biol. Conserv.* **33**, 133–145.

Mitchell, P. L. (1989). Repollarding large neglected pollards: a review of current practice and results. *Arbor. J.* **13**, 125–142.

Mitchell, P. L. and Kirby, K. J. (1989). "Ecological effects of forestry practices in long-established woodland and their implications for nature conservation". Oxford Forestry Institute Occasional Paper No. 39, University of Oxford.

Morris, M. G. (1974). Oak as a habitat for insect life. *In* "The British Oak; its history and natural history" (M. G. Morris and F. H. Perring, eds), pp. 276–297, E. W. Classey, Farringdon.

Nature Conservancy Council (1984). "Nature Conservation in Great Britain". NCC, Peterborough.

Nature Conservancy Council (1986). "Nature Conservation and Afforestation in Britain". NCC, Peterborough.

Nature Conservancy Council (1989). "Guidelines for the Selection of Biological SSSIs". NCC, Peterborough.

Neff, D. J. (1957). Ecological effects of beaver habitat abandonment in the Colorado Rockies. *J. Wildl. Manage.* **21**, 80–84.

Niemelä, J., Haila, Y., Halme, E., Lahti, Pajumen, T. and Punttila, P. (1988). The distribution of carabid beetles in fragments of old coniferous taiga and adjacent managed forest. *Ann Zool. Fennici*, **25**, 107–119.

Oates, M. R. and Warren, M. S. (1990). "A Review of Butterfly Introductions in Britain and Ireland". World Wide Fund for Nature, Godalming.

Osborne, P. J. (1964). "The Effect of Forest Clearance on the British Insect Fauna". Proceedings of the 12th International Congress of Entomologists, pp. 456–457, London.

Owen, J. A. (1988). Destruction of habitat at Windsor. Letter to the editor. *Coleopt. Newsl.* **31**, 11–13.

Owen, J. A. (1989). The 1987 Presidential Address — Part 2. A preliminary account of the beetles of the RSPB Loch Garten reserve. *Br. J. Entomol. Nat. Hist.* **2**, 17–28.

Paviour-Smith, K. (1964). Habitats, headquarters and distribution of *Tetratoma fungorum* F. (Col. Tetratomidae). *Entomol. Monthly Mag.* **100**, 71–80.

Peterken, G. F. (1981). "Woodland Conservation and Management". Chapman & Hall, London.

Peterken, G. F. and Allison, H. (1989). "Woods, Trees and Hedges: a Review in the British Countryside". Focus on Nature Conservation No. 22. Nature Conservancy Council, Peterborough.

Pollard, E. (1979). Population ecology and change in range of the white admiral butterfly *Ladoga camilla* in England. *Ecol. Entomol.* **4**, 61–74.

Pollard, E. (1982). Monitoring butterfly abundance in relation to the management of a nature reserve. *Biol. Conserv.* **24**, 317–328.

Pollard, E., Hall, M. L. and Bibby, T. J. (1986). "Monitoring the Abundance of Butterflies 1976–1985". Research and Survey in Nature Conservation No. 2, Nature Conservancy Council, Peterborough.

Pollard, E., Hooper M. D. and Moore, N. W. (1974). "Hedges". Collins, London.

Rackham, O. (1980). "Ancient Woodland: Its History, Vegetation and Uses in England". Arnold, London.

Rackham, O. (1986). "The History of the British Countryside". Dent, London.

Rackham, O. (1989). "The Last Forest: The Story of Hatfield Forest". Dent, London.

Ranson, C. E. (1978). Nature conservation in Epping Forest. *In* "Epping Forest – the Natural Aspect" (D. Corke, ed), pp. 58–79. Essex Naturalist No. 2, Essex Field Club, London.

Read, R. W. J. (1982). *Mesites tardii* (Curtis) (Coleoptera: Curculionidae) new to west Cumbria, with notes on the species in Britain. *Entomol. Gaz.* **33**, 233–242.

Remmert, H. (1987). Natural successions – an important goal of nature conservation. *In* "Council of Europe. Workshop on the situation and protection of ancient natural and seminatural woodlands in Europe". Report of meeting at Bayerischen Wald, F.R.G., pp. 27–28. Council of Europe, Strasbourg.

Roskam, H. C., de Gier, H., Persoon, B., de Bruijn, G. J. and ter-Keurs, W. J. (1991). Secondary succession of gall inducers in coppice. *In* "Evolutionary Ecology of Gall-forming Insects" Proceedings 18th International Congress of Entomological Variation, 1988, in press.

Russell, L. K. (1979). Beetles associated with slime moulds (Mycetozoa) in Oregon and California (Coleoptera: Leiodidae, Sphindidae, Lathridiidae). *Pan-Pacific Entomol.* **55**, 1–9.

Salisbury, E. J. (1924). The effects of coppicing as illustrated by the woods of Hertfordshire. *Trans. Hert. Nat. Hist. Soc.* **18**, 1–21.

Sanderson, H. R. (1975). Den tree management for grey squirrels. *Wildl. Soc. Bull.* **3**, 125–131.

Savely, H. E. (1939). Ecological relations of certain animals in dead pine and oak logs. *Ecol. Monogr.* **9**, 321–385.

Scheffer, T. C. and Cowling, E. B. (1966). Natural resistance of wood to microbial deterioration. *Annu. Rev. Phytopathol.* **4**, 147–170.

Shirt, D. B. (ed.) (1987). "British Red Data Books: 2. Insects". Nature Conservancy Council, Peterborough.

Silverborg, S. B. (1959). Rate of decay in northern hardwoods following artificial inoculation with some common heart rot fungi. *For. Sci.* **5**, 223–228.

Sollins, P. (1982). Input and decay of coarse woody debris in coniferous stands in western Oregon and Washington. *Can. J. For. Res.* **12**, 18–28.

Speight, M. C. D. (1985). The extinction of indigenous *Pinus sylvestris* in Ireland: relevant faunal data. *Ir. Nat. J.* **21**, 449–453.

Speight, M. C. D. (1986). "Criteria for the selection of insects to be used as bio-indicators in nature conservation research". Proceedings of the 3rd European Congress of Entomology, Amsterdam, Part 3, pp. 485–488.

Speight, M. C. D. (1989). "Saproxylic Invertebrates and Their Conservation". Nature & Environment Series No. 42, Council of Europe, Strasburg.

Squance, M. A. and Harris, P. (1988) Sherwood Forest Country Park. A Draft Strategy to Protect and Promote its Natural Heritage. Nature Conservancy Council. East Midlands Region. Grantham (unpublished).

Steel, D. and Mills, N. (1988). A study of plants and invertebrates in an actively coppiced woodland (Brasenose Wood, Oxfordshire). *In* "Woodland Conservation and Research in the Clay Vale of Oxfordshire and Buckinghamshire" (K. J. Kirby and F. J. Wright, eds), pp. 116–122. Research and Survey in Nature Conservation No. 15, Nature Conservancy Council, Peterborough.

Stephens, J. F. (1839). "A Manual of British Coleoptera". Longman, Orme Brown, Green & Longmans.
Sterling, P. H. and Hambler, C. (1988). Coppicing for conservation: do hazel communities benefit? *In* "Woodland Conservation and Research in the Clay Vale of Oxfordshire and Buckinghamshire". (K. J. Kirby and F. J. Wright, eds). pp. 69–80. Research and Survey in Nature Conservation No. 15, Nature Conservancy Council, Peterborough.
Stubbs, A. E. (1972). Wildlife conservation and dead wood. Supplement to the Journal of the Devon Trust for Nature Conservation, pp. 1–18.
Stubbs, A. E. (1982). Hoverflies as primary woodland indicators with reference to Wharncliffe Wood. *Sorby Rec* **20**, 62–67.
Stubbs, A. E. and Falk, S. J. (1983). "British Hoverflies. An Illustrated Identification Guide". British Entomological and Natural History Society, London.
Styskel, E. W. (1983). Problems in snag management implication — a case study. *In* "Snag Habitat Management — Proceedings of the Symposium" (J. W. Davis, G. A. Goodwin and R. A. Ockenfels, eds), pp. 24–27. USDA Forest Service General Report RM-99 Rocky Mountains Forest and Range Experimental Station, Fort Collins.
Sutton, S. L. and Beaumont, H. E. (1989). "The Butterflies and Moths of Yorkshire: Distribution and Conservation". Yorkshire Naturalists' Union, Leeds.
Swift, M. J. (1973) The estimation of mycelial biomass by determination of the hexoamine content of wood tissue decayed by fungi. *Soil Biol. Biochem.* **5**, 321–332.
Swift, M. J. (1977) The ecology of wood decomposition. *Sci. Prog.* **64**, 175–199.
Sylvester, W. D., Sollins, P., Verhoeven, T. and Cline, S. P. (1982). Nitrogen fixation and acetylene reduction in decaying conifer boles: effects of incubation time, aeration and moisture content. *Can. J. For. Res.* **12**, 646–652.
Szujecki, A. (1979). Some problems of zoocoenosis restoration in degraded forest habitats. *Mem. Zool.* **32**, 57–67.
Tabor, R. C. C. (1989). A role for hazel in woodland conservation. *Q. J. For.* **83**, 177–182.
Thomas, J. A. (1974). Ecological studies of hairstreak butterflies. Ph.D. thesis, University of Leicester.
Thomas, J. A. (1989). Ecological lessons from the reintroduction of Lepidoptera. *The Entomologist* **108**, 56–68.
Thomas, J. A. (1991). Rare species conservation: Case studies of European butterflies. *In*: "The Scientific Management of Temperate Communities for Conservation". pp. 149–197. 29th symposium of the British Ecological Society, Blackwell Scientific publications, Oxford.
Thomas, G. (1980). "The Butterflies of Scotland". Croom Helm, London.
Toole, E. R. (1965). Inoculation of bottmland red oaks with *Poria fissilis* and *Polyporus hispidus*. *Plant Dis. Rep.* **49**, 81–83.
Toole, E. R. (1966). Comparison of two methods for inoculating nuttal oaks with *Pleurotus ostreatus*. *Plant Dis. Rep.* **50**, 552–553.
Trisca, F. J. and Cromack, K. (1980). The role of wood debris in forests and streams. *In* "Forests: Fresh Perspectives from Ecosystems Analysis" (R. H. Waring, ed.), pp. 171–190. Proceedings of the 40th Biology Colloquium (1979). Oregon State University Press. Corvallis.
Tubbs, C. R. (1968). "The New Forest: An Ecological History". David & Charles, Newton Abbot.
Van Wagner, C. E. (1968). The line intersect method in forest fuel sampling. *For. Sci.* **14**, 20–26.
Walter, J. -M. (1987). Remarks on the dynamics of old growth forests as irreplaceable elements of the European natural heritage. *In* "Workshop on the Situation and Protection

of Ancient Semi-natural Woodlands in Europe". pp. 17–26. Council of Europe Environmental Encounters Series No. 3, Strasbourg.

Waring, P. (1988). Responses of moth populations to coppicing and the planting of conifers. *In* "Woodland Conservation and Research in the Clay Vale of Oxfordshire and Buckinghamshire" (K. J. Kirby and F. J. Wright, eds), pp. 82–113. Research and Survey in Nature Conservation No. 15, Nature Conservancy Council, Peterborough.

Waring, P. (1989). General requirements of moths in woodlands. NCC Moth Conservation Project News Bulletin. Unpublished report. Nature Conservancy Council. Peterborough.

Warren, M. S. (1984). The biology and status of the wood white butterfly, *Leptidea sinapis* (L.), in the British Isles. *Entomol. Gaz.* **35**, 207–223.

Warren, M. S. (1985a). The influence of shade on butterfly numbers in woodland rides, with special reference to the wood white, *Leptidea sinapis*. *Biol. Conserv.* **33**, 147–164.

Warren, M. S. (1985b). Habitat utilization by larvae of the heath fritillary, *Mellicta athalia*, and other related species in S. E. France. *Bull. Brit. Ecol. Soc.* **16**, 24–26.

Warren, M. S. (1987a). The ecology and conservation of the heath fritillary butterfly, *Mellicta athalia*. I. Host selection and phenology. *J. Anim. Ecol.* **24**, 467–482.

Warren, M. S. (1987b). The ecology and conservation of the heath fritillary butterfly, *Mellicta athalia*. II. Adult population structure and mobility. *J. Anim. Ecol.* **24**, 483–498.

Warren, M. S. (1987c). The ecology and conservation of the heath fritillary butterfly, *Mellicta athalia*. III. Population dynamics and the effects of habitat management. *J. Appl. Ecol.* **24**, 499–513.

Warren, M. S. (1989), National heath fritillary survey and conservation report, 1989. Unpublished report to the Joint Committee for the Conservation of British Insects.

Warren, M. S. and Fuller, R. J. (1990). "Woodland Rides and Glades: their Management for Wildlife". Nature Conservancy Council, Peterborough.

Warren, M. S., Pollard E., and Bibby, T. J. (1986). Annual and long-term changes in a population of the wood white butterfly. *Leptidea sinapis*. *J. Anim. Ecol.* **55**, 707–719.

Warren, M. S., Thomas, C. D. and Thomas, J. A. (1984). The status of the heath fritillary butterfly, *Mellicta athalia* Rott., in Britain. *Biol. Conserv.* **29**, 287–305.

Warren, W. G. and Olsen, P. F. (1964) A line intersect method for measuring logging waste. *For. Sci.* **10**, 267–276.

Welch, R. C. (1969). Coppicing and its effect on woodland invertebrates. *Q. J. Devon Trust* **22**, 969–973.

Welch, R. C. (1978). Changes in the distribution of the nests of Formica rufa L. (Hymenoptera: Formicidae) at Blean Woods National Nature Reserve, Kent, during the decade following coppicing. *Insectes Sociaux* **25**, 173–186.

Welch, R. C. (1986). What do we know about insects in Scottish woods? *In* "Trees and Wildlife in the Scottish Uplands" (D. Jenkins ed.) pp. 95–100. ITE Symposium No. 17, Institute of Terrestrial Ecology, Cambridge.

White, I. M. (1982). An assessment of the entomological value of some Highland birch woods. Invertebrate Site Register Report No. 9. Nature Conservancy Council, London (unpublished).

Wilde, S. A., Youngberg, C. T. and Hovind, J. H. 1950. Changes in the composition of ground water, soil fertility and forest growth produced by the construction of beaver dams. *J. Wildl. Manage.* **14**, 123–128.

Willmott, K. J. (1988) An appraisal of Chidding fold forest and its current butterfly population. Unpublished report to The Nature Conservancy Council.

Worsley, P. (1978). On the ecology of the beaver and some speculative applications. *Quatern. Newsl.* **25**, 3–4.

9

Lepidoptera as Indicators of Change in the Semi-natural Grasslands of Lowland and Upland Europe

A. ERHARDT AND J. A. THOMAS

I. Introduction 213
II. Intensively Cultivated Grassland 215
III. Abandoned Grassland 217
IV. Changes in Species and Families of Lepidoptera 218
A. Subalpine Meadows 218
B. British Lowland Grassland 220
V. Reasons for Rapid Change in Insect Populations 222
VI. Discussion and Conclusions 231
References 234

I. INTRODUCTION

Wild or traditionally farmed grassland is of prime importance for Lepidoptera in Europe. About 50% of the butterfly species of the British Isles breed wholly or largely in this biotope (Thomas, 1984), and a similar dependency exists in other European countries (e.g. Blab and Kudrna, 1982; Tax, 1990). Over most of the continent, the large majority of these species breed mainly or entirely in "secondary" grassland which has been created and maintained by agriculture. Primary grasslands – in which grassland is the climax vegetation – have been destroyed by man in many parts of Europe, and now account for an extremely small proportion of this biotope. Examples include certain undercliffs, dunes, parts of the floodplains of a few rivers, steep gorges and rocks, areas cleared by avalanches and, in the alpine region, areas above the timberline. In Britain, only *Thymelicus acteon* Poda and *Melitaea cinxia* L. of the "grassland" butterflies have more than 25% of their colonies breeding on primary grassland (Heath *et al.*, 1984; Thomas, 1983a; Thomas and Simcox 1982).

Until 50 years ago, most secondary grassland in Europe had been managed in a similar manner for hundreds or thousands of years. There were shifts in emphasis between grazing, hay making, arable farming and abandonment in different nations as techniques improved and as demand and populations changed, but in general the same traditional methods of farming prevailed. Since the war there has been an industrialization of agriculture. This occurred first in the richer European nations, was hastened in some poorer countries when they joined the EEC, and is now spreading through Eastern Europe.

This revolution has involved the use of modern machinery, chemical fertilizers and pesticides, and new breeds of plants. It has increased the yields from flat landscapes, particularly where the soil was already nutrient rich or moist, to such an extent that it is frequently unnecessary or uneconomic to farm less productive soils, especially those on steep slopes. There has thus been a double change in the management of grasslands, with intensification in some and abandonment in others.

The extent to which either process has occurred varies with soil type, aspect, topography, regional farming subsidies and the prosperity of each nation. Wells (1989) and Mellanby (1981) list changes in the more vulnerable categories of lowland grassland caused by agricultural intensification in Britain: only 0.1% of Huntingdonshire clay soils remain as unimproved pasture and just 1% of East Anglian fen marshes now survive. Serious, but less drastic, was the destruction of 82% of the coastal marshlands of Essex in 1938–1981, mainly as a result of war-time conversion to arable land, and the loss of 80% of unimproved lowland calcareous grassland since 1940.

Abandonment has affected many of the remaining unimproved areas of grassland. In southern Britain, almost all unfertilized calcareous or neutral grassland was traditionally cropped short by sheep, cattle or rabbits. However in 1974–1982, an analysis of 208 such hillsides revealed that 70% of sites were receiving no appreciable grazing, and only 14% were being grazed to a mean turf height of 2 cm or less (Thomas, 1983a,b, unpublished; Thomas *et al.*, 1986). Abandonment is an even greater problem in the Central Alps (Erz, 1973; Surber *et al.*, 1973). Today only about 5% of all unfertilized cultivated grassland is managed in the traditional way of mowing or light grazing, and in the Jura mountains the figure is below 1% (Erhardt, 1985a; Zoller and Bischoff, 1980).

In this chapter we examine the effects of both the intensification of agriculture and abandonment on the diurnal Lepidoptera of European grasslands. We compare these with the changes occurring in populations of plants and other invertebrates, and conclude that insects react quickly to apparently minor environmental changes. Because they are also conspicuous, diurnal Lepidoptera are therefore good indicators of changes that may be occurring over a longer period in longer lived organisms. The main sources of data are a study made in the Central Alps of Switzerland, and intensive autecological studies made in Britain.

The detailed methodology and results of the Swiss research are given by Erhardt (1985a,b,c). It was carried out on the north-facing and south-facing slopes of one subalpine valley. A comparison was made between the densities of Lepidoptera along modified transects in 1977–1979 in traditionally managed meadows, in others that had been abandoned and had reached different stages in a succession to woodland, and in yet others that had received varying intensities of fertilization. Several species, notably most in the genus *Erebia*, occur near the centre of their geographical ranges in this locality. In theory, these should consist predominantly of eurytopic species (Kühnelt, 1948; Schwerdtferger, 1975), which complete their entire life cycle in the same vegetation type (autochtonous species) but which are not restricted to specialized habitats. These, again in theory, should be more resistent to small-scale change than the stenotopic species (autochthonous species that are restricted to a few narrow niches) but less resistant to change than xenotopic species, which comprise migrants and visitors that commute between different types of vegetation.

The British studies, on the other hand, were entirely of stenotopic species that lived near the edges of their geographic ranges, and also on lowland sites (Thomas, 1991). Although the number of species studied has been fewer than those in Switzerland, the mechanisms responsible for population change have been examined in greater detail.

II. INTENSIVELY CULTIVATED GRASSLAND

Figure 1(a) shows the ecological status of all butterfly species recorded in different types of cultivated grassland in a subalpine Swiss valley (excluding rare species that were recorded fewer than five times per flight period). The numbers of species of plants and autochthonous butterflies in these grasslands are compared in Fig. 1b. All meadows used for these figures are south-facing and comparable, except for the most heavily fertilized example which is in a valley bottom.

This pattern is similar to, but more detailed than, figures compiled for the number of butterflies breeding in fields in lowland Britain that are under different kinds and intensities of cultivation (Thomas, 1984). In the

(a)

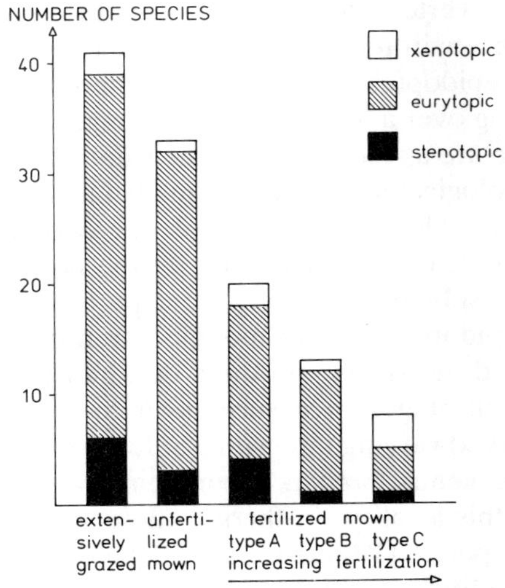

(b)

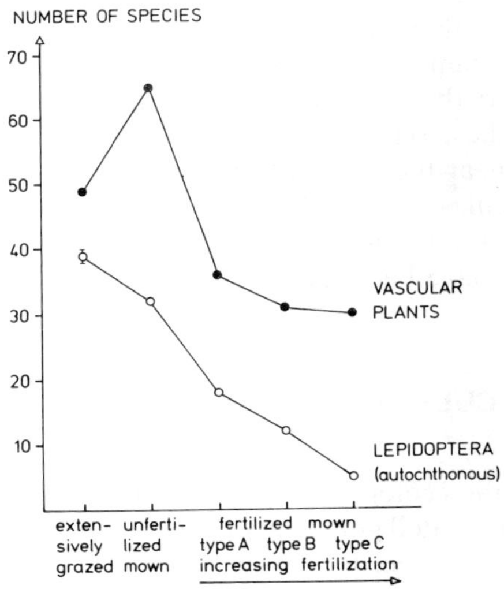

Fig. 1. (a) Numbers and ecological status of Lepidoptera species in cultivated meadows: (□), xenotopic; (▨), eurytopic; (■), stenotopic. (b) Comparison of plant and Lepidoptera species in cultivated meadows; (●——●), vascular plants: (○——○), autochthonous Lepidoptera; (Φ), range of species numbers; type A: fertilization moderate, mown once/season, south-facing slope; type B: fertilization strong, mown twice/season, south-facing slope; type C: fertilization very strong, mown twice/season, valley bottom; unfertilized mown meadows are mown half to once/season.

Swiss example, extensively grazed meadows were inhabited by the most Lepidoptera species. Unfertilized mown meadows were also rich in Lepidoptera, but increasing fertilization caused a severe decline in the species richness of Lepidoptera.

In general, the diversity of Lepidoptera is correlated with plant diversity in all investigated vegetational types (Fig. 1(b); $p < 0.05$, Spearman's rank correlation). This correlation is clearly due to the dependence of the phytophagous larvae on vascular plants. However, there is a discrepancy in the case of unfertilized mown meadows. This is because the act of mowing produces well-balanced conditions for competition between growing herbs, and results in a sward that is especially rich in herb species. On the other hand, the sudden and uniform breakdown of the structure of the vegetation by mowing reduces butterfly populations by destroying the flowers used by adult Lepidoptera and, more harmfully, by killing many of the young stages. This disturbance is more harmful to Lepidoptera in subalpine meadows than is light grazing and trampling by cattle, because cattle graze selectively and leave patches of sward undisturbed. As a consequence, butterfly diversity is lower in unfertilized mown meadows, although plant diversity is higher there than in lightly grazed meadows. The bare hollows of ground caused by the hooves of grazing animals may also be essential for certain species of Lepidoptera that require an unusually warm microclimate (Thomas, 1990, 1991).

Another difference between the figures for Lepidoptera and plants is that the fall in species richness among insects is three times greater than that among plants over the three categories of mown grassland that experienced increasing fertilization in Switzerland (Fig. 1(b)). This does not necessarily reflect a greater robustness among plant populations in the long term. The results of intensive British studies, described later, suggest that insect populations may merely be reacting more quickly to a change in their environment; plant populations may eventually experience comparable declines to Lepidoptera.

III. ABANDONED GRASSLAND

When cultivated grassland is abandoned, a succession of different vegetation types occurs, each with its own fauna. The plant succession is complex and not fully explained in many parts of Europe (Bischof, 1980). However, in the subalpine region of the Swiss Central Alps, the principal trends can be summarized as follows (Bischof, 1980, 1984): on unfertilized mown meadows early successional stages are dominated by grasses and herbs. These are then invaded either by dwarf shrubs (*Vaccinium myrtillus* L., *V. uliginosum* L., *Calluna vulgaris* L.) or by young trees (*Betula pendula* Roth, *Alnus viridis* Chaix). Later stages may eventually revert to forest climax vegetation dominated by spruce (*Picea abies* L.).

It is still widely believed, as a result of studies in central Europe, that the abandonment of grassland is wholly beneficial for Lepidoptera (Reichholf, 1973; Bierhals, 1976; Gerlach, 1976; Ulrich, 1982). Our results from a subalpine valley (Fig. 2(a),(b)) do not support this conclusion, nor do those from the lowland grasslands of Britain (Lipscomb and Jackson, 1964; Frazer, 1965; Frazer and Hyde, 1965; Dempster, 1971; Morris, 1971, 1978; Thomas, 1983a,b, 1984, 1991; Anon, 1984).

In Switzerland, there was again a good correlation between the number of species of plants and Lepidoptera over the range of successional stages sampled ($p < 0.05$, Spearman's rank correlation), although the deviation in Lepidoptera in unfertilized mown meadows was again evident (Fig. 2(b)). Thus although the diversity of plant species was lower in abandoned areas that were still dominated by grass and herbs, that of Lepidoptera was higher. In later successional stages dominated by dwarf shrubs, reduced plant diversity leads to a decrease in butterfly species. Even so, slightly more species occur in these stages than in the unfertilized mown meadows.

Species richness of Lepidoptera falls much more rapidly than that of plants with the arrival of shrubs and trees. Indeed, successional stages dominated by *Alnus viridis* are practically devoid of Lepidoptera. The climax vegetation of woodland dominated by spruce is inhabited by a few, but characteristic, species of Lepidoptera.

Similar trends were recorded on north-facing subalpine slopes as on the south-facing slopes represented in Fig. 2, although species richness was generally lower on the former. The species composition also differed on the two aspects. Many species were confined to south-facing slopes but a few were restricted to northerly aspects. The latter include *Erebia pharte* Hbn., *E. eriphyle* Frr., and a rare endemic of the European alps, *Euphydryas intermedia wolfensbergeri* Frey.

IV. CHANGES IN SPECIES AND FAMILIES OF LEPIDOPTERA

A. Subalpine Meadows

The trends in total species richness described in the previous two sections mask a more rapid turnover in the presence of individual species and

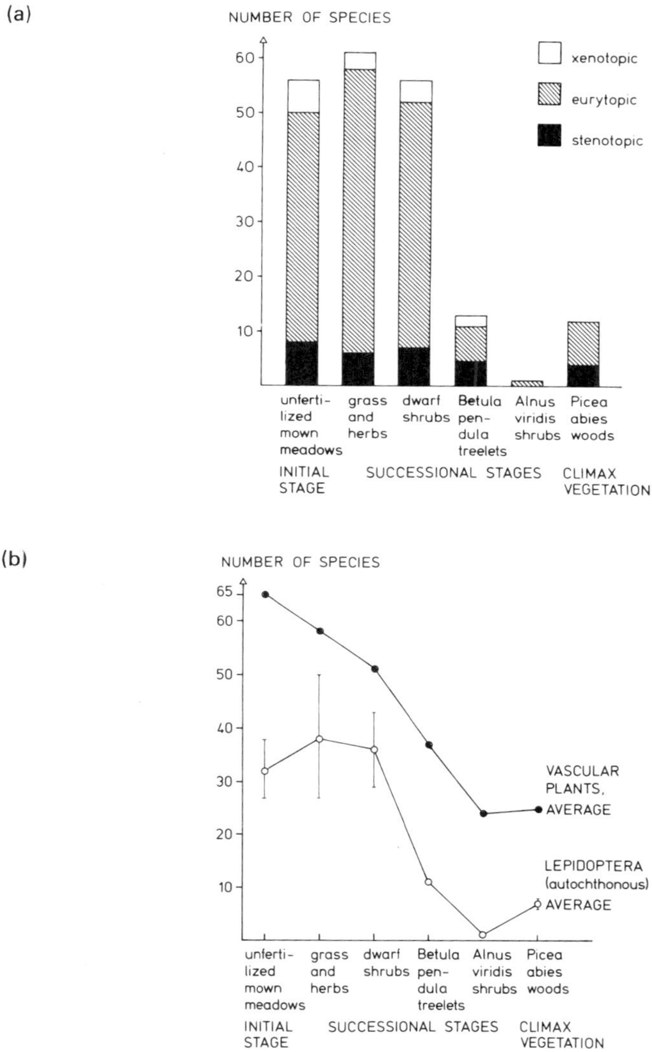

Fig. 2. (a) Numbers and ecological status of Lepidoptera species in different successional stages, south-facing slope: (□), xenotopic; (▨), eurytopic (■), stenotopic. (b) Comparison of plant and Lepidoptera species in different successional stages, south-facing slope: (●——●), vascular plants; (○——○), autochthonous Lepidoptera; (Φ), range of species numbers.

families of Lepidoptera in fertilized or abandoned grassland (Fig. 3). In general, not only does species diversity fall with increasing fertilization, but the proportion of xenotopic species shows a marked increase. Thus in the most heavily fertilized meadow, 38% of Lepidoptera are xenotopic species whereas only 3% and 5% of those in unfertilized mown and lightly grazed meadows, respectively, are visitors or vagrants. There are, however, exceptions: *Paleochrysophanus hippothoe eurydame* Hffmgg. is an endangered subspecies that breeds mainly in moist valley bottoms. Although a stenotopic species, it has not been harmed by fertilization and may well have profited, for fertilization favours its larval foodplant, *Polygonum bistorta* L.

At a family or generic level, Papilionids (*Papilio machaon* L., *Parnassius appollo* L.) were only recorded regularly in early successional stages (Fig. 3). *Erebia* species were mainly eurytopic, as expected of a genus near its centre of distribution, and showed the greatest species diversity in unfertilized mown meadows and recently abandoned grassland. Lycaenidae, on the other hand, have a high percentage of stenotopic species in the Alps. They occurred mainly in lightly cultivated meadows and decreased steeply with increasing fertilization and completely in the later stages of abandonment. In contrast to other Lepidoptera families, the species number of Lycaenidae closely paralleled that of vascular plants in the different categories of vegetation. Zygaenidae were again different: all were eurytopic and their representation extended to successional stages containing dwarf shrubs, albeit in reduced abundance (Erhardt, 1985c). Finally, Geometridae show the opposite trend to that observed in Lycaenidae: the number of stenotopic species (including rarities) increases in later successional stages and reaches a maximum in woods, with the exception of those dominated by *Alnus viridis*. The fact that a few stenotopic species of Geometridae also occur in cultivated meadows (Fig. 3) enhances the value of this family as an indicator group.

B. British Lowland Grassland

Thomas (1991) has reviewed autecological research on seven species of butterfly that inhabit dry neutral or calcareous grassland in Britain, and on one species (*Papilio machaon*) of wet meadow or fen. Also reviewed were studies in Poland and France of two species of *Maculinea* that inhabit moist lowland hay meadows.

Populations of all these butterflies fluctuated considerably from one generation to the next, due partly to short-term variation in the weather. But overriding this "noise", populations also changed very rapidly and directionally in response to successional change: in the cases of *Lysandra*

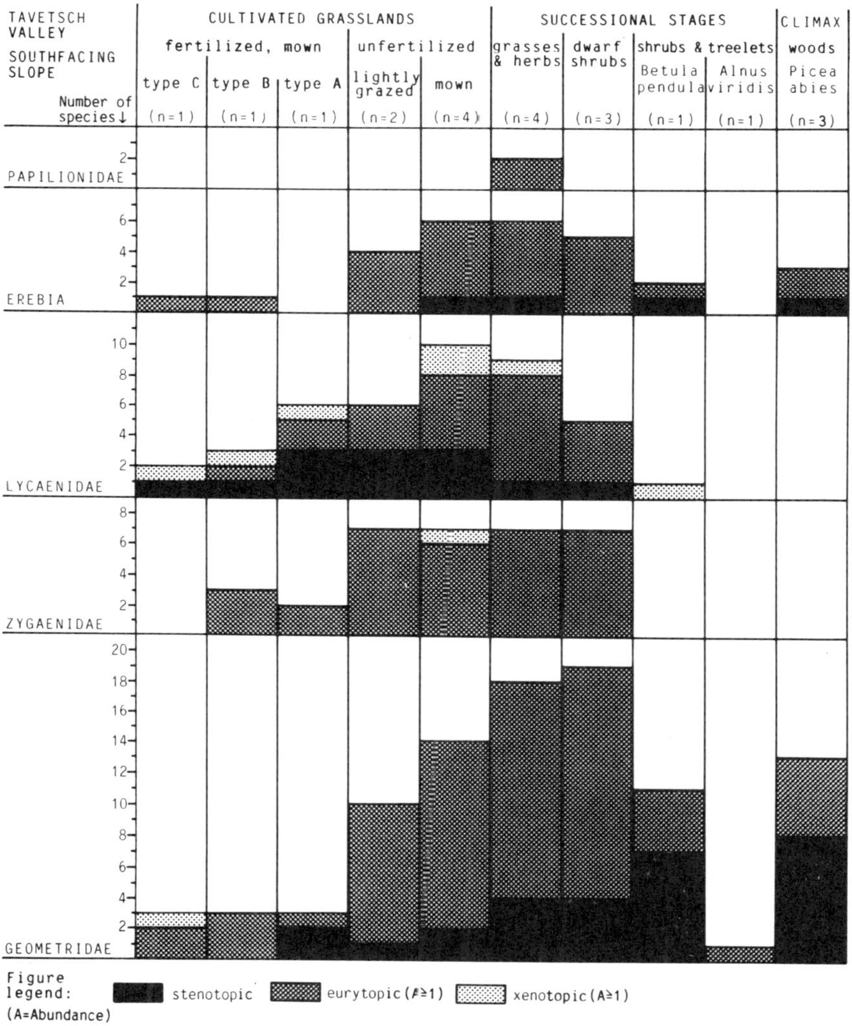

Fig. 3. Species number and ecological status of different Lepidoptera families in cultivated meadows and in different stages of abandoned grassland.

bellargus Rott., *Hesperia comma* L. and *Maculinea arion* L., there are examples where populations of several thousand adults became extinct in the first 2 or 3 years after grazing diminished (Thomas, 1976, 1983b, 1990, 1991; Thomas *et al.*, 1986). With one species, *M. arion*, which preys on the brood of *Myrmica sabuleti* Meinert ants for most of its larval period, the ant host also showed considerable population changes in response to ecological succession (Thomas, 1984).

It was noted in all studies that populations of Lepidoptera and ant species changed much more rapidly than did those of the former's foodplants. Indeed on some sites, the local extinction of *Hesperia comma* coincided with an increase in its foodplant, *Festuca ovina* L. This, however, was exceptional. With every other species, the populations of hostplants eventually experienced the same kind of changes as those of the Lepidoptera. In four studies this was measured with sufficient accuracy for these comparative trends to be quantified, excluding "noise" (Table I)

All examples of insect and plant population changes in Table I are attributable to successional change. In three cases the habitat improved for the species in question and in three cases it deteriorated. On average, the increase or decrease of the four Lepidoptera populations was 12 times more rapid than that of their foodplants, although anecdotal accounts indicate that, in several examples, foodplants also eventually became extinct on some deteriorating sites (Thomas, 1983a; Thomas *et al.*, 1986; Thomas, 1991). The ant *Myrmica sabuleti* also changed quite rapidly, but on average at only three times the rate of the equivalent foodplant (Table I).

It follows that many populations of these butterflies disappeared from sites where their foodplants were still comparatively common, at least for a few years. This is one of the general conclusions from conservation research into butterfly species in a wide range of biotopes in The Netherlands (Anon, 1990; Tax, 1990) and in Britain (Thomas, 1984, 1991). For example in the English county of Suffolk, 42% of butterfly species have become extinct in the past 140 years, but all the foodplants of these species are still present and only 5% of all vascular plants have disappeared over the same period (Thomas, 1991). Comparable figures for other groups of wildlife in Suffolk are the loss of 12% of amphibia species and 3% of mammal species, while the number of bird species has increased by 14% during this period. There has even been a high rate of extinction among Lepidoptera populations on British grassland nature reserves, although again most foodplants remain (Thomas, 1984). The implications of Table I are that these plants may also disappear, if the same factors responsible for Lepidoptera declines continue over a long enough period.

V. REASONS FOR RAPID CHANGES IN INSECT POPULATIONS

There is nothing new in the concept that populations of phytophagous insects often react more rapidly and track minor changes in their environment more closely than the plants on which they feed. Almost all the

TABLE I. A comparison between changes in the population size of Lepidoptera and their foodplants after the habitat changed on their breeding sites. In the case of *Maculinea arion*, changes are given for the initial foodplant *Thymus praecox* and the ant host *Myrmica sabuleti*. For populations that were sampled more than twice, the factor by which populations changed is calculated from a log regression of the data. Sources of data: *Papilio machaon*, Dempster and Hall (1980); other species, J. A. Thomas (1991).

Species	Habitat change	No. samples	Period (years)	Mean factor by which populations changed	Factor by which change in insect population exceeded that of the foodplant
Papilio machaon	Deterioration:	5	4	0.10	6.9
Peucedanum palustre	drying fenland	5	4	0.69	
Maculinea arion	Deterioration:	3	3	0.03	27.3
Myrmica sabuleti	reduced grazing	3	3	0.42	2.0
Thymus praecox		3	3	0.82	
Myrmica sabuleti	Deterioration:	4	1	0.39	2.2
Thymus praecox	reduced grazing	4	1	0.86	
Myrmica sabuleti	Improvement:	4	3.5	18.8	4.5
Thymus praecox	increased grazing	4	3.5	4.2	
Thymelicus acteon	Improvement:	2	3	4.4	3.4
Brachypodium pinnatum	reduced grazing	2	3	1.3	
Lysandra bellargus	Improvement:	2	7	16.6	9.2
Hippocrepis comosa	increased grazing	2	7	1.8	

Lepidoptera studied show characteristics closer to the *r* end of the *r*–K spectrum (Southwood, 1977) than do those of their foodplants, most notably in their shorter generation times. What is surprising, however, is the extreme rapidity with which a number of Lepidoptera have changed in abundance, when the change in their foodplant populations was scarcely noticeable (Thomas, 1991). In Britain, these species include *Plebejus argus* L., *Lysandra bellargus, Maculinea arion, Aricia agestis* Denis & Schiff. (Bourn, 1989), *Melitea cinxia, Hipparchia semele* L., *Thymelicus acteon* and *Hesperia comma* of the "grassland species" of butterfly. This is about 28% of the species that are characteristic of this biotope; a similar figure is obtained when the butterflies of other British biotopes are considered (Thomas, 1991).

The rapidity with which many species of butterfly disappeared, or in a few cases increased, when there was no obvious habitat change caused many scientists to believe that factors such as physical or genetic isolation, butterfly collectors, pesticides, air pollution or a changing climate were responsible. These and other theories formed the rationale for most attitudes towards insect conservation in the 1950s and 1960s, often with disastrous results (Thomas, 1984). These factors can now largely be dismissed for reasons given by Thomas (1984), although it is recognized that genuine climate change would indeed be a major factor; the fears that predominated in the 1960s and early 1970s erroneously assumed that the climate was cooling.

It is now clear, from the British studies of butterflies, that those grassland species that react very quickly to abandonment or intensification occupy extremely narrow niches during their egg and larval stages. For example, egglaying by *Lysandra bellargus* in Britain is almost entirely restricted to *Hippocrepis comosa* plants that are growing in a sward less than 3 cm tall, whereas its close relative, *L. coridon* Poda, lays on almost any *H. comosa* plant that is available (Fig. 4).

British colonies of *Maculinea arion, Hesperia comma, Melitaea cinxia*, *Plebejus argus* and *Hipparchia semele* occupy a similar niche to that of *L. bellargus*, being restricted as larvae to examples of their foodplant that grow in very early successional (short or sparse) swards; so does the ant *Myrmica sabuleti* (Thomas, 1984, 1991; Thomas *et al.*, 1986; Thomas, 1985a,b). *Thymelicus acteon*, on the other hand, restricts breeding to an equally narrow niche at the opposite end of the grassland succession: it lays only in mature clumps of *Brachypodium pinnatum* Beauv. that are 15–50 cm tall (Thomas, 1983a, 1990). *Aricia agestis* restricts egglaying to examples of its foodplant, *Helianthemum chamaecistus* Mill., that have both unusually thick mesophylls and an unusually high nitrogen content, a growth form so rare that it was not recorded at all by random sampling on some sites (Bourn, 1989).

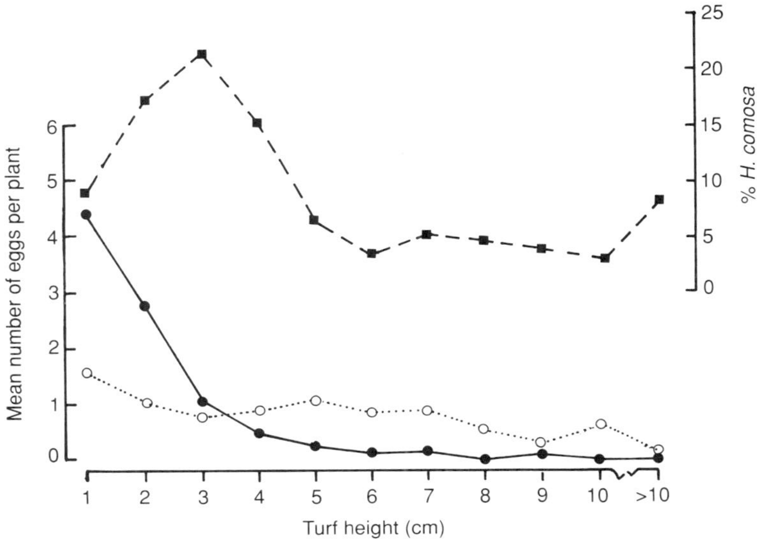

Fig. 4. The distribution of the eggs of *Lysandra bellargus* (●) and *L. coridon* (○), and of their foodplant *Hippocrepis comosa* (■) in turf of varying heights on a south-facing lowland grassland in Britain (from Thomas, 1991).

Any minor change in the growth form of the foodplants of the above species, such as results from a small and temporary relaxation in grazing, causes their populations to crash or – if this increases the proportion of plants that are suitable for egglaying – dramatically to increase (Thomas, 1991). One reason why foodplant populations also eventually decline (or increase) is that their reproduction is affected by the same habitat changes. Thus *Peucedanum palustre* L. produces less seed and fewer seedlings on drying fens (Dempster and Hall, 1980), as does *Sanguisorba officinalis* L., the initial foodplant of *Maculinea nausithous* Bgstr. and *M. teleius* Bgstr., in abandoned wet meadows (Thomas, unpublished). *Hippocrepis comosa, Thymus praecox* L. and *Plantago lanceolata* L., the foodplants respectively of *L. bellargus*, *M. arion* and *Melitea cinxia*, are also perennial, but not immortal, and since their seedlings establish only in short open turf, each plant eventually follows its herbivore into decline if the sward becomes tall and dense (Thomas, 1983b, 1991). In contrast, when *Brachypodium* grassland is abandoned, there is an immediate increase in *Thymelicus acteon* because the *B. pinnatum* is released to grow into the mature clumps selected for oviposition. But these plants also then set much seed, and the grass slowly increases (Thomas, 1983a, 1990).

The reason why most quick-reacting species of British butterfly are restricted to much narrower niches than those occupied by their foodplants appears to be a need to live in comparatively warm microclimates for the region they inhabit. For example, not only is *L. bellargus* restricted, in Britain, to south-facing slopes in the southern counties, but the 1–2 cm-tall *Hippocrepis comosa* plants used by its larvae are about 4–7°C warmer than those growing in a 7 cm-tall sward during the main period of spring feeding (Fig. 5). Similar measurements have been made for the niches occupied by *M. arion* (Thomas, 1984), *Melitea cinxia* (Thomas and Simcox, unpublished) and *P. argus* (Thomas, 1985a), and undoubtedly apply to *H. semele* and *Hesperia comma*. A similar situation exists among several other species of butterfly that breed in British woodlands and heathlands: indeed about 20% of all British butterfly species are restricted to very warm, early successional stages within their biotopes. Of these species, all but *M. cinxia* depend wholly or largely on man to generate their habitats (Thomas, 1991).

Maculinea arion, and possibly others among these stenotopic species, has almost as narrow a niche throughout its European range, although

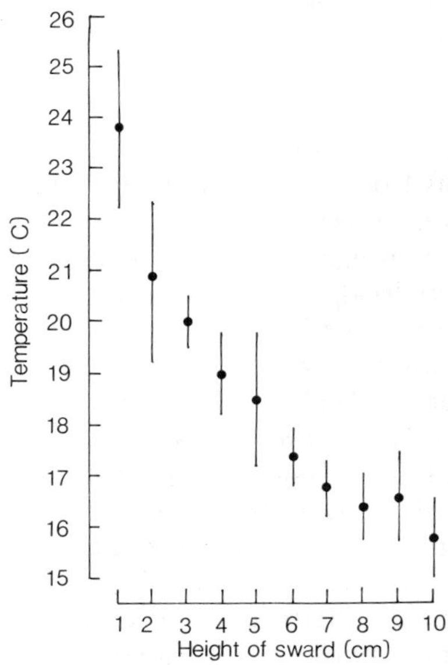

Fig. 5. The mean temperature ± SD at the soil surface beneath *Hippocrepis comosa* plants growing in different heights of turf on a *Lysandra bellargus* site. Measured on 18 April at 12.00–12.15 h in hazy–bright conditions (from Thomas, 1983b).

the exact nature of this niche differs. Thus in the lowlands of central France, where summer temperatures are roughly 3°C warmer than on its British sites, *M. arion* and its ant host abound in a narrow band of taller grassland on flat and north-facing aspects (Thomas, 1991). *M. arion* is thus equally vulnerable to successional change throughout its range, although at present it happens to be less threatened in warm climates than in cool ones because abandoned unfertilized grassland is currently far commoner than heavily grazed unfertilized grassland in central and northern Europe.

The other British species exhibit "regional stenotopism" (Kuhnelt, 1943), due to the ecological stresses of living on the borders of their geographical ranges. *Papilio machaon*, for example, has a greater range of foodplants and occurs more widely over much of mainland Europe than in England (Wiklund, 1974), and appears to be much less vulnerable to minor successional changes. Other British species, such as *L. bellargus* and *H. comma*, are widespread and common in the warmer climates of central and southern Europe, where they breed on slopes of all aspects and in swards up to about 30 cm tall (Thomas, 1991).

The greater vulnerability of most species that are restricted to warm microclimates at high altitudes or at the north of their ranges can be demonstrated by using two measurements of the British habitat of *Plebejus argus*, recorded in north Wales (Thomas, 1985a) and in the warmer climate of south Devonshire (Read, 1985). The figures refer to heathland, but the principle is the same for true grassland.

In north Wales, *P. argus* is virtually restricted not only to south-facing slopes which, on average, are 7°C warmer in spring than the north-facing aspects in this region, but also to very short swards within these slopes, which are 8–13°C hotter than nearby tall vegetation. In other words, breeding is confined to south-facing areas that have usually been burnt or disturbed within the previous 5 years (Thomas, 1985a,b). But in Devonshire, *P. argus* is not restricted to southerly aspects, and breeds in the building phase as well as the pioneer phase of heathland (Read, 1985); this extends the longevity of individual patches of its habitat from roughly 5 to 15 years.

Many physical and botanical characteristics of the heaths in a third region, Dorset, were measured in both 1978 and 1987 (N. R. Webb, personal communication). These can be used to plot the distribution of both the northern and southern habitats of *P. argus* on particular heaths in this region in each year (Thomas 1991; Fig. 6). The distribution of the butterfly's foodplants corresponds to that of the heathland boundaries (Fig. 6a). It can be seen that in 1987, *P. argus*' exacting northern habitat existed as seven small islands in just four of these 84 heaths

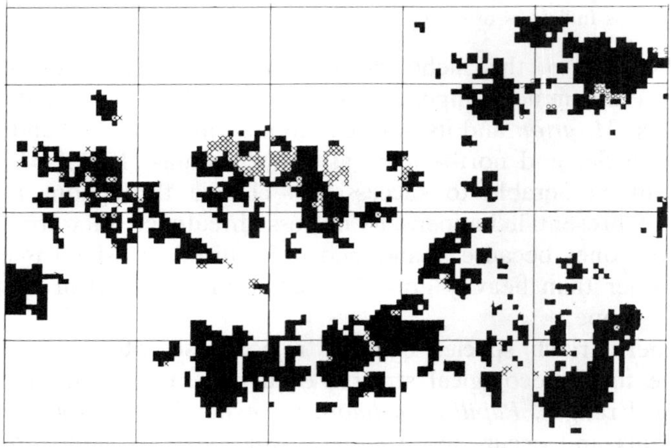
(a)
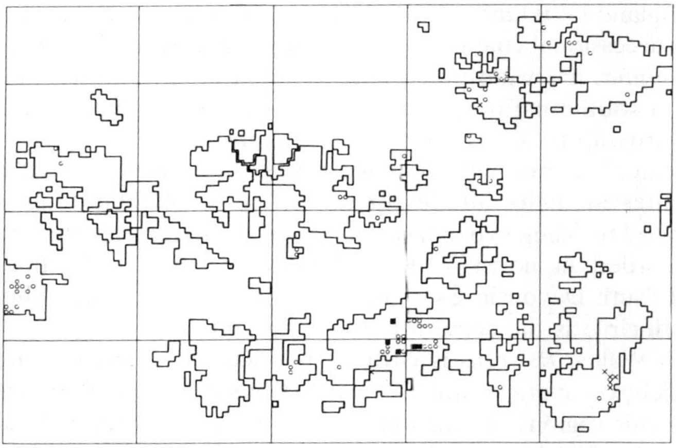
(b)
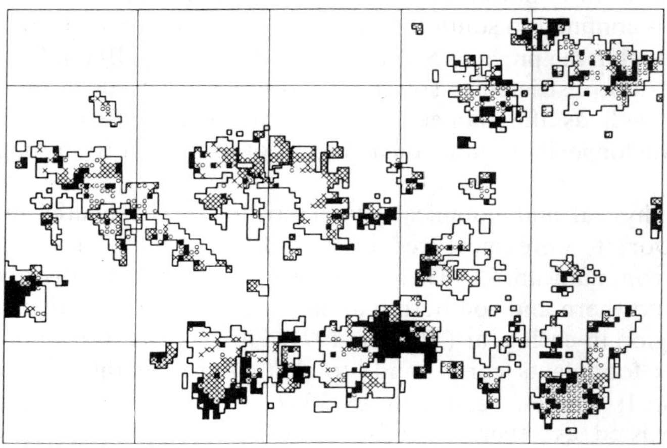
(c)

(Fig. 6(b)), whereas its southern habitat was 70 times larger, and existed as 112 islands within 45 heaths (Fig. 6(c)). In addition, there was much less continuity of breeding habitat using the northern criteria for *P. argus*. This is an important factor, for all students of this species have concluded that it is extremely sedentary (Thomas, 1985b; Read, 1985; Ravenscroft, 1990): indeed, when introduced to an extensive area of unoccupied habitat in 1942, *P. argus* increased during the following 40 years, but its mean rate of spread was no more than 1 km per decade (Thomas 1985b).

Lowland heath was traditionally farmed by burning, cutting and grazing in a regular but piecemeal way, which would have produced a continuous replenishment of *P. argus'* ephemeral habitat in many small patches, all very close to one another. Nearly all British lowland heath has been abandoned by farmers during the past 50 years, and the succession is now rekindled at long and irregular intervals, usually as a result of large-scale accidental fires. It is scarcely surprising, therefore, that this specialized butterfly has experienced dramatic declines on most northern sites (Thomas, 1985b) but, due to its broader niche in a warmer climate, has survived very much better in the south (Thomas and Webb, 1984).

Severe declines have occurred among all British species of butterfly that depend on early successional grassland habitats, even in the south of the country (Thomas, 1991). The problem, as with *P. argus*, is exacerbated by the sedentary behaviour of most species that have been studied (Thomas, 1991). Thus not only have their specialized habitats diminished in recent years, but these are often beyond the colonizing range of the species if they recur (Thomas, 1991). This again explains why populations of sedentary species of Lepidoptera are frequently more vulnerable to changing management than their foodplants. Many plants can avoid unfavourable periods in a particular patch either through dormancy, for example as buried seed, or as long-lived perennials that can withstand long periods without reproducing.

Thomas (1991) reviewed autecological research into 20 species of European butterfly, and concluded that all were limited in both distribution and abundance by the specialized habitats of the young stages, whereas the adult butterflies tended to be generalists. This is also the conclusion of

---

Fig. 6. The distribution and continuity of the foodplants and habitat of *Plebejus argus* on Dorset heathland, using Thomas's (1985a,b) and Read's (1985) definitions of its requirements, respectively, in north-west and southern Britain (from Thomas, 1991). Heathland data for 1978 (○), 1987 (X) and present in both years (■) from N. R. Webb and R. T. Clarke (personal communication). (a) All heathland (= distribution of foodplants); (b) the north-western habitat of *P. argus* (— = boundaries of heath); (c) the southern habitat of *P. argus* (— = boundaries of heath).

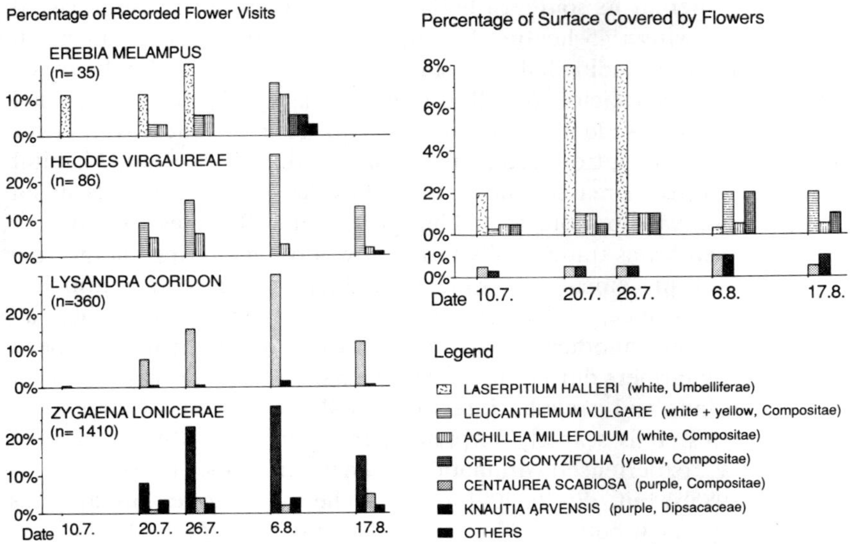

Fig. 7. Flower visits of selected Lepidoptera species and the abundance of flowers that were visited on south-facing, early successional subalpine grassland.

Erhardt's (1985a,b,c) research in subalpine meadows, although here there is evidence that adult nectar resources may also be important.

Knowledge of adult feeding preferences in different species is surprisingly limited, despite the demonstration that this can affect longevity and fecundity (Gilbert, 1972, 1984; Dunlap-Pianka *et al.*, 1977; Murphy *et al.*, 1983), as well as population density (Gilbert, 1984) and even choice of oviposition sites (Murphy *et al.*, 1984). However, records made on selected species in a south-facing early successional stage of grassland show clear-cut differences when faced with the same choice of nectar (Fig. 7).

Erebia melampus Fuessl. was the only species that visited the very abundant Umbellifer *Laserpitium halleri* Cr. It became more opportunistic when this plant had faded, but still showed a clear preference for white flowers, such as those of *Leucanthemum vulgare* Lam. and *Achillea millefolium* L., which were also almost the only flowers visited by *Heodes virgaureae* L.. In contrast, the blue Lycaenid *Lysandra coridon* restricted its flower visits to the purple flowers of *Centaurea scabiosa* L. *Zygaena lonicerae* Schev. showed yet another preference, restricting its visits mainly to *Knautia arvensis* (L.) Coult. (Fig. 7).

Factors responsible for these preferences include the proboscis lengths and energy requirements of Lepidoptera, and corolla tube lengths, colour, scent and the composition of the nectar of different flowers (Erhardt, in preparation). The ability of butterflies to produce saliva in order to dilute nectar, which can be highly viscous — even crystalline — in certain Umbellifers, may also be an important factor.

VI. DISCUSSION AND CONCLUSIONS

Similar results were obtained from the studies of Lepidoptera in the lowland grasslands of Britain and central Europe, and from subalpine meadows in Switzerland, even though many species live near the limits of their ranges in Britain. In all situations, a much greater diversity or abundance of Lepidoptera was found in secondary grassland that was still being managed in the traditional way, although a very few species benefited from increased fertilization and rather more species from prolonged abandonment. Since shrubs invade abandoned montane meadows more rapidly than subalpine ones (Kienzle, 1982), the problem is likely to be greater at lower altitude.

The rate at which individual species or families of Lepidoptera changed when meadows were fertilized or abandoned was surprisingly rapid, though probably no more so than in other invertebrate populations (e.g. Morris, 1971, 1978). Temperate species of ant may be less sensitive to changed management, due to the longevity of workers and queens, and to their ability to withstand temporary disasters by eating their young and resuming breeding in the following year (G. W. Elmes, personal communication). Nonetheless, colonies of the ant *Myrmica sabuleti* reacted about three times more quickly than the perennial plant *Thymus praecox* when grazing was reduced (Table I). But, on the one site where this was accurately measured, a population of the parasitic butterfly, *Maculinea arion*, changed 10 times more quickly than its ant host.

Diurnal Lepidoptera can therefore be used as sensitive indicators of apparently minor changes in the structure of grassland habitats, in the knowledge that the same factors will eventually affect populations of longer lived organisms, although not necessarily to so severe an extent. The short-lived, specific Hymenopteran parasitoids of Lepidoptera are likely to show even greater and faster changes, but these are less conspicuous and too unfamiliar to most conservationists to be of practical use as indicators.

The extent to which many species of Lepidoptera depend wholly or largely on secondary grassland that is managed by man in traditional ways begs the question as to where these species lived before the land was cultivated. This, of course, applies to other groups of wildlife and to other biotopes. In the case of grassland, species were probably more localized before man began large-scale clearances of trees; some were probably confined to rare primary grassland habitats (which have since been destroyed), and spread once the new habitats were created.

None the less, it is remarkable how many species are restricted to narrow man-made niches, most involving a very warm microclimate. In Britain, this applies to about 20% of butterfly species when all biotopes are considered. Thomas (1991) has suggested that this group may be a relic from 4000–8000 years ago, when the summer climate was about 2–3°C warmer than today. Under those conditions, species that are now confined to exceptionally warm microclimates would probably have been much more widespread due to their ability to inhabit cooler habitats such as tall grassland, partly shaded woodland and woodland pasture. When the climate cooled, 3500–4500 years ago, these species might have been expected to disappear from the British Isles, and from northern and upland sites throughout Europe, retreating to present centres of distribution in central and southern lowland Europe. But before this could occur, much of the land had already been cultivated, providing warm man-made refugia into which these species could have moved. They have been trapped in these ever since (Thomas, 1991).

Whether this is the whole explanation or not, there is no doubting the rapidity and the scale of extinctions among many invertebrate species due to the recent demise of these traditional forms of grassland management. Since they have lived largely under this system for a few thousand generations, it would be surprising if natural selection had not produced local races of Lepidoptera that were adapted to particular types of management, differing from those that now live in primary habitats (Schremmer, 1949). For example, this might explain the extreme sedentariness of several species of butterfly that are restricted to ephemeral, early successional stages of grassland and other biotopes. Far from contradicting Southwood's (1977) hypothesis that these otherwise r-selected species should be highly mobile, they provide an exception to help prove the rule that the characteristics of species are determined by the habitat templets in which they evolved. For one characteristic of traditional farming is the creation of a continuous supply of early successional habitats, each very close to the next in both space and time (Thomas, 1991). There would have been strong selective forces on the species living in these to become sedentary.

There is a need for considerably more research into the ecology of grassland Lepidoptera if conservationists are to stem current declines. This should include such basic research as identifying both the species and required structure of larval foodplants, the food requirements of adults, the minimum area needed to support a population, and the effect of different types of management regime (Erhardt, in press).

Despite large gaps in current knowledge, we conclude that nature conservationists would be considerably more successful if they paid more attention to the Lepidoptera (and invertebrates) of European grasslands. A few western nations, such as The Netherlands and Great Britain, are already well endowed with grassland nature reserves, but here the management of many sites is inadequate. So many species of Lepidoptera have disappeared from British grassland reserves that there is a strong case for switching future resources towards rectifying this, for example by organizing rotational management schemes, rather than use funds to obtain more reserves (Thomas, 1991).

Although there is a close correlation between plant and Lepidoptera diversity on most sites, evaluations based on invertebrates are likely to produce different results to those based solely on plants, and are likely to identify the minority of sites that have a more or less unbroken history of traditional management. There is an urgent need to establish more grassland nature reserves in most nations, representing as wide a range as possible of the diverse but characteristic assemblages of Lepidoptera that still survive. This need is particularly acute in the uplands of Europe, especially in the Alps and Pyrenees, where unfertilized mown and lightly grazed meadows have become extremely rare in many regions. Equally pressing are the lowland grasslands of east Europe, where intensification of agriculture is the greatest threat. Here, until recently, existed vast areas still under traditional cultivation that supported assemblages of wildlife that made many western nature reserves seem insignificant. Much has already been fertilized or abandoned, and the remainder will inevitably disappear if western governments give agricultural aid to these new democracies. Future generations would surely be grateful if agricultural aid was linked to subsidies to ensure that equal areas are managed in the traditional way. Diurnal Lepidoptera would form a good basis for selecting these conservation areas, and for monitoring the continuing richness of their wildlife.

Acknowledgements

We wish to thank Professors H. Zoller and W. Sauter for their helpful advice and discussions, and Professor H. A. Moser, N. Bischof, E. de Bros, J. Hartmann, O. Hoeg-Guldberg, J. de Marmels, L. Rezbanyai, P. Sonderegger and St. Whitebread for help with literature searches, identification, unpublished information and stimulating discussions. We also thank M. Wall for statistical advice and are particularly grateful to Silvia Engeli for typing the manuscript. Research in Switzerland was supported by the Swiss National Science Foundation (Project No. 3.643−0.75).

REFERENCES

Anonymous (1984). "The Management of Chalk Grassland for Butterflies". Nature Conservancy Council, Peterborough.

Anonymous (1990). "Beschermingsplan Dagvliners", Ministerie van Landbouw en Visserij, Gravenhage, Netherlands.

Bierhals, E. (1976). Oekologische Folgen der Vegetationsebtwicklung und des Wegfalls der Bewirtschafungsmassnahmen. "Brachflachen in der Landschaft", pp. 1−156. KTBL-Schrift 195, Munster-Hiltrup.

Bischof, N. (1980). Pflanzensociologische Untersuchungen von Sukzessionen aus gemahten Magerrasen in der subalpinen Stufe der Zentralalpen. Ph.D. thesis, Universitat Basel.

Bischof, N. (1984). Pflanzensoziologische Untersuchungen von Sukzessionen aus gemahten Magerrasen in der subalpinen Stufe der Zentralalpen. *Beit. geobotan. Landesauf. Schweiz*, **60**, 1−128.

Blab, J. and Kudrna, O. (1982). "Hilfsprogramm fur Schmetterlinge". Kilda-Verlag, Steinfurt.

Bourn, N. A. D. (1989). The ecology and conservation of *Aricia agestis* in southern England. M.Sc. thesis, University College, London.

Dempster, J. P. (1971). Some effects of grazing on the population ecology of the Cinnabar Moth (*Tyria jacobaeae* L.). *In* "The Scientific Management of Animal and Plant Communities for Conservation" (E. Duffey and A. S. Watt, eds), pp. 517−526. Symposium of the British Ecological Society 11. Blackwell Scientific Publications, Oxford.

Dempster, J. P. and Hall, M. L. (1980). An attempt at re-establishing the swallowtail butterfly at Wicken Fen. *Ecol. Entomol.* **5**, 327−334.

Dunlap-Pianka, H., Boggs, C. L. and Gilbert, L. E. (1977). Ovarian dynamics in *Heliconiine* butterflies: Programmed senescence versus eternal youth. *Science* **197**, 487−490.

Erhardt, A. (1985a). Lepidopterafauna in cultivated and abandoned grassland in the subalpine region of Central Switzerland. "Proceedings of the 3rd Congress of European Lepidopterology", Cambridge 1982, pp. 63−73. Societas Europaea Lepidopterologica.

Erhardt, A. (1985b). Diurnal Lepidoptera: sensitive indicators of cultivated and abandoned grassland. *J. Appl. Ecol.* **22**, 849−861.

Erhardt, A. (1985c). Wiesen und Brachland als Lebensraum für Schmetterlinge: eine Feldstudie im Tavetsch. *Denkschr. Schweiz. Naturforsch. Ges.* **98**, 154S.

Erhardt, A. (1991). Zum Schutz der Schmetterlinge in der Schweiz: Die Notwendigkeit eines grösseren, wissenschaftlich fundierten Engagements. *Nota Lepidopterologica*, in press.

Erz, W. (1973). Das Brachflächenproblem aus der Sicht von Naturschutz und Landschaftspflege. *Jahrb. Nat. Landschaft.* **22**, 1−95.

Frazer, J. F. D. (1965). Butterflies of chalk grassland: A conservation problem. "Proceedings of the 12th International Congress of Entomology, London 8−16 July 1964" (P. Freeman, ed.), p. 355. London.

Frazer, J. F. D. and Hyde, G. E. (1965). The decline of the chalk grassland butterflies. *Wildlife* **7**, 212−215.

Gerlach, P. (1976). 40 Thesen zur Problematik der Brachflächen. "Brachflächen in der Landschaft", pp. 1−42. KTBL-Schrift, 195, Münster-Hiltrup.

Gilbert, L. E. (1972). Pollen feeding and reproductive biology of Heliconius butterflies.

Proc. Nat. Acad. Sci. USA, **69**, 1403–1407.
Gilbert, L. E. (1984). The biology of butterfly communities. *In* "The Biology of Butterflies" (R. I. Vane-Wright and P. R. Ackery, eds), pp. 41–54. Academic Press, London.
Heath, J., Pollard, E. and Thomas, J. A. (1984). "Atlas of Butterflies in Britain and Ireland" Viking, Harmondsworth.
Kienzle, U. (1982). Vegetationswechsel (Sukzession) in brachliegenden Streuwiesen und Magerweiden des Napfgebietes. *Mitt. Naturforsch. Ges. Luzern* **27**, 112–166.
Kühnelt, W. (1943). Die Leitformenmethode in der Ökologie der Landtiere. *Biol. Gen.* **17**, 106–146.
Lipscomb, C. G. & Jackson, R. A. (1964). Some considerations on some present day conditions as they affect the continued existence of certain butterflies. *Entomol. Rec.* **76**, 63–68.
Mellanby, K. (1981). "Farming and Wildlife". Collins, London.
Morris, M. G. (1971). Differences between the invertebrate faunas of grazed and ungrazed chalk grassland. IV. Abundance and diversity of homoptera-auchenorhyncha. *J. App. Ecol.* **8**, 37–57.
Morris, M. G. (1978). Grassland management and invertebrate animals — a selective review. *Sci. Proc. R. Dublin Soc. Ser. A* **6**, 247–257.
Murphy, D. D., Launer, A. E. and Ehrlich, P. R. (1983). The role of adult feeding in egg production and population dynamics of the checkerspot butterfly *Euphydryas editha*. *Oecologia* (*Berlin*) **56**, 257–263.
Murphy, D. D., Menninger, M. S. and Ehrlich, P. R. (1984). Nectar source distribution as a determinant of oviposition host species in *Euphydryas chalcedona*. *Oecologia* (*Berlin*) **62**, 269–271.
Ravenscroft, N. O. M. (1990). The ecology and conservation of the silver-studded blue butterfly *Plebejus argus* L. on the Sandlings of East Anglia, England. *Biol. Conserv.*, **53**, 213b.
Read, M. (1985). The silver-studded blue conservation report. M.Sc. thesis, Imperial College.
Reichholf, J. (1973). Die Bedeutung nicht bewirtschafteter Wiesen für unsere Tagfalter. *Nat. Landsch.* **48**(3), 80–81.
Schremmer, F. (1949). Die Wiese als Lebensgemeinschaft. "Sammlung 'Bios'" Band 7, pp. 1–107. Wien.
Schwerdtfeger, F. (1975). "Oekologie der Tiere", Bd. III Synökologie, pp. 1–451. Parey, Hamburg.
Southwood, T. R. E. (1977). Habitat, the templet for ecological strategies? *J. Anim. Ecol.* **46**, 337–365.
Surber, E., Amiet, R. and Kobert, H. (1973). Das Brachlandproblem in der Schweiz. "Bericht der Eidg. Anstalt für das forstl. Versuchswesen", Birmensdorf, No. 112, pp. 1–138.
Tax, M. H. (1990). "Atlas van de Nederlandse Dagvlinders". P. J. L. Schoenmakers, Wageningen.
Thomas, C. D. (1985a). Specializations and polyphagy of *Plebejus argus* L. (Lepidoptera: Lycaenidae) in North Wales. *Ecol. Entomol.* **10**, 325–340.
Thomas, C. D. (1985b). The status and conservation of the butterfly *Plebejus argus* L. (Lepidoptera: Lycaenidae) in North West Britain. *Biol. Conserv.* **33**, 29–51.
Thomas, J. A. (1976). "The Ecology and Conservation of the Large Blue Butterfly". ITE, UK.
Thomas, J. A. (1983a). The ecology and status of *Thymelicus acteon* in Britain. *Ecol. Entomol.* **8**, 427–435.
Thomas, J. A. (1983b). The ecology and conservation of *Lysandra bellargus* in Britain.

J. Appl. Ecol. **20**, 59–83.

Thomas, J. A. (1984). The conservation of butterflies in temperate countries: past efforts and lessons for the future. *In* "The Biology of Butterflies" (R. I. Vane-Wright and P. R. Ackery, eds), pp. 333–353. Academic Press, London.

Thomas, J. A. (1990). The conservation of Adonis blue and Lulworth skipper butterflies — two sides of the same coin. *In* "Calcareous Grassland Ecology and Management" (S. H. Hillier, D. W. H. Walton and D. A. Wells, eds,) pp. 112–117. Bluntisham Books, Huntingdon.

Thomas, J. A. (1991). Rare species conservation: case studies of European butterflies. *In* "Scientific Management of Temperate Communities" (I. F. Spellerberg, F. B. Goldsmith and M. G. Morris, eds), pp. 149–197. 29th Symposium of the British Ecological Society. Blackwell Scientific Publications, Oxford.

Thomas, J. A. and Simcox, D. J. (1982). A quick method for estimating larval populations of *Melitea cinxia* L. during surveys. *Biol. Conserv.* **22**, 315–322.

Thomas, J. A. and Webb, N. R. (1984). "Butterflies of Dorset" Dorset Natural history and Archaeological Society, Dorchester.

Thomas, J. A., Thomas, C. D., Simcox, D. J. and Clarke, R. T. (1986). Ecology and declining status of the silver-spotted skipper butterfly (*Hesperia comma*) in Britain. *J. Appl. Ecol.* **23**, 365–380.

Ulrich R. (1982). Vergleich von bewirtschafteten Wiesen und Brachen hinsichtlich des Wertes für unsere Tagfalter. *Nat. Landsch.* **57**, 378–382.

Wells, T. C. E. (1989). The re-creation of grassland habitats. *Entomologists* **108**, 97–108.

Wiklund, C. (1974). The concept of oligophagy and the natural habitats and hostplants of *Papilio machaon* L. in Fennoscandia. *Entomol. Scand.* **5**, 151–160.

Zoller, H. and Bischof, N. (1980). Stufen der Kulturintensität und ihr Einfluss auf Artenzahl und Artengefüge der Vegetation. *Phytocoenologia* **7**, 35–51.

10

Conserving Insects of Aquatic and Wetland Habitats, with Special Reference to Beetles

G. N. FOSTER

I. Introduction: Sites, Habitat, Communities and Species 238
II. International Species Protection 238
III. The British Red Data Book — Insects 240
A. Species Selection 240
B. Threats and Causes of Loss reported in the Red Data Book 242
C. Threats not mentioned in the Red Data Book 245
D. Conservation Measures Specified in the Red Data Book 247
IV. Species Rarity and Site Evaluation 249
A. Nationally Scarce Species and National Notability 249
B. Species Rarity and Quality Scores 249
C. Lists, Assemblages, Communities and Guilds 250
D. Site Quality Criteria 251
E. Water Beetle Site List Evaluation 252
V. The Low Profile of Insect Conservation 255
A. The World Conservation Strategy 255
B. Ignorance 256
VI. Acceptance of the Need for Insect Conservation by Non-entomologists 257
A. A Non-entomologist's Perception 257
B. Colourful Species 257
C. Large Size 257
D. Surface-active Insects 258
E. Food for Other Animals 258
F. Environmental Education 258
G. New and Unstable Habitats 259
H. Five Tactics 259
References 260

I. INTRODUCTION: SITES, HABITATS, COMMUNITIES AND SPECIES

The conservation of insects can be considered on nine levels:

1. Sites containing habitats conserved for reasons not primarily concerned with insects.
2. Sites containing habitats known to support insects with intrinsic appeal.
3. Sites containing habitats known to support rare insects.
4. Habitats supporting requirements specific to insects.
5. Entire macro-invertebrate communities.
6. Particular insect groups.
7. Species endangered nationally.
8. Species endangered throughout their range.
9. Endangered races.

This can be reduced to a site–habitat–communities–species hierarchy, in which most conservation effort has been directed to level 1 by non-entomologists and to level 2 by entomologists. With aquatic and wetland habitats, the ecology and biology of individual insect species is so poorly understood that it is difficult to command attention beyond level 2. If one accepts this hierarchy, it seems illogical that concern has been devoted to the plight of a few endangered species whereas little has been done that recognizes the need to conserve endangered assemblages.

The rest of this review runs roughly in reverse order to the hierarchy proposed above, and exposes the problems with special reference to aquatic Coleoptera.

II. INTERNATIONAL SPECIES PROTECTION

Legal protection of European insects has been reviewed by Collins (1987). The Bern and Ramsar Conventions are the most relevant of the international initiatives to aquatic insect conservation. The Convention on the Conservation of European Wildlife and Natural Habitats (the Bern Convention) was signed in 1979 to conserve European flora and fauna, with emphasis on habitat protection, education and planning. The Convention's Appendix 2 now includes 51 species of insect. There is an obligation on all countries party to the Convention, which includes most EC and EFTA countries, to take strict measures to conserve such species.

In March 1986 a request was made for the selection of candidate species for Appendix 2 (N. M. Collins, personal communication). A literature review was undertaken and 10 specialists representing 10 countries were asked to produce short lists of endangered species. The criteria suggested by the International Union for Conservation of Nature and Natural Resources (IUCN, now known as the World Conservation Union) were:

1. That the species should be under serious threat in Europe as a whole, though not necessarily in every place.
2. That the species should be reasonably easy to identify.
3. That the species' distribution should be mainly in Europe.

The review (Anon., 1986), which included material from the provisional British red data list, indicated that 192 species of water beetle were considered to be threatened in at least one country in Europe. The panel's "short" list ran to 117 species. *Dytiscus latissimus* L., a non-British species, and *Hydroporus elongatulus* Sturm were each considered endangered in five countries. The latter species was considered too small and difficult to identify to justify inclusion. *D. latissimus* and another larger species, *Graphoderus bilineatus* (DeGeer) (noted as endangered in France, Denmark and Italy, and found in the Norfolk Broads in the 1900s), were therefore put forward and listed in Appendix 2, along with 16 other aquatic insects, all of which were dragonflies.

D. latissimus was already protected by law in Austria, parts of Belgium, the Federal Republic of Germany and Hungary. This species has declined or even become extinct throughout most of its western European range, but still occurs in Denmark (Holmen, 1987), where protection measures taken as a result of the Bern Convention may hamper ecological studies and, *inter alia*, successful introductions elsewhere.

The species most in danger of extinction are the endemics of the Iberian Peninsula and of some Mediterranean islands, but none fitted the criteria in use, and the status of most is not yet established. The rest of those listed are more widely distributed species which are still well established either in the south of Europe or Scandinavia, despite being endangered or extinct in western and central Europe.

The case for international protection is weakened by the evolutionary stability of insects. Some populations separated since the end of the Pleistocene (e.g. Swedish and Spanish *Helophorus lapponicus* Thomson; Angus, 1983) are reproductively compatible. It can be argued that if such glacial relict populations are not in special need of protection, then species that were widespread until recently can survive in underdeveloped parts of their ranges. But this is naive. It overlooks the fact that all parts of Europe are subject to development pressure, which is often less hindered by conservation considerations in the lesser known areas. Each glacial relict population should be conserved intact, at least long enough to be available for DNA and enzyme studies, which may yet establish new grounds for conservation (and for or against introductions as a conservation measure).

An opportunity for community protection, as opposed to the protection of individual species, arose with the Convention on Wetlands of International Importance especially as Wildfowl Habitat (the Ramsar Convention). This came into force in 1975, and has 40 parties as signatories who act by conserving all wetlands on a "List of Wetlands of International Importance". The first of the "Cagliari Criteria" for identifying such sites is concerned with quantitative data for waterfowl. Thus much of the listing has been strongly biased towards birds, and it is still largely a matter of chance whether sites that are internationally important for insects are included. There was change of emphasis at the conference in Regina, Canada in 1987, such that internationally important sites could be designated on the basis of general criteria relating to rare plant and animal species (criteria 2a–d), or because they are either unusual or good representatives of a regionally restricted type of wetland (criterion 3). In the absence of attempts to identify such types on the basis of their invertebrate fauna, the signatories can hardly be blamed for deficiencies in implementing the convention.

III. THE BRITISH RED DATA BOOK — INSECTS

A. Species Selection

The British Red Data Book (RDB) for Insects (Shirt, 1987) is the main reference source for information on endangered species in Great Britain. Its remit extended to the Isle of Man but not to the Republic of Ireland, Northern Ireland or the Channel Islands. The list of 1786 species is divided into "endangered", "vulnerable" and "rare" species and subspecies, plus some minor categories. Endangered taxa (RDB 1) were originally defined by IUCN as those in danger of extinction and whose survival was considered unlikely if the causal factors of decline continued to operate. This includes taxa considered to be extinct, those whose numbers have been reduced to a critical level, or whose habitats have been so dramatically reduced that the taxa are deemed to be in immediate danger of extinction. For the purposes of the British Red Data Book, the definition is fulfilled if any of the following criteria are met: species (or subspecies) known only as a single population within one 10-km square;

species occurring in habitats known to be specially vulnerable; species undergoing rapid and continuous decline over the past 20 years and now confined to fewer than five 10-km squares; those species considered extinct which, if rediscovered, would need protection.

The next category, "vulnerable" (RDB 2), was defined for taxa likely to move into the endangered category in the near future if the causal factors continued to operate. This included taxa of which most or all populations were decreasing because of overexploitation, extensive destruction of habitat or other environmental disturbance; taxa with populations that had been seriously depleted and whose ultimate security was not assured; and taxa that were still locally abundant but under serious threat throughout their range. This therefore includes species that are declining throughout their range, or that live in vulnerable habitats or that have small populations.

The third main category (RDB 3) of IUCN and the British Red Data Book is "rare". This covers taxa with small populations not considered at present to be either endangered or vulnerable, but at risk. Most of these taxa occur within restricted geographical areas or habitats, or are thinly scattered over a wider range. The RDB definition includes taxa which are believed to be rare but which are too recently discovered or recognized to be certain of placement. Despite the strong emphasis laid on the degree of threat, as opposed to the degree of rarity, for all RDB categories, the only criterion given for RDB 3 is that the species should exist (or be known to exist) in 15 or fewer 10-km squares. This overlaps with the criteria for selection of nationally notable species (see below).

A candidate list of RDB species was hurriedly drawn together for water beetles (Foster, 1978) a year before the launch of the National Recording Scheme. Publication of the Red Data Book took 9 years, of which 5 were spent in collating draft entries from a large number of entomologists. The original list of 29 species of Hydradephaga, Spercheidae, Hydrophilidae, Hydraenidae and Elmidae increased during this period to 52 species; two species were deleted but 10 species were added to the British checklist as a result of intensified recording and generic revisions. Nevertheless all nine of the RDB 1 species were in the first list as were eight of the final list of 12 RDB 2 species. It was possible to guess at the species most likely to be endangered without reference to a modern recording scheme, because Professor F. Balfour-Browne had operated a national method of recording for this group, based on vice-counties, for about 60 years. This ended in 1962, when his journals and card indices were deposited in the Royal Scottish Museum, now the National Museum of Scotland. There was thus a large and accessible data base, with maps published for most British species of Hydradephaga and those palpicorn taxa recognized as either species or aquatic by Balfour-Browne (1940, 1950, 1958). The records for the rarer species could quickly be converted into 10-km records whereas, even now, generation of maps of "ubiquitous" species is time consuming and unedifying.

It is difficult to draw comparisons between insect orders in the Red Data Book, because the cover is not comprehensive. About 14.5% of British insects are listed in the book, with relatively high percentages for some groups (e.g. 21.4% of butterflies and 22.0% of Odonata), no listings for some major groups (e.g. Homoptera and Hymenoptera: Parasitica) and "near average" representation for the Heteroptera (14.6%), Coleoptera (14.0%) and Diptera (13.8%).

The preponderance of dipteran species of unknown biology listed in the Red Data Book has attracted comment, but can perhaps be justified by the size of the order; the proportion of species at risk is the same as for the Coleoptera.

No comparable analysis can be made of the percentage of threatened dipteran species that are associated with wetlands (Table I), because accounts are given for only 82 of the 496 species listed as RDB 1 or 2. However, 45% of the species with accounts are aquatic. Assuming that a similar percentage of the other 414 species is aquatic, this gives a figure of 3.7% of the British checklist being associated with water and considered either endangered or vulnerable. The equivalent value for beetles of 1.3% can perhaps be accounted for by differences in the morphology and ecology of the larvae of the two orders. This low percentage also indicates that there has been no overstatement of the case that aquatic Coleoptera are threatened, in comparison to other groups.

The main use to which RDB listing has been put is in the scheduling of sites of special scientific interest (SSSI) (Nature Conservancy Council, 1989). At the very least this has resulted in more effort to record the continued presence of RDB species on known sites. A disadvantage has been that the idea, although sound, is undermined by the overenthusiastic listing of certain groups within the RDB. Some Nature Conservancy Council (NCC) officers have difficulty in believing that a site has any entomological interest if RDB species are absent — and *vice versa*, if RDB species occur frequently, this suggests that they are not really endangered. A lack of confidence also arises from our inability to give clear advice on the requirements of RDB species.

B. Threats and Causes of Loss reported in the Red Data Book

When the threats indicated in each species account of the RDB are categorized (Table II), it is obvious that the main pressure is construed to

TABLE I. Wetland insects listed in the British Red Data Book — Insects (Shirt, 1987).

Order Subdivisions	Species numbers RDB 1	RDB 2	RDB 3	RDB 1 and 2 spp. as % of British list
Odonata	4	2	3	14.6
Orthoptera	0	1	0	3.3
Heteroptera	1	0	–	0.2
Trichoptera	9	4	18	6.5
Lepidoptera of reeds and sedges	2	1	–	0.1
Coleoptera	30	21	–	1.3
Hymenoptera: Aculeata	0	0	0	0
Diptera with accounts in RDB	(10)	(27)	–	(3.7)

TABLE II. Threats to and causes of loss of endangered and vulnerable insects listed in the British Red Data Book — Insects (Shirt, 1987).

Threats

Agricultural development (67 references) including: drainage (38); agricultural improvement/intensification (9); canalization, drainage of the East Anglian meres, machine clearance of ponds and ditches (3); mixed to arable farming, pollution by fertilizers, pollution by pesticides (2); conversion of brackish to fresh water, loss of foodplant, loss of heathland, pollution by farm effluent, reclamation, reed-cutting, robust clearance to steep-sided ponds (1).

Natural changes (14) including: scrub and tree invasion (7); succession (4); drying out or up, loss of reed/sedge cutting (2); marine incursion (1).

Recreational (18) including: disturbance (6); recreational pressure (3); holiday development (2); bank clearance for angling, caravan sites, landscaping, powerboat wash and bank reinforcement, sailing, trampling, yacht moorings (1).

Urban and industrial development (26) including: oil-related developments (3); barrage schemes, commercial development, gravel extraction, lowering of water table, nuclear power stations (2); ball-clay extraction, colliery tipping, factory development, industrial development, infilling of gravel pits, landscaping, loss of heathland, pollution from housing development, pollution by sewage plant, road construction, rubbish tipping, siltation from road development, urbanization (1).

Causes of loss

Drainage (52) including: canalization, drainage for afforestation, drainage of the East Anglian meres (3); drainage for agricultural improvements, drainage for peat harvesting, drying out or up, lowering of water table (2); reclamation (1).

Pollution (29) including: eutrophication (4); long-term decline in water quality, fertilizers, pesticides (2); farm effluent, housing development, sewage plant (1).

be from agricultural improvement, followed by urbanization and industrial developments, then by recreational pressures and lastly by natural changes, including those associated with neglect. Drainage is mentioned more frequently than pollution as a damaging activity.

Changes in the balance of fenland habitats, other than natural hydroseral developments, clearly began with drainage systems developed from Roman times and then intensified in the Elizabethan age (Summers, 1976). Drainage of the East Anglian meres was specifically mentioned in connection with two water beetles (*Rhantus aberratus* Gemminger & von Harold and *Graphoderus bilineatus*) which barely survived into the twentieth century (both being last reported in 1904 and 1906, respectively) from east Norfolk). The evidence that drainage caused these extinctions is circumstantial, but compelling in view of the large proportion of still water habitats that were lost. Whittlesea Mere, drained in the 1850s (Godwin, 1978), was the second largest of the English lakes. Stephens (1828) records a specimen of *R. aberratus* from Whittlesea Mere and later (Stephens, 1829) the record " 'Plentiful near Cambridge in 1829.' – C. Darwin, Esq.". In the Red Data Book, Whittlesea Mere is also mentioned in connection with a leaf beetle, *Galeruca interrupta* Illiger, last found (in Dorset) in 1919.

An unusual threat is associated with nuclear power stations. These are usually sited in remote coastal areas and, almost by definition, site clearance must result in loss of important habitats. Also, the water issuing from plants is warm and unsuitable for the previous occupants of neighbouring water bodies.

Perhaps the most idiosyncratic threat listed is "disturbance". Some RDB entries specifically refer to disturbance associated with increased public pressure on dune systems, but it must be admitted that several references to disturbance threatening water beetles are nebulous and perhaps false.

The specialized insect faunas of saltmarshes and tidal lagoons are not included in this analysis. Clearly, improvements in coastal defence systems are perceived as the main threat. Of the two references to oil-related development, one concerns loss of Dorset heathland (for *Chrysops sepulcralis* (F.)). The other relates to *Ochthebius lenensis* Poppius, a beetle confined within Britain to north-east Scotland. A recent survey (Foster, 1988) confirmed its RDB 2 status on the basis that it is associated with saltmarsh pools for only a limited period in their cycle of formation and degradation. Consequently large tracts of saltmarsh are required in order to protect this species; these are threatened by the infrastructure of improved roads, platform yards and a pipe construction factory.

C. Threats not mentioned in the Red Data Book

Climatic change is not perceived as a cause of loss of aquatic species in the entries for aquatic insects in the Red Data Book, although the reduction of species such as *Emus hirtus* (L.) is noted as being part of a general decline in Europe. A revised version of the Red Data Book would presumably recognize that thermophilous species are likely to expand in distribution again. Global warming would result in contraction of the ranges of boreal species and, more importantly, major losses of

Fig. 1. Areas of Great Britain vulnerable to a rising sea level (reproduced from Boorman *et al.*, 1989).

fenland habitat through marine incursions (Boorman *et al.* 1989). The map of areas vulnerable to a rising sea level in Britain (Fig. 1) corresponds to the combined distribution of most of our RDB fenland insects. This emphasizes the value of relict fenland on slightly higher ground, such as the pingo mires of west Norfolk and the fens of Anglesey.

The rapidity with which peat is being lost was not perceived as a threat when the Red Data Book was written. This loss is well known as a fenland phenomenon, but drainage, associated with agricultural intensification, means that the thinner peat soils will soon be gone. Burton and Hodgson (1987), in a survey of English and Welsh lowland peat soils, found that 56% of the remaining 240 km^2 of Fenland peat soils (*cf.* 1480 km^2 in 1630 AD) is less than 1 m deep (Fig. 2). If wastage of cultivated peats progresses at current rates then two-thirds of the remaining area will be lost by 2050 AD. The 10 500 ha of peats thicker than 1 m are largely confined to river valleys, deep basins, recently drained areas and to nature reserves and flood relief washlands. The latter two types of area are protected from wastage, but much of the deep peat habitat could be lost as a result of permanent flooding.

Another unrecorded threat is acidification. Given that the diversity of invertebrates is lower in acid streams than in alkaline ones (Sutcliffe and

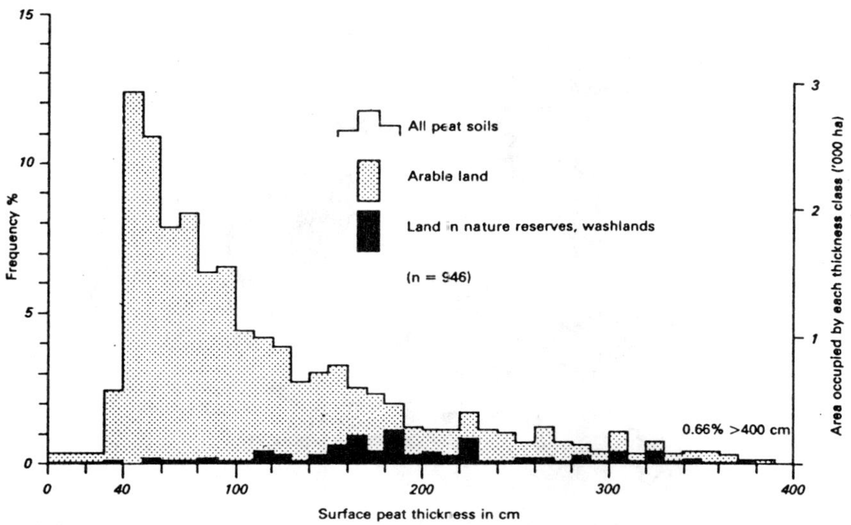

Fig. 2. Frequency and area of peat soils in East Anglian fenland in 1985 by dm thickness classes (reproduced from Burton and Hodgson, 1987).

Hildrew, 1989), it is odd that no species are considered at risk from such changes. The nearest to a reference is the long-term decline in water quality, which is indicated as a threat to the elmid beetles *Normandia nitens* (Müller) and *Stenelmis canaliculata* (Gyllenhal). Mentions of afforestation are primarily concerned with loss of habitat through drainage, not with siltation or acidification of catchments.

The extraction of ground water was also overlooked in the Red Data Book. The wetlands associated with the fluctuating meres of the Breckland are well known for their entomological interest, but the international importance of Irish turloughs and East Anglian pingo fens for wetland Coleoptera has only recently been established. The interest of these sites depends partly on their relict status, but also on the high quality of the ground water and on the fluctuating water table in chalk or limestone. Nilsson (1984) has established that changing water levels contribute to a high diversity of water beetles in northern Swedish kettleholes, through the maintenance of both a high habitat complexity (by lack of floral conformity) and a high production that is associated with recurrent nutrient pulses. Population pressures in areas associated with fluctuating meres will result in a greater demand for artesian water and a consequent lowering of water table. This cannot be prevented by orthodox site protection measures.

Finally, the rapidity with which investment priorities can change almost caused the loss of "The Flow Country" before there had been a proper evaluation of its conservation values, or any recording at all for many groups of invertebrates. The Caithness Biological Records Centre survey (Spirit, 1987) revealed that the caddis *Nemotaulius punctatolineatus* (Retzius) bred in dubh lochans there and that the distinctive beetle *Oreodytes alpinus* (Paykull) was common in some sandy lochans. The latter was a new species for Britain, having previously been known only as a Pleistocene fossil in southern England. No plant or other animal has such a restricted status in Britain (Lindsay *et al.*, 1988). New gaps in our knowledge will almost certainly be revealed by similar unexpected demands for development.

A better appreciation of the need to conserve communities, as opposed to species, would have shown that general environmental changes are just as important as habitat destruction.

D. Conservation Measures Specified in the Red Data Book

The main measure specified for aquatic insects is site protection. Protected sites rank from National Nature Reserves (NNR) and sites of special scientific interest (SSSI) (about 20 references each), then county trust reserves, National Trust land (mainly Wicken Fen), reserves of the Royal Society for the Protection of Birds, and lastly local nature reserves. The conservation areas of the New Forest are regularly mentioned, as are sites owned by the Ministry of Defence, such as Woolmer Bog (for *Graphoderus zonatus* (Hoppe)) and the Stanford Training Area on the Breckland; local conservation management groups should be aware of their responsibilities. The need for the notification of more SSSIs and for negotiations with land owners appears to be equally important, according to the number of references in site accounts.

Only one species, *Aeshna isosceles* (Müller), is listed on Schedule 5 of the Wildlife and Countryside Act 1981, and no account suggests that further listing of aquatic insects is desirable. A later proposal to include *Coenagrion mercuriale* (Charpentier) was resisted by the NCC even though it is listed on Appendix 2 of the Bern Convention. This species is clearly better established in Britain than elsewhere in its range. There are more references to a need for increased knowledge about modern distribution than about biology and ecology, although it is implicit in many accounts that life cycle and habitat requirements are no better known now than when the species were first described. Ignorance of modern records hampers ecological studies. In a few cases, localities have not been disclosed to those preparing the accounts. This creates a further possible threat to the species concerned through the loss of information when the record holder dies. Another problem concerns proprietorial rights to records. Not all those kept by government bodies are in the public domain, and private recorders increasingly expect compensation for their efforts other than the satisfaction of contributing to a national data base.

Only one case of an introduction is cited, the large marsh grasshopper (*Stethophyma grossum* (L.)) to an NNR in Surrey. Examples of failed introductions (e.g. *Agabus undulatus* Schrank to East Norfolk in 1906 by F. Balfour-Browne (1950)) could have been included. Introduction is proposed for only one species, the lesser silver water beetle, *Hydrochara caraboides* (L.). A proposal for its re-establishment in Woodwalton Fen NNR from its sole surviving British location on the Somerset Levels met with opposition, no doubt because of experience with the work needed to maintain the Dutch form of the Large copper butterfly (*Lycaena dispar* Haworth) there.

About 11 other management suggestions are included in the Red Data Book. These may be summarized as control of carr, changes in ditch management, maintenance of water quality standards and exclusion of pleasure boats. There are also many suggestions about the need for monitoring.

IV. SPECIES RARITY AND SITE EVALUATION

A. Nationally Scarce Species and National Notability

RDB species should be too rare for their presence to be used to evaluate the conservation value of most sites. It is therefore necessary to categorize the scarce species not presently considered to be under threat. In Great Britain this has been achieved by "national notability", with species ratings developed in connection with NCC's Invertebrate Site Register (Ball, 1986). List A species (Na) were those known to occur in 30 or fewer 10-km squares of the national grid, and List B species (Nb) those known from 100 or fewer squares. The system was applied to species not subject to active recording schemes by a process of consultation, and by using coarser measurements, such as the number of vice-counties from which the species is known (seven or fewer for Na, 20 or fewer for Nb). Ball (1986) also lists other approaches to rating species, such as regional notability and habitat indicators. A point-scoring system based on Red Data Book and notability status was described by Ball (1986). In this a certain number of points based on the presence of two RDB species, four Na or five Nb species, or some equivalent mixture) is taken to indicate suitability for scheduling as a SSSI. Ball criticizes this method because it fails to take into account the diligence of the recorder. This system has been modified (Nature Conservancy Council, 1989) so that now one refers to "nationally scarce species" which occur in 16–100 10-km squares (all rarer species thus acquiring RDB status). Although points are still awarded, including some for "regionally scarce species", no threshold for SSSI selection is presently defined.

In addition to Ball's criticisms, one might also query whether a site found, in the course of a general survey, to contain a nationally scarce beetle, moth and bug, should be accorded the same status as one from which three nationally scarce beetles were found during a specialist survey. The next question to consider is whether the proportion of the total number of species found to be scarce is more important than the number of scarce species present.

B. Species Rarity and Quality Scores

A system of point scoring was proposed by Foster (1987), in which lists of water beetle species were used to generate an aggregate quality score. Each species in the list was scored on a geometric progression from 1 for the commonest species to 32 for the rarest. Scores could be national, regional or local, and were originally based on the relative rarity of each species in an area, with a modification to account for those that are under-recorded because of difficulties in identification or capture. The aggregate of geometric scores should reflect the hollow curve nature of the distribution of populations within communities, and should accentuate differences between site lists. Extra points were scored for associations of rarities. The quality scores were also standardized by dividing by the number of species present, to yield a "species quality score" (SQS). This mean score per species should minimize the problems imposed by diligent collection, by variation in collecting abilities and by seasonality.

This approach has been tested by Eyre and Rushton (1989), who confirm that a standardized score based on a geometric scale of points results in a clear interpretation of site rarity. They found that the weighting for rarity association improved the detection of sites of the highest conservation potential. Indirect evidence for the robustness of SQS came from a study in 1986 of East Anglian ditches (Foster *et al.*, 1989): SQS could be related to vegetation management and ditch clearing, and highlighted the need for regular clearings along larger drains combined with neglect for the smaller ditches.

C. Lists, Assemblages, Communities and Guilds

The word "assemblage" is often used to avoid drawing conclusions about the true status of site species lists, especially the inference that they constitute communities. The word "community" should be acceptable if one uses the definition of James and Boecklen (1984): "a collection of species that happen to exist together because of converging accidents of space, time and environmental needs".

Root (1967) defined a guild as "a group of species that exploit the same class of environmental resources in a similar way". Hawkins and MacMahon (1989) described this as "a seductively simple definition" and as a "Gordian knot of ecological thinking". They decided that it was "a useful but artificial construct of the minds of ecologists" with the resource exploited being "investigator-defined". But if the concept can be used to understand the ground plan of communities and so improve comparisons between communities, then it must have value in conservation even if it provokes disdain among theoretical ecologists.

Water beetle communities are evidently made up of a limited number of guilds, mainly predatory, with few interlopers from other taxonomic groups. Nilsson (1986) identified two guilds based on 10 "core species" out of 33 species of dytiscid water beetles occurring in a northern Swedish pond. Guild I comprised five *Hydroporus* species and guild II three species of *Agabus*, with another species, *Agabus labiatus* Brahm, being treated separately. Guild I species were small, with larvae that fed in June on Cladocera and chironomid larvae. Guild II species were larger, with larvae that fed on culicid larvae and mayfly nymphs in May and early June. *A. labiatus* is a small member of the genus; its larvae develop at the same time as guild I species and are unusual in that they swim rather than crawl. Interspecific competition between water beetles appears to be very weak according to Larson (1985) and Nilsson (1986), with guilds based on small (3–5 mm body length) and moderate-sized (8–14 mm) beetles within a few genera dominating standing water sites. A proper development of the idea of functional groupings within aquatic insect communities is hampered by a lack of information on feeding range, feeding behaviour, life cycle strategies, and by a failure to identify spatial relationships within the habitat. Differences between bottom-crawling and swimming insects, and between those of the edge and the open substratum are well known, but most species of aquatic insect live in the complex matrix of debris and vegetation at the water's edge, where spatial relationships are destroyed in any attempt to study them.

A better concept of guild structure might improve methods of site evaluation. Guild numbers and species packing within guilds would indicate the complexity of trophic relationships far better than crude indices of diversity, such as species numbers. It should also permit comparisons to be made over greater geographical ranges.

D. Site Quality Criteria

Ratcliffe (1977) specified 10 criteria for assessing comparative site quality in the British "Nature Conservation Review": size, diversity, naturalness, rarity, fragility, typicalness, recorded history, position relative to other high quality sites, potential value and intrinsic appeal. Usher (1986) suggested 14 further criteria. Those relevant to aquatic insects are: education value, representativeness, scientific value, uniqueness, availability, management considerations, replaceability, wildlife reservoir potential, threat of human interference, and lastly, because of natural enrichment, importance for migratory wildfowl.

Given that much site information comprises little more than a listing of species, it is important to consider what can be achieved with such lists rather than waste time on discussing the relative importance of the criteria. Usher (1986) emphasizes that evaluation is based on attributes, criteria and values. A site's attributes would be the species list, what we know of the site's history and a site description in terms of water flow, pH, brackishness, vegetation, shading, etc. The attribute of a list might generate quantifiable criteria such as species richness (the length of the list), rarity (based on the number of rare species) and diversity (as an index calculable if numbers of each species have been estimated). The history of the site might yield a non-quantifiable estimate of "naturalness". The age of a standing water site, if known, may allow quantification of naturalness, but it should be remembered that the fauna of rivers and natural lakes, although dating back to the end of the Pleistocene, may only be as "old" as the last major pollution incident.

Some criteria, such as "representativeness", can only be used to group sites into comparable series rather than to evaluate them. On the other hand, a site may be most "typical" of a group, and this typicalness can be used to evaluate sites if an attribute can be developed. Eyre and Rushton (1989) demonstrated that typicalness, based on deviation from the "centroids" of ordination scores for site groups (Eyre *et al.*, 1986), could be measured separately from rarity for aquatic and ground beetles.

Usher (1986) points out that values have to be placed on the criteria. Many scientists rank diversity as a more important criterion than rarity, but the value judgement of society may well give rarity a pre-eminence.

E. Water Beetle Site List Evaluation

It is easier to compile meaningful lists of species from aquatic habitats than it is from terrestrial ones. Apart from the obvious boundaries of each site, all species can be collected in a net whereas an attempt to record terrestrial species, particularly the nocturnal ones, involves a variety of techniques, some of which require several visits.

It is wasteful if the species list data base is used solely to generate maps of individual species. The water beetle mapping scheme, initiated in 1979, has produced at least 4000 lists of five or more species associated with particular sites. Adult water beetles tend to be present at all times of year, so it is possible to undertake multivariate analysis without needing to take seasonality into account. This should be true for other aquatic macro-invertebrates which exist as recognizable stages throughout the year.

A rationale has therefore been proposed and tested whereby species lists are subjected to multivariate analyses, detrended correspondence analysis (DECORANA; Hill, 1979a) and two-way species indicator analysis (TWINSPAN; Hill 1979b). DECORANA, by ordinating sites and species, allows the major ecological factors that dictate variation to be identified (by "indirect gradient analysis", as opposed to direct gradient analysis by the programme CANOCO; Ter Braak, 1987). TWINSPAN produces a classification of site lists based on an ordered two-way table of sites and species with each division into subgroups characterized by indicator species.

Most "gradient analyses" of species lists indicate that the main factor determining community type is water flow (Table III). Other major factors are pH, brackishness, vegetation (fen conditions versus open substratum) and seasonal permanence. Sometimes these gradients are "confused" (for the human observer, if not for the computer), with gradient extremes being intermittent calcareous water and acid flushes, or permanent acid water and (temporary) saltmarshes. DECORANA permits objective ranking of these variables. Similarly TWINSPAN permits objective grouping of sites, though the point at which the divisions should stop is best judged by the field experience of the recorder. Jeffries (1988) extended the permanence concept by noting that the distance of a temporary pond from a large, permanent water body could be identified by TWINSPAN as a variable influencing the invertebrate fauna. The importance of the presence of a large, open water body in dictating the assemblage of fens had also been detected in an analysis of sites in southern Scotland (Foster, 1987). Factors such as these were overlooked when authors analysed data subjectively (see a comparison of factors listed by six authors in Larson, 1985). It is difficult to see how one can offer constructive suggestions about the management of wetland sites without being aware of the main factors dictating community types.

The objective classification of lists into "end-groups" can be followed by ranking sites on the basis of a series of factors derived from the single attribute of a species list. These are the number of species, the number which are either RDB species or nationally notable, the species quality score (SQS), the aggregate quality score, and a typicalness rating. Choice of a hierarchy for these factors is as difficult as is the choice between the criteria listed by Usher (1986). In practice the topmost sites within each end-group are easily recognized by any three factors (other than typicalness). It was originally thought that it would be logical to give highest conservation priority to the best sites within each end-group. In practice this is not so. Some communities are associated with polluted or severely disturbed sites; others are found only in relict sites, all of which appear to require conservation. This reflects the way in which the classification of TWINSPAN is achieved, with the species present in most assemblages being retained in the centre of the matrix. This central mass of data from low quality ponds and ditches cannot easily be interpreted,

TABLE III. Ecological gradients recognized by DECORANA (without downweighting) of water beetle species lists.

Area and number of sites	Axis	Eigen-value	Extremities of axis
Lincolnshire and	1	0.441	Running to acid
northern East Anglia,	2	0.307	Halophil to permanent and base-rich
547 sites	3	0.242	Acid to permanently brackish
Cheshire, Staffordshire,	1	0.489	Fast running to stagnant
Derbyshire, Nottinghamshire,	2	0.386	Stagnant fen to bare substratum
208 sites	3	0.242	Not interpretable
Wales	1	0.595	Running to saltmarsh
230 sites	2	0.403	Open to acid
	3	0.289	Not interpretable
Southern Scotland,	1	0.423	Riverine to brackish pools
193 sites	2	0.403	Upland flushes to fens
	3	0.289	Open acid to brackish
North-east England	1	0.448	Upland mires to limestone ponds
(without rivers),	2	0.290	Lakes to saltmarshes
349 sites	3	0.235	Not interpretable
South-west England,	1	0.669	Running to saltpans
172 sites	2	0.362	Open over mud to intermittent streams
	3	0.240	Calcareous running to acid running
Weald,	1	0.447	Shaded running to tidal pools
395 sites	2	0.375	Chalk stream to shaded acid streams
	3	0.228	Not interpretable

but the conservation message from TWINSPAN is usually clear. The main value of this attempt to analyse conservation priorities objectively is that the entire species list is utilized, the common species as indicators for the main community types, the rare species because of their RDB or notable status, and all species in contributing to SQS.

The identification of communities by TWINSPAN opens up other possibilities. First, invertebrate lists can be analysed independently of site descriptions. It is not necessary to force-fit a relationship between the invertebrates found and the vegetation or "spot" records of physicochemical variables of a site. Some of the worst examples of previous misuse are associated with pollution studies in which invertebrate counts are reduced to an index for treatment on the same level as any one of a number of physicochemical variables. For example, in a study of invertebrates as indicators of acid mine pollution, Dills and Rogers (1974) converted invertebrate data into the Shannon–Wiener index of diversity, and were thus able to report on the possibility of using a macro-invertebrate index whilst naming only one fly and one crustacean. The Biological Monitoring Working Party score (BMWP; Chesters, 1980) comes close to this extreme in Britain. This is an index calculated by assigning a score of 1–10 to each invertebrate family present in a sample, and summing these numbers to obtain a score. The value for each family is dependent on its sensitivity to organic pollution, the highest values being given to the most sensitive. This oversimplified approach is undoubtedly better than ignoring invertebrates, but it fails to utilize the variation in sensitivity within families and the possibility for more sensitive recording. Wright *et al.* (1984) classified running water sites by a TWINSPAN analysis of macroinvertebrates, and were able to demonstrate the possibility of predicting community types by using environmental data; pollution could be identified by the absence of sensitive species.

V. THE LOW PROFILE OF INSECT CONSERVATION

A. The World Conservation Strategy

A report on the British contribution to IUCN's conservation programme for 1985–1987 (UK Committee for IUCN, 1987), names nine species of mammal, 21 birds, two reptiles, three amphibians and one crustacean. Only three insects are specifically mentioned, the tsetse fly and the chequered skipper and large blue butterflies. On what basis can such an uneven coverage be justified in implementation of the world conservation strategy?

Are we to blame as entomologists for achieving such a low profile in conservation matters? The immensity of the insect fauna has rightly caused emphasis to be placed on habitat protection rather than on intensive studies of threatened individual species. But so often we lose the confidence of those attempting to conserve habitats when we are unable to offer more by way of advice than a list of latin names. This immensity has caused a fragmentation of effort and knowledge. The IUCN Committee's report mentions at least 60 groups associated with conservation. There are probably as many again concerned with British insects.

Perhaps most entomologists are not interested in conservation. Amateur entomologists have a reputation for the ruthless pursuit of the most recently rediscovered rarity, and many professional entomologists look down upon practical conservation as being undisciplined, unscientific and irrelevant to the long-term survival of insects. Another problem would appear to be that the gap between the amateur insect collector and the professional insect ecologist is more evident than in other disciplines. This is associated with the difficulties of taxonomy and identification; insect ecologists are notorious for their failure to follow taxonomic change.

B. Ignorance

The main threat to the conservation of aquatic insects is ignorance. This falls into four categories: a lack of enlightenment on the part of the land owner or developer; a lack of direction on the part of the conservation manager; low public awareness; and forms of ignorance on the part of the entomologists. The last include a lack of knowledge of the distribution of species within the UK and over entire ranges; a lack of knowledge of life cycle stages, such as the larvae of some caddis and beetles (recognition of larval morphobiotic features may provide insight into guild selection and guild packing); a lack of knowledge of feeding requirements, such as the range of host plants of many phytophagous species, the extent of opportunistic prey responses in predators, and the susceptibility to predation; a failure to understand critical parts of the development and behaviour of some species, such as the demand for pupation sites, the ability to survive drying out and the significance of movement by flight; and an inability to predict the consequences of environmental changes, other than habitat loss.

VI. ACCEPTANCE OF THE NEED FOR INSECT CONSERVATION BY NON-ENTOMOLOGISTS

A. The Non-entomologist's Perception

It is unlikely that the problem of ignorance will be overcome in time to be of use in conservation. We must therefore build on present knowledge, which means knowledge of small species groups. The non-entomologist's perception can be divided into six areas: aesthetically appealing species; large species; surface-active species; food for large animals; "pond dipping" as a first exposure to insect communities; and pioneer species as the first sign of life in a newly constructed pond. Some of these may have more fundamental appeal.

B. Colourful Species

Dragonflies and damselflies command almost as much interest as butterflies. The number of species in the UK (44) is too small to permit the manipulation of species lists, but regional thresholds (such as 15 species in south-east England and nine in most of Scotland) are available for the scheduling of SSSI (Nature Conservancy Council, 1989). Such a simplistic approach is justifiable for species with an intrinsic appeal, but it is an inappropriate attribute for use with other groups.

Odonate nymphs are long lived and water breathing, so many species occupy permanent, deep water habitats and could be considered as a group of predators complementary to the dytiscid beetles, which also occupy shallower and less permanent habitats.

C. Large Size

Large aquatic insects, such as dragonflies, a few of the larger bugs, the great diving beetles (*Dytiscus* spp.) and the great silver water beetle (*Hydrophilus piceus* L.), have some interest for the average conservationist. Few small wetland insects, other than those that bite, generate interest in lay conservationists, except when the presence of a rare species can be used to support a case.

Is it possible to relate site quality, using an objective evaluation, to the presence of larger species? Large aquatic insect species have proportionately higher biomasses than small ones (Smock, 1980), and such insects could be expected to be of the k-strategist type associated with stable and complex habitats. This argument is unfortunately flawed by the fact that many large insects are strong fliers and tend to be among the pioneer species associated with temporary or new habitats, and are "satellite" species in the sense of Nilsson (1986), with no constant association with diverse and stable communities.

D. Surface-active Insects

Water skaters, crickets, striders and whirligig beetles attract interest from members of the public and are thus worth conserving on that basis alone. Surface-active insects are specialized predators, and merit consideration as the "iceberg tip" indicators of high quality habitats. This is particularly true for whirligig beetles, whose larvae have external gills and therefore also require highly oxygenated water, even though most live in stagnant water. Unfortunately *Gyrinus* spp. cannot easily be either caught or identified. Moreover, one species, *G. substriatus* L., is almost ubiquitous and has far less demanding requirements than other members of the genus.

E. Food for Other Animals

The mayflies and caddis have an appeal to anglers because knowledge of their phenology and behaviour improves the chances of catching fish. It is part of angling lore that entomologists know less about insects than fly fishermen. Perhaps this can be turned to advantage for conservation? Ornithologists regard an abundance of insects as beneficial because their presence attracts birds. This may be undesirable if it encourages further development of overpopulated, fouled and trampled mudflats in otherwise natural habitats. However the survival of many old ponds, and their associated insect faunas, must owe much to angling and bird watching.

F. Environmental Education

The fact that the expression "pond dipping" is self-explanatory underlines the importance of this activity in environmental education. Most beginners' guides to pond life stress the great variety of life to be found. An average pond dip is more likely to reveal the great diversity *within* a particular group such as crustaceans or bugs or beetles. Perhaps this is a lost opportunity to emphasize the importance of diversity within groups as well as within the animal kingdom?

G. New and Unstable Habitats

One of the first activities in habitat creation is either cleaning out the pond or digging a new one. Insects are the earliest colonists of such new habitats and attain a high profile until vegetation develops. Pond construction may contribute to the survival of some insect species that cannot be conserved in any other way. The main pioneer insects, such as corixids, dytiscids and *Helophorus* spp., could not possibly be in need of habitat protection, yet some of the rarest water palpicorn beetles could be termed "relict pioneers", because they fleetingly colonize new sites in extremely small numbers and within a restricted area. Examples include *Helophorus longitarsis* Wollaston, *Laccobius simulatrix* d'Orchymont and *Limnebius crinifer* Rey.

The value of insects in selecting unstable, unvegetated sites for conservation has been recognized. Bird diversity and population densities are used to evaluate coastal mudflats, but not shingle beds in river valleys. An awareness of the need to conserve such habitats may focus attention on the highly sedentary behaviour of many of the "pioneer" species that are associated with them.

H. Five Tactics

Although one cannot deny the instant appeal of conserving the last known site for a great rarity, this approach is easily undermined, usually by discovering it elsewhere! It appeals mainly to the collecting instincts of entomologists. Rarity is a forceful argument, but can more safely be applied to whole communities of insects, perhaps identified by subgroups such as the pioneers, the surface feeders and the "showy" species. A strategy is therefore proposed that highlights the community as the basis of conservation. There are five main tactics.

1. Heightened awareness of the need to maintain complex communities by reference to "high profile" taxa.
2. Heightened awareness of the value of wetland insects in identifying sites of conservation value that cannot easily be identified by reference to other groups.
3. A rationale for community analysis developed in terms understood and used by conservation decision makers.

4. Community studies generating plausible management suggestions.
5. An inventory of wetland sites supporting internationally important species and communities for potential listing in the Ramsar Convention — a key site approach.

Acknowledgements

I am grateful to Angus Laing and to Mrs Margaret Palmer, both of the Nature Conservancy Council, for their comments on an early draft of this paper.

REFERENCES

Angus, R. B. (1983). Evolutionary stability since the Pleistocene illustrated by reproductive compatibility between Swedish and Spanish *Helophorus lapponicus* Thomson (Coleoptera, Hydrophilidae). *Biol. J. Linn. Soc.* **19**, 17–25.

Anonymous (1986). Rare and endangered water beetles in Europe. *Balfour-Browne Club Newsl.* **37**, 1–12.

Balfour-Browne, F. (1940, 1950, 1958). "British Water Beetles", **Vols 1–3**. Ray Society, London.

Ball, S. G. (1986). Terrestrial and freshwater invertebrates with Red Data Book, Notable or Habitat Indicator Status. "Invertebrate Site Register 66". Nature Conservancy Council, Peterborough.

Boorman, L. A., Goss-Custard, J. D. and McGrorty, S. (1989). "Climatic Change, Rising Sea Level and the British Coast", Institute of Terrestrial Ecology, HMSO Publication.

Burton, J. G. O. and Hodgson, J. M. (1987). Lowland peats in England and Wales. "Soil Survey, Special Survey 15", Harpenden.

Chesters, R. K. (1980). Biological Monitoring Working Party. The 1978 national testing exercise. Department of the Environment, Water Data Unit Technical Memorandum **19**, 1–37.

Collins, N. M. (1987). The protection of insects by law in the countries of the European Community. *Cah. Liaison O.P.I.E.* **21**, 17–30.

Dills, G. and Rogers, D. T. (1974). Macroinvertebrate community structure as an indicator of acid mine pollution. *Environ. Pollut.* **6**, 239–262.

Eyre, M. D. and Rushton, S. P. (1989). Quantification of conservation criteria using invertebrates. *J. Appl. Ecol.* **26**, 159–171.

Eyre, M. D., Ball, S. G. and Foster, G. N. (1986). An initial classification of the habitats of aquatic Coleoptera in north-east England. *J. Appl. Ecol.* **23**, 841–852.

Foster, G. N. (1978). The British Red Data Book. *Balfour-Browne Club Newsl.* **9**, 1–5.

Foster, G. N. (1987). The use of Coleoptera records in assessing the conservation value of wetlands. *In* "The Use of Invertebrates in Site Assessment for Conservation" (M. L. Luff, ed.), pp. 8–17. University of Newcastle upon Tyne.

Foster, G. N. (1988). The status of *Ochthebius lenensis* Poppius (Coleoptera: Hydraenidae) in the Moray Firth area. *Entomol. Gaz.* **39**, 77–81.

Foster, G. N., Foster, A. P., Eyre, M. D. and Bilton, D. T. (1989). Classification of water beetle assemblages in arable fenland and ranking of sites in relation to conservation value. *Freshwater Biol.* **22**, 243–354.

Godwin, H. (1978). "Fenland: its Ancient Past and Uncertain Future". Cambridge University Press, Cambridge.
Hawkins, C. P. and MacMahon, J. A. (1989). Guilds: the multiple meanings of a concept. *Annu. Rev. Entomol.* **34**, 423–451.
Hill, M. O. (1979a). "DECORANA – a FORTRAN Program for Detrended Correspondence Analysis and Reciprocal Averaging". Cornell University, Ithaca, New York.
Hill, M. O. (1979b). "TWINSPAN – a FORTRAN Program for Arranging Multivariate Data in a Two-way Table by Classification of the Individuals and Attributes". Cornell University, Ithaca, New York.
Holmen, M. (1987). *Dytiscus latissimus* L. – en truet vandkalv (Coleoptera: Dytiscidae). *Entomol. Medd.* **55**, 29–30.
James, F. C. and Boecklen, W. J. (1984). Interspecific morphological relationships and the densities of birds. *In* "Ecological Communities. Conceptual Issues and the Evidence" (D. R. Strong, D. Simberloeff, L. G. Abele and A. B. Thistle, eds), pp. 458–477. Princeton University Press, New Jersey.
Jeffries, M. (1988). Do water beetle communities reflect the wider freshwater community? *Balfour-Browne Club Newsl.* **42**, 14–17.
Larson, D. J. (1985). Structure in temperate predaceous diving beetle communities (Coleoptera: Dytiscidae). *Holarctic Ecol.* **8**, 18–32.
Lindsay, R. A., Charman, D. J., Everingham, F., O'Reilly, R. M., Palmer, M. A., Rowell, T. A. and Stroud, D. A. (1988). "The Flow Country. The Peatlands of Caithness and Sutherland". Nature Conservancy Council, Peterborough.
Nature Conservancy Council (1989). "Guidelines for Selection of Biological SSSIs". Nature Conservancy Council, Peterborough.
Nilsson, A. N. (1984). Species richness and succession of aquatic beetles in some kettlehole ponds in northern Sweden. *Holarctic Ecol.* **7**, 149–156.
Nilsson, A. N. (1986). Community structure in the Dytiscidae (Coleoptera) of a northern Swedish seasonal pond. *Ann. Zool. Fenn.* **23**, 39–47.
Ratcliffe, D. (1977). "A Nature Conservation Review", Vols 1 and 2. Cambridge University Press, Cambridge.
Root, R. B. (1967). The niche exploitation pattern of the blue-gray gnatcatcher. *Ecol. Monogr.* **37**, 317–350.
Shirt, D. B. (ed.) (1987). "British Red Data Books: 2 Insects". Nature Conservancy Council, Peterborough.
Smock, L. A. (1980). Relationships between body size and biomass of aquatic insects. *Freshwater Biol.* **10**, 375–383.
Spirit, M. G. (1987). "Freshwater Invertebrates of Caithness". Wick, Caithness and Sutherland District Councils Community Programme Agency.
Stephens, J. C. (1828). Illustrations of British entomology. *Mandibulata* **2**, 1–112.
Stephens, J. C. (1829). Illustrations of British entomology. *Mandibulata* **2**, 113–200.
Summers, D. (1976). "The Great Level". David & Charles, Newton Abbot.
Sutcliffe, D. W. and Hildrew, A. G. (1989). Invertebrate communities in acid streams. *In* "Acid Toxicity and Aquatic Animals". (R. Morris, E. W. Taylor, D. J. A. Brown and J. A. Brown, eds), pp. 13–29. Society for Experimental Biology, Seminar Series **34**.
Ter Braak, C. J. F. (1987). "CANOCO – a FORTRAN Program for Canonical Community Ordination by Partial Detrended Correspondence Analysis, Principal Components Analysis and Redundancy Analysis (Version 2.1)". TNO Institute of Applied Computer Science, Wageningen.
Usher, M. B. (1986). Wildlife conservation evaluation: attributes, criteria and values. *In* "Wildlife Conservation Evaluation" (M. B. Usher, ed.), pp. 4–44. Chapman & Hall, London.

UK Committee for IUCN (1987). "The British Contribution to IUCN's Conservation Programme". Nature Conservancy Council, Peterborough.

Wright, J. F., Moss, D., Armitage, P. D. and Furse, M. T. (1984). A preliminary classification of running-water sites in Great Britain based on macro-invertebrate species and the prediction of community type using environmental data. *Freshwater Biol.* **14**, 221–256.

11

Creating New and Successional Habitats for Arthropods

M. B. USHER AND R. G. JEFFERSON

I. Introduction 263
II. Arthropods in Ecological Successions 264
A. Review 264
B. Arthropod Succession at Wharram Quarry Nature Reserve 266
III. Habitat Creation 269
A. Review 269
B. Pond Creation: the Development of Freshwater Communities 271
IV. Habitat Re-creation 272
A. Review 272
B. Re-creation of Butterfly habitats by the Nature Conservancy Council 273
V. Habitat Mosaics of Different Successional Age 275
A. Review 275
B. Spider Communities of Upland Heather Moorland 276
VI. Discussion 278
References 280
Appendix I 283

I. INTRODUCTION

In preparing this review, it became more and more apparent that the conservation of insects cannot be divorced from the conservation of the other biota of the habitats in which those insects live. Although it has often been assumed that conservation of a plant community will lead to conservation of all of the consumer species (discussed by Usher, 1980), it is now clear that such a generalization is not always true. One of the features of this review will, therefore, be to investigate the extra conservation effort needed for the conservation of arthropod species and communities.

Another sweeping generalization is that all natural communities are dynamic, responding to year-to-year variation in climate, the occasional effects of disease, etc. However, in considering new habitats for insects, the process of ecological succession is perhaps the single most important factor causing change in the arthropod community. This review is therefore set within the context of ecological succession.

The aim of the review is to explore three facets of the creation of new habitats for insects. First, what happens when totally new habitats are created? It is perhaps possible to generalize and say that conservationists seldom think of the creation of totally new habitats, except for bodies of freshwater. Hence, second, what happens when habitats that used to exist are re-created on a site? With an increasing awareness of the conservation importance of traditional forms of management, habitat re-creation is receiving considerable attention from conservationists. However, both habitat creation and re-creation tend to be "one-off" events, whereas in some ecosystems, especially those managed by fire, there are cyclical forms of management, with an ecological succession allowed to proceed only to a defined stage before the system is rejuvenated. Third, what are the effects of such periodic management regimes? Examples in this review will tend to be drawn from three sources; our own recent experience in Yorkshire, a survey of arthropod management in national nature reserves (NNRs), as shown in Appendix I and the Lepidoptera (especially butterflies), the group of insects that has received most conservation attention. Appendix I relates specifically to the work of the Nature Conservancy Council on NNRs; no attempt has been made to survey the management for arthropods on the reserves of other bodies (RSNC, RSPB, etc). or more widely on SSSIs. Many entries in Appendix I, although listed for a particular species or group of species, have a wider conservation function, relating also to botanical management, provision of breeding sites for amphibians, etc.

II. ARTHROPODS IN ECOLOGICAL SUCCESSION

A. Review

The majority of studies of succession have related to plants (Usher, 1979), though there is an increasing body of information about the complex interrelationships between plant succession and herbivorous arthropods (Brown, 1982, 1985).

One of the first studies of arthropod succession was conducted by Richards (1926) on a lowland heath fauna in Surrey. A more recent study, by Southwood *et al.* (1979), analysed a secondary succession on abandoned arable land, using three stages which they called "young field" (1–2 years abandoned), "old field" (6–7 years) and the presumed later stage, a birch woodland with a successional age of about 60 years. Taxonomic diversity of the plants peaked at 16 months after spring initiation of the succession; taxonomic diversity of the Heteroptera and adult Coleoptera increased during this period, but subsequently remained high. Southwood *et al.* argued that the arthropod diversity is maintained by the increasing structural diversity of the green plants, compensating for their falling taxonomic diversity. The continuation of this experiment, annually creating new "young field" sites, has extended the data set to include Orthoptera, Dermaptera, Dictyoptera, Psocoptera, Homoptera and Thysanoptera (Brown and Southwood, 1983). Both taxonomic and trophic diversity of the insects increased with successional age (for example in the young field site, 93.0% of species were herbivores, 5.5% predators and 1.5% fungivores; in the woodland site, 74.6% were herbivores, 10.0% predators, 14.6% fungivores and 0.8% scavengers). Besides these changes in diversity, the niche breadth of the sap-feeding herbivores decreased with successional age, whilst generation times became greater.

Such studies are descriptive, implicitly suggesting that the plant species (represented taxonomically or by their structure) influence the invertebrate species and communities. Although such information is essential for the conservation of targeted insect species, another interesting facet of such studies is the question of whether the arthropods themselves affect the course of the plant succession. Experimental approaches to this question are discussed by Brown (1982), though she draws attention to the need for carefully controlled experiments to ascertain that the methods used to exclude insects (e.g. insecticides) do not themselves affect plant growth (Brown *et al.*, 1987). The interrelationships between insects and plants in succession become even more complex if a vertebrate herbivore is also considered (Gibson *et al.*, 1987a,b). The possibility exists that manipulation of both the mammal and insect herbivores may be important tools for the conservation manager wishing to direct changes in plant community composition (Gibson *et al.*, 1987a).

For herbivorous insects, succession does not only relate to the presence in the community of an appropriate species of food plant or to an appropriate environmental architecture; there may be other more subtle facets of the succession that are of importance. A case study is of the two butterfly species *Euphydryas chalcedona* and *E. editha*. Larvae of *E. chalcedona* were commonest where the host plant (*Diplacus aurantiacus*) grew near to the main nectaring species (*Eriodictyon californicum*) (Murphy, 1983; Murphy *et al.*, 1984). Similarly, for the rare *E. editha bayensis*, intake of both carbohydrate and amino acids by pre-oviposition females leads to more and heavier eggs, with an increased probability of larval survival (Murphy *et al.*, 1983). The plants are not, however, only of importance to the butterflies; the butterflies are themselves pollen vectors for the plants (Murphy, 1984). These examples of *Euphydryas* illustrate another facet of insects in succession; the different developmental stages (here adults and larvae) have different environmental requirements.

For conservation management, the interactions between plants and insects have several implications. First, it must be realized that all newly created or re-created ecosystems are dynamic, and hence subject to change in their species complement. Second, it may be possible to determine the direction of change of a plant community by managing either the insect or the mammalian herbivores. Third, if management is targeted at a specific insect species, it may not be sufficient to maintain large populations of a larval food plant because other species, such as a nectar plant, may also be essential. Finally, all of these considerations imply that if the conservation objectives of any newly created or re-created habitat are to be achieved, management of that habitat is essential. A *laissez-faire* approach, which may be appropriate in a late successional habitat, e.g. the conservation of insects associated with dead wood (Speight, 1989), is unlikely to be successful.

B. Arthropod Succession at Wharram Quarry Nature Reserve

Unlike the studies reviewed, which investigated the insects of the above-ground plant communities, Parr (1980) investigated the below-ground succession at Wharram Quarry. The quarry, worked through the first four decades of the twentieth century, had been successively abandoned. The oldest successional stage was a closed sward of grasses, particularly *Dactylis glomerata* and *Arrhenatherum elatius*, with frequent bushes of *Crataegus monogyna* and *Rosa* spp. The youngest successional stage was an open community, with bare chalk patches, largely of legumes (*Lotus corniculatus* and *Medicago lupulina*) and rosette herbs (*Thymus praecox*, *Hieracium pilosella*, etc.), described by Jefferson (1984).

Data for the Collembola and Acari, shown in Fig. 1, are related to an ordination axis from the youngest to the oldest successional age. As with

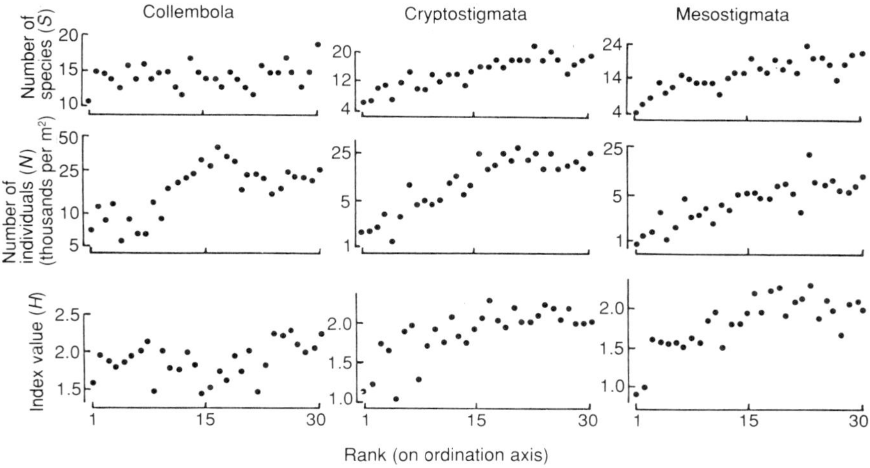

Fig. 1. Changes in the number of species (*S*), the number of individuals (*N*) and the Brillouin diversity index (H) of the Collembola and Acari (Cryptostigmata and Mesostigmata) during an ecological succession at Wharram Quarry Nature Reserve. The successional sequence, ranked from 1 (youngest) to 30 (oldest), is based on an ordination of the complete soil arthropod community, and correlates closely with botanical ordinations. The illustration, reproduced by permission of Chapman and Hall Ltd from Jefferson and Usher (1986), is based on the work of Parr (1980).

Brown and Southwood's (1983) study, the taxonomic diversity of the soil arthropod community increased (from about 25 to 55 species). This increase is solely attributable to the mites since the mean number of species of Collembola remained at about 15 throughout the sequence. However, although the Collembola mean remained more or less constant, the species complement changed, as shown for some of the more abundant species in Fig. 2. The differences show that it would be possible to label some Collembola as early successional, whilst others are late successional, and the abundance of a few species (e.g. *Tullbergia affinis* and *Isotoma viridis*) cannot be related to successional age. The abundance of all groups increased with succession (Fig. 1), as for the above-ground arthropod community studied by Brown and Southwood (1983). The diversity also increased during the succession, at least as far as the stage of grassland being invaded by shrubs (it is not known how diversity would change if the succession proceeded further towards mixed deciduous woodland).

The above-ground fauna at Wharram Quarry has not been studied in the same detail, although Compton (1982) surveyed the arthropod fauna,

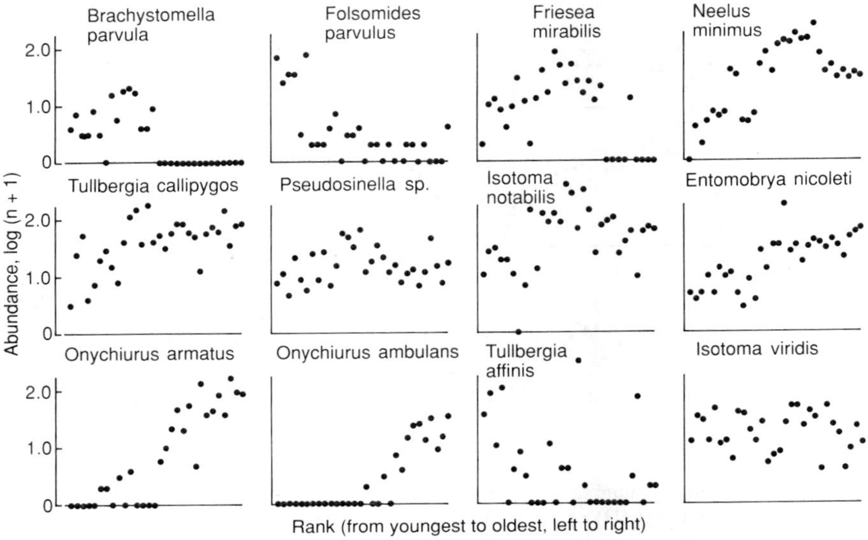

Fig. 2. Changes in the abundance of 12 species of Collembola during an ecological succession at Wharram Quarry Nature Reserve. The species selected are the 10 most abundant, together with *B. parvula* and *O. ambulans* (the two showing the greatest extremes in successional distribution). The successional sequence is the same as in Fig. 1; data from Parr (1980).

largely with an aim of determining which were the potential herbivores of *Lotus corniculatus* (see also Compton *et al.*, 1988). It is reasonably certain that the reserve was colonized by *Melanargia galathea* (marbled white butterfly) in 1981 (Jefferson and Usher, 1986). In the Yorkshire Wolds, where this species has possibly increased its distribution during the last decade, *M. galathea* is a species of grassy valley sides (Rafe and Jefferson, 1983) and hence could not have colonized Wharram Quarry until a grassland community had become established. There is no documentation of when other insect species colonized the reserve, but Compton (1982) and the unpublished reserve management plan list the species presently recorded.

Besides the general concepts discussed at the end of the review section, the example of Wharram Quarry raises one unresolved, and probably unresolvable, question: should insect species be introduced to such habitats? Dempster (1989) started a discussion of the topic by saying "there are few subjects which generate more heated debate amongst ecologists and naturalists than the question of the advisability of translocating species from one place to another". Resolution of this dilemma for individual sites is important if habitats are to be created or re-created for the conservation of insects and other arthropods.

III. HABITAT CREATION

A. Review

The review in this section concentrates on habitats created purposely or partially for the benefit of invertebrates (or wildlife more generally). However, natural processes are also creating new habitats; riverine shingle, sand banks and landslips are examples. A range of species is associated with such naturally disturbed habitats. For example, species particularly in orders Coleoptera, Diptera and Hymenoptera are associated with fresh cliff faces resulting from coastal landslips in softer geological strata such as boulder clay (a number of these species is rare, as listed by Shirt (1987)). *Melitaea cinxia* (Glanville fritillary butterfly) is associated in Britain with the unstable undercliff on the south coast of the Isle of Wight.

There are, however, rather few examples of the creation of completely new habitats specifically for terrestrial arthropods. One example is the recent development of managing expanded field margins (Deane, 1989), although, as pointed out by Morris and Webb (1987), these are essentially enlargements of existing habitats. Further research will be required to elucidate many aspects of the value of such newly created features in arable farmland.

Because of the popularity of the group (Ratcliffe, 1977), some attention has been given to creating habitats for butterflies. Butterflies can be a very obvious feature of newly created habitats, even if no planning for insects was part of the original habitat creation scheme. For example, when the colliery spoil of Mitchell Main Colliery, Dearne Valley, South Yorkshire, was reclaimed, sown with grass and planted with groves of trees in the mid 1960s, there were no plans to conserve insects. Within 4–6 years of reclamation, butterflies breeding on the site included *Polyommatus icarus* (common blue), *Aglais urticae* (small tortoiseshell), *Inachis io* (peacock), *Lasiommata megera* (wall), *Maniola jurtina* (meadow brown) and *Coenonympha pamphilus* (small heath), and probably species of *Pieris* (M. B. Usher, personal observations). This represents virtually all of the species that could be expected in such a habitat in northern England, without any planning to attract butterflies either by planting appropriate food plants or nectar sources.

This list of butterflies compares favourably with a list given by Davis (1989) for a newly completed landfill site in Essex, where grassland, including wildflower species, was sown in autumn 1983. The importance of Davis's study is that it is experimental, not just observational, with replicated treatments, routine monitoring of the butterflies and vegetation recording. By the fourth summer after sowing, at least six species were breeding (*P. icarus*, *A. urticae*, *M. jurtina* and *C. pamphilus* from the Yorkshire list, together with *Thymelicus sylvestris* (small skipper) and *T. lineola* (Essex skipper)), with a total of 18 species recorded. Interestingly, a decision was taken in 1987 to introduce two further species, *M. galathea* (marbled white) and *Hipparchia semele* (grayling), neither of which was thought to be able to colonize the site naturally due to its isolation from extant colonies. It is still too early to assess the success of these introductions.

There is now a considerable literature on the creation and management of flower-rich grassland (Wells *et al.*, 1982, 1989; Wells, 1989). Having established the techniques for creating such habitats, attention is being focused on the whole of the community (e.g. the scheme to create a herb-rich grassland, specifically for butterflies, on a ryegrass ley, outlined by Warren and Stephens, 1989). Creating a total environment is clearly a logical progression from creating a habitat with the appropriate mixture of plant species. However, it is important that one understands which species need to be introduced and which will colonize naturally. The landfill site described by Davis (1989) was sown with 31 species of plants, of which 27 established, but a further 140 species or so have colonized naturally. To re-create wildflower meadows there are now guidelines on the appropriate species mix (with two or three dozen species of herbs) to use; will it be possible to develop similar guidelines for the key insect species to introduce?

It may not be sufficient just to create the appropriate botanical mixture and introduce a selection of insects. Weiss and Murphy (undated) have discussed the importance of topography and microclimate in habitat restoration projects designed to conserve four endangered butterfly species in the San Francisco Bay area. In that area, however, desiccation is probably of greater importance for many insects than it would be in the moister climate of the UK.

The examples reviewed all involve the creation of terrestrial ecosystems. Reference to Appendix I shows that perhaps the greatest effort in habitat creation is in ponds for Odonata. The case studies, described below, are therefore based on the creation of freshwater habitats.

B. Pond Creation: the Development of Freshwater Communities

Two brief case studies will be used to demonstrate the development of invertebrate communities following the creation of ponds.

The first relates to a pond created in February 1981 on Hopewell House Farm, Knaresborough, North Yorkshire. The pond was designed to be small (24 m^2), being a bulbous end to a drainage ditch, taking mostly spring water but also some field drainage. The aim was to create a wetland feature; insects were not a primary reason for its creation. Irregular monitoring by dip net occurred during the 5-year period following the pond's creation (see Fig. 3). Damselflies (Odonata) were observed at the pond in summer 1981, but nymphs were not obtained in the dip net sample till 1984 (after which they were present in all samples, three sets obtained by different sampling techniques not being included in Fig. 3). The Corixidae (Hemiptera) also first appeared in 1984 and have been in all subsequent samples. Dytiscidae (Coleoptera) have been recorded in all samples, though fly larvae (Diptera; at least three species of Chironomidae, *Chaoborus*, and species of Culcididae, Dixidae and Ceratopogonidae) seem to be less predictable (e.g. Chironomidae were present in the 1983 and 1986 samples, but not in the 1984 sample). The presence of *Caenis horaria* in 1986 indicates that the water in the pond is likely to be relatively free from pollution by artificial fertilizers; the only other species of mayfly (Ephemeroptera), in all samples, was *Chloeon dipterum*.

Natural colonization of this pond has clearly produced a diverse invertebrate community within a 5-year period. The origin of these colonists remains a mystery; sampling of a nearby, permanently flowing ditch indicated a community with fewer species than the newly created pond.

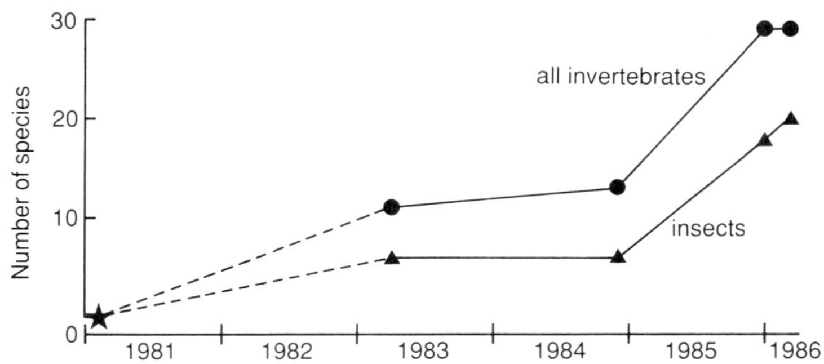

Fig. 3. Colonization of a newly created farm pond in North Yorkshire by insects and other invertebrate animals. Data from Usher (1987).

The presence of platyhelminths (*Polycelis nigra*), molluscs (*Planorbis leucostoma*) and crustacea (*Asellus aquaticus*, daphnids and copepods) in the 1983 sample indicates that it is not only flying insects that can colonize new water bodies.

The second case study relates to the creation of pools for Odonata (dragonflies and damselflies), a recurrent feature of NCC management (Appendix I). An example of what has been achieved at Loch Lomond NNR is the creation of an L-shaped pool (north–south length *ca* 23 m, east–west length *ca* 15 m, water width 6–8 m, maximum depth *ca* 0.9 m) in the lee of a shelterbelt in September 1986. Within 1 year five species of Odonata were observed breeding in the pool (Table I); of particular interest was *C. puella*, a species previously unrecorded from the Loch Lomond area, which is at the northern edge of its British distribution. Two additional species were observed breeding in the second and third years after pool creation (J. Mitchell, personal communication). The Odonata are clearly rapid colonists of newly created pools! Wherever data have been collected, it does appear that many species are extremely mobile, reaching new water bodies relatively rapidly. The question of whether to introduce freshwater species will probably not arise unless conservation is aimed at a species currently beyond its present biogeographical range.

IV. HABITAT RE-CREATION

A. Review

The majority of conservation management activity listed in Appendix I concerns the re-creation of habitats. Methods of re-creating grassland habitats have received considerable attention recently (e.g. Wells, 1989), but targeted research will be required to elucidate the intricate relationships with (a) vertebrate grazers, (b) the structure of neighbouring metapopulations, and (c) microclimate.

The study of sheep grazing a base-rich grassland by Gibson *et al.* (1987c) is important in understanding how grazers can change vegetation succession away from dominance of a grassland sward by a few species to a more diverse sward with more herbaceous species. They concluded that any grazing was better than none, and that more intensive grazing resulted in swards with more of the locally occurring calcicolous species. Amongst the butterflies arriving as breeding species (on a field that had been arable only 5 years previously), were *Argynnis aglaia* (dark green fritillary),

TABLE I. Colonization of a newly created pond in Loch Lomond NNR by Odonata. The pool was created in September 1986; B implies breeding in the year indicated, O implies that the species was observed, – that there was no observation of the species. Data provided by J. Mitchell.

Species	1987	1988	1989
Aeshna juncea	B	B	B
Coenagrion puella	B	B	B
Cordulegaster bottonii	O	O	O
Enallagma cyathigerum	–	B	B
Ischnura elegans	B	B	B
Lestes sponsa	–	B	B
Libellula quadrimaculata	O	O	O
Pyrrhosoma nymphula	B	B	B
Sympetrum danae	B	B	B
No. of breeding species	5	7	7

Aricia agestis (brown argus) and *M. galathea* (marbled white). Research is now required to elucidate the effect of grazing on other re-created habitats.

The study of metapopulation dynamics is still relatively new, although Ehrlich and Murphy (1987) have applied the concept in planning the conservation of *Euphydryas* (checkerspot) species. They considered that it is essential to conserve what they term a "reservoir population" ("a demographic unit which can be expected to persist through essentially all natural environmental perturbations") of fragmented populations. This can act as a source of colonists to habitats that contain "satellite populations", populations that are small and will probably not maintain themselves in perpetuity. Again, there are research requirements for understanding both the demography and genetics of such metapopulations; the studies of Bengtsson (1988) on *Daphnia* rockpool species on the Swedish Baltic coast show how rich a field of study this could be.

B. Re-creation of Butterfly Habitats by the Nature Conservancy Council

The majority of target species of insects for conservation management are butterflies (*cf.* the list in Appendix I). In many cases the management is aimed at re-creating a habitat that has altered under changing socio-economic conditions. The demise of a market for coppice produce led to that system of woodland management becoming rare; *Mellicta athalia* (heath fritillary) requires sunny sheltered woodland in an early stage of succession or regeneration (Warren *et al.*, 1984), and hence became rare when coppice management declined. The death of many rabbits following the introduction of myxomatosis led to downlands becoming overgrown; *Hesperia comma* (silver-spotted skipper) will only lay its eggs in short *Festuca ovina* turf surrounded by bare ground or scree (recent sheep stocking on abandoned downs has led to a slight expansion in the restricted range of this butterfly) (Thomas *et al.*, 1986). *Plebejus argus* (silver-studded blue) requires early successional habitat as eggs are laid on food-plants along the margins between vegetation and bare ground (Thomas, 1985a), with a warm microclimate (i.e. sheltered, south-facing slopes). The possible ways of managing habitats for *P. argus* are described by Thomas (1985b). Summaries of possible management activities to favour chalk grassland butterfly species in Britain are given in BUTT (1986).

A particular example of management for butterflies is Gait Barrows NNR, purchased by the NCC in 1976, which includes about 30 ha of old coppice woodland. This had not been coppiced regularly for a long time (though there had been some clearance for firewood, pig keeping and limestone pavement removal). Attempts to re-create the coppice woodland environment started in the 1982–1983 winter; then, and each winter subsequently, *ca* 400 m of ride have been widened and 0.33 ha of woodland coppiced (A. C. Aldridge, personal communication). This form of conser-

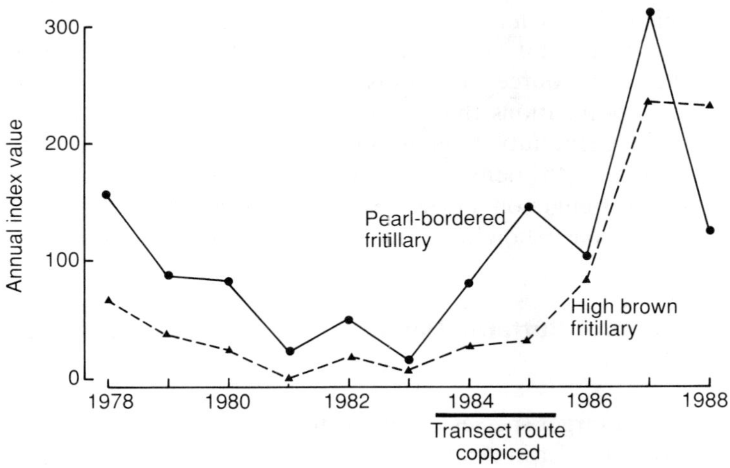

Fig. 4. Annual index values for two species of butterfly observed during transect walks at Gait Barrow NNR (data collected by Mrs Gillian Barker, provided by A. C. Aldridge). There had been extensive clearings by a previous owner about 1970; the whole transect route was coppiced between the 1983–1984 and 1985–1986 winters.

vation management has been targeted at three species of butterfly, *Hamearis lucina* (Duke of Burgundy), *Boloria euphrosyne* (pearl-bordered fritillary) and *Argynnis adippe* (high brown fritillary). Whereas the former has remained scarce, with only occasional sightings in the NNR, there is circumstantial evidence to suggest that the other two species have responded positively to the habitat re-creation (see Fig. 4).

These examples, partly the results of research and partly unreplicated demonstrations, illustrate how much knowledge has to be gained to re-create the habitat suitable for single insect species. Much of the research in Britain is summarized in BUTT (1986). To extend such conservation activities to even a handful of the British insect fauna would require resources of both research and management that are currently inconceivable. Inevitably, then, the question will be asked about whether there are short cuts, ways of creating or re-creating habitats that will support viable populations of many conservationally important arthropod species.

V. HABITAT MOSAICS OF DIFFERENT SUCCESSIONAL AGE

A. Review

A mosaic of patches of different successional (or developmental) age is often associated with ecosystems managed by fire. In many instances the managed blocks are large, as in the case of Mediterranean-type shrubland (e.g. minimum diameters of burnt areas of 200–500 m described by Schlettwein and Giliomee (1987) in their comparison of the Collembola, Cicadellidae and Formicidae of South African fynbos of different successional age). In a more intensive study of ants, Donnelly and Giliomee (1985) showed that species richness declined from fynbos 3 months after burning (29 species) to 4 years after burning (20 species; 16 species common to both sites), with further declines to 39 year fynbos (14 species) and in pine stands (9 species, none of which were exclusively found in the pine). With a community of 36 ant species, it becomes apparent that a mosaic of all successional ages is required to conserve this biodiversity.

Smaller patches are burnt in the British upland *Calluna* heaths, although management is endeavouring to favour grouse production, red deer or sheep grazing etc., rather than arthropod conservation. Although relatively little is known about the effects of such management on arthropod biodiversity (Usher and Gardner, 1988), there is some evidence to suggest that it does maintain maximum biodiversity of spiders (Usher and Smart, 1988) and beetles (Gardner and Usher, 1989). Considering the conservation of the early successional plants and soil fauna of a chalk quarry, Jefferson and Usher (1987) suggested that small scrapes, to create a mosaic of areas of different successional age from newly exposed chalk to communities of 40–50 years development, would be an appropriate management aim. Conservation of early successional arthropod communities requires active management intervention, often re-creating conditions for the colonizing species, though because of restricted ranges of colonization small-scale mosaics may be more appropriate than larger scale mosaics.

B. Spider Communities of Upland Heather Moorland

Usher and Smart (1988) analysed counts of spiders collected in transects of pitfall traps running from unburnt old *Calluna vulgaris* (*ca* 20–25 years old), across recently burnt or cut areas, and back into unburnt old heather. Their study was extended by a programme of pitfall trapping during 1988 (late April to early October), in the centres of eight burnt areas, six cut areas and in 10 heather stands of different ages, which yielded 3997 individuals. Summaries of the distribution of 12 of the more common species are shown in Fig. 5, where the first row of four species could be considered to be early successional, with few individuals appearing in pitfall traps set in unburnt heather (465 trap weeks) but many individuals in the areas burnt or cut in 1987 (370 and 287 trap weeks, respectively), The second row illustrates four species that show relatively little successional relationship, whilst the bottom row illustrates four species that could be described as later successional since they occur predominantly in the pitfall traps set in heather.

Such an interpretation, however, assumes only two successional states, early and later. However, as the heather community develops after a fire, it is possible that the arthropod community continues to develop. Although historical records of when each small area of the heather mosaic was burnt have not been kept, the height of the heather over this limited moorland area (all sites were within 300 m of each other, on reasonably flat ground) can be an indicator of age, at least until the *Calluna* is about 40 cm tall. The abundances of the four later successional species in Fig. 5 are plotted against the height of the heather in Fig. 6. Here it can be seen that *Lepthyphantes ericaeus* tends to become slightly less abundant as the heather becomes taller (except for one anomalous point). Both *Lepthyphantes tenuis* and *Pelecopsis mengei* become more abundant in traps with taller heather. Such interpretations need to be made with care since the effectiveness of a pitfall trap may be influenced by the changing vegetation structure. There is no clear relationship between abundance and heather

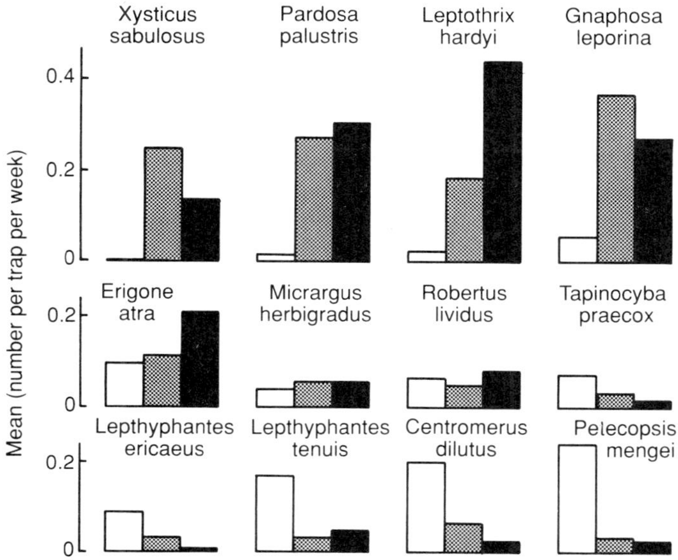

Fig. 5. The frequency of 12 of the more abundant species of spiders in pitfall traps on Danby Low Moor, North York Moors, in 1988. Unshaded histograms indicate the number per trap per week in stands of *Calluna*, the shaded histograms the number in areas burnt in April 1987, and the filled histograms the number in areas cut in April 1987.

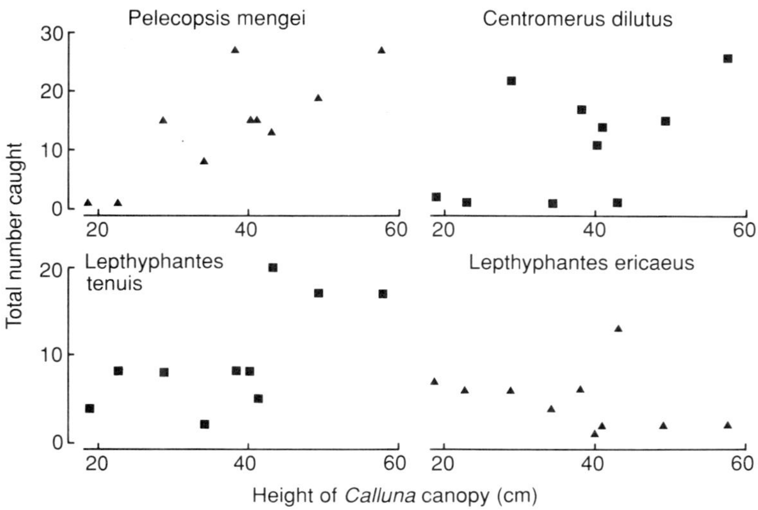

Fig. 6. The relationship between the abundance of four spider species in pitfall traps and the height of the *Calluna* in which those traps were located. The four species selected are those in the bottom row to Fig. 5.

height for *Centromerus dilutus*, nor for *Walckenaeria acuminata*, a species that Usher and Smart (1988) found to be characteristic of the unburnt heather.

The 1988 study on Danby Low Moor, North Yorkshire, yielded 92 species. Except for some uncommon species, all species occurred in all three of the management habitat types (unburnt, burnt and cut heather). That fact might indicate that management is unimportant in this community for the existence of all species. However, the data in Figs 5 and 6 clearly indicate species preferences, and hence the intricate mosaic of heather ages is perhaps the best guarantee that the diversity of this spider community will be conserved.

VI. DISCUSSION

In attempting to review the creation and re-creation of habitats for arthropod conservation, it is apparent that there are very few examples of new habitats designed specifically for arthropods. The exception is of pools for Odonata, especially on heathland reserves. Although this locally increases biodiversity, the creation of ponds in farmland can have spectacular effects on biodiversity in otherwise species-poor environments. Farm ponds have rarely been created specifically for arthropods, just as many other facets of conservation management are not primarily for arthropods (e.g. coppicing, grazing), but often the arthropod community can benefit from management targeted at other groups or species.

The question as to whether the creation of completely new habitats for invertebrates is justified on sites of existing high value for nature conservation (including nature reserves) needs to be carefully considered by conservation managers. Clearly the creation of a new feature for a target invertebrate species, or community, which can subsequently destroy or damage existing species or habitats is undesirable. A common example of this, particularly in unimproved farmland outside protected areas, is the creation of a pond in an area of species-rich fen vegetation. Rigorous examination of each habitat creation project is clearly required and knowledge of the autecology of the target species is necessary. Such examination would be assisted if there were more published case studies of habitat creation, recording not only the scientific objectives and scientific achievements but also the management inputs of resources, the techniques used and subsequently a long-term evaluation of success or failure.

Habitat re-creation, on the other hand, is often related to the re-establishment of a traditional form of land use which has declined due to changing socio-economic conditions. Species of arthropods dependent upon such habitats tend to have become rare or local, and hence their conservation has become a matter of priority. As opposed to such "local" species are the "ubiquitous" species that appear to be able to benefit from any kind of habitat creation — conservation headlands, flower-rich swards, farm ponds, new woodlands or hedgerows etc. The study by Davis (1989) of a landfill site seems particularly relevant in this context. A number of more ubiquitous butterfly species rapidly colonized the site; the more local species had to be introduced by translocating individuals from colonies in the geographical area. Almost certainly, effective arthropod conservation of newly created habitats will rely largely upon natural colonization, with supplementation by the planned release of a few local species that can be expected to persist under the management regime of the newly created habitat.

There are two other features of this review that seem to be important, relating first to the amount of management intervention (disturbance) required and second to the availability of information available to the manager.

First, by whatever mechanism (changing plant species composition, changing microclimate, interaction between the insects and plants), succession affects the arthropod community. Many arthropods are characteristic of the early and middle stages of an ecological succession. These species will either become scarce, or even extinct, under *laissez-faire* management policies, as demonstrated for a number of butterfly species by Warren *et al.* (1984), Thomas (1985a,b) and Thomas *et al.* (1986). The important point is to define the conservation goal, and then find management techniques to achieve that goal. When considering newly created and re-created habitats, management may need either to "hold" a succession at an intermediate stage (e.g. by sheep grazing) or to restart the succession locally by designing systems of periodic disturbance (e.g. by fire). In the latter case, research will clearly be required to define the intensity of disturbance required, an appropriate periodicity, tolerable or required stochasticity in both these aspects of disturbance, and an appropriate spatial scale for the disturbance.

Second, much research aimed at conservation management is not research in the traditional way that research is carried out; it is often not possible to experiment with endangered populations nor is it possible to plan appropriate controls. The butterfly data for Gait Barrow NNR (Fig. 4) is a case in point. Here there is essentially only one annual observation, suggestive of an increase in butterfly abundance following the restoration of a coppicing regime. There is, however, no uncoppiced control, and it is possible that the species would have increased anyway whether or not the coppicing work had been carried out. Many conservation experiments cannot be designed with independent replicates, appropriate controls etc., and hence it becomes increasingly important to find national (or regional) data against which to compare the management results of an individual site.

Such an approach was adopted by Davis (1989), who compared the increase in population of several common butterflies on his landfill site with trends established widely in the geographical area (East Midlands/ East Anglia). This comparison showed that the increases on his site were real, being larger than increases within the region as a whole. Such a comparison highlights the importance of biological recording, assembling national series of statistics not only of distribution but also of year-to-year abundance (Berry, 1988; Crawford *et al.*, 1990). If the records can be regionally or nationally co-ordinated, as for common birds (e.g. Fuller *et al.*, 1985) or for butterflies (referred to in Heath *et al.*, 1984), with appropriate methodologies for recording and calculating index values, they then form an important "control" against which to compare many facets of conservation management. Such background data should be a priority area of insect conservation.

Acknowledgements

We are indebted to the help that we have received from the Regional Officers and Chief Wardens of the 15 NCC regions in collecting data for Appendix I. In particular we should like to thank Adrian Fowles and Dr Roger Key for their overviews of insect conservation in the NCC, and Mrs Jean Waite for typing the MS.

REFERENCES

Bengtsson, J. (1988). Life histories, interspecific competition and regional distribution of three rockpool *Daphnia* species. Doctoral thesis, University of Uppsala.

Berry, R. J. (1988). "Biological Survey: Needs and Network. A Report of a Working Party set up by the Linnean Society of London". PNL Press, London.

Brown, V. K. (1982). The phytophagous insect community and its impact on early successional habitats. *Proc. 5th Int. Symp. Insect–Plant Relationships* 205–213.

Brown, V. K. (1985). Insect herbivores and plant succession. *Oikos* **44**, 17–22.

Brown, V. K. and Southwood, T. R. E. (1983). Trophic diversity, niche breadth and generation times of exopterygote insects in a secondary succession. *Oecologia* **56**, 220–225.

Brown, V. K. Leijn, M. and Stinson, C. S. A. (1987). The experimental manipulation of insect herbivore load by the use of an insecticide (malathion): the effect of application on plant growth. *Oecologia* **72**, 377–381.

BUTT (Butterflies under Threat Team) (1986). The management of chalk grassland for butterflies. *Focus Nat. Conserv.* **17**, 1–80.

Compton, S. G. (1982). Studies of insects associated with *Lotus corniculatus* L. PhD thesis, University of Hull.

Compton, S. G., Beesley, S. G. and Jones, D. A. (1988). Variation in the colour of the keel petals in *Lotus corniculatus* L. 5. Successional differences in the distribution of dark-keeled plants. *Heredity* **61**, 235–245.

Crawford, T. J., Toy, R. and Usher, M. B. (1990). Key indicators for British wildlife: stage 1. *In* "National Perspectives in Biological Recording in the UK" (G. Stansfield and P. T. Harding, eds). National Federation for Biological Recording, Cambridge, pp. 47–61.

Davis, B. N. K. (1989). Habitat creation for butterflies on a landfill site. *Entomologist* **108**. 109–122.

Deane, R. J. L. (1989). "Expanded Field Margins: their Costs to the Farmer and Benefits to Wildlife". Nature Conservancy Council, Peterborough.

Dempster, J. P. (1989). Insect introductions: natural dispersal and population persistence in insects. *Entomologist* **108**, 5–13.

Donnelly, D. and Giliomee, J. H. (1985). Community structure of epigaeic ants in a pine plantation and in newly burnt fynbos. *J. Entomol. Soc. S. Africa* **48**, 259–265.

Ehrlich, P. R. and Murphy, D. D. (1987). Conservation lessons from long-term studies of checkerspot butterflies. *Conserv. Biol.* **1**, 122–131.

Fuller, R. J., Marchant, J. H. and Morgan, R. A. (1985). How representative of agricultural practice in Britain are Common Bird Census farmland plots? *Bird Study* **32**, 56–70.

Gardner, S. M. and Usher, M. B. (1989). Insect abundance on burned and cut upland *Calluna* heath. *Entomologist* **108**, 147–157.

Gibson, C. W. D., Brown, V. K. and Jepson, M. (1987a). Relationships between the effects of insect herbivory and sheep grazing on seasonal changes in an early successional plant community. *Oecologia* **71**, 245–253.

Gibson, C. W. D., Dawkins, H. G., Brown, V. K. and Jepsen, M. (1987b). Spring grazing by sheep: effects on seasonal changes during early old field succession. *Vegetation* **70**, 33–43.

Gibson, C. W. D., Watt, T. A. and Brown, V. K. (1987c). The use of sheep grazing to recreate species-rich grassland from abandoned arable land. *Biol. Conserv.* **42**, 165–183.

Heath, J., Pollard, E. and Thomas, J. A. (1984). "Atlas of Butterflies in Britain and Ireland". Viking, Middlesex.

Jefferson, R. G. (1984). Quarries and wildlife conservation in the Yorkshire Wolds, England. *Biol. Conserv.* **29**, 363–380.

Jefferson, R. G. and Usher, M. B. (1986). Ecological succession and the evaluation of non-climax communities. *In* "Wildlife Conservation Evaluation" (M. B. Usher, ed.), pp. 69–91. Chapman and Hall, London.

Jefferson, R. G. and Usher, M. B. (1987). The seed bank in soils of disused chalk quarries in the Yorkshire Wolds, England: implications for conservation management. *Biol. Conserv.* **42**, 287–302.

Morris, M. G. and Webb, N. R. (1987). The importance of field margins for the conservation of insects. *BCPC Monogr.* **35**, 53–65.

Murphy, D. D. (1983). Nectar sources as constraints on the distribution of egg masses by the checkerspot butterfly, *Euphydryas chalcedona* (Lepidoptera: Nymphalidae). *Environ. Entomol.* **12**, 463–466.

Murphy, D. D. (1984). Butterflies and their nectar plants: the role of the checkerspot butterfly *Euphydryas editha* as a pollen vector. *Oikos* **43**, 113–117.

Murphy, D. D., Launer, A. E. and Ehrlich, P. R. (1983). The role of adult feeding in egg production and population dynamics of the checkerspot butterfly *Euphydryas editha*. *Oecologia* **56**, 257–263.

Murphy, D. D., Menninger, M. S. and Ehrlich, P. R. (1984). Nectar source distribution as a determinant of oviposition host species in *Euphydryas chalcedona*. *Oecologia* **62**, 269–271.

Parr, T. W. (1980). The structure of soil microarthropod communities, with particular reference to ecological succession D Phil thesis, University of York.

Rafe, R. W. and Jefferson, R. G. (1983). The status of *Melanargia galathea* (Lepidoptera: Satyridae) on the Yorkshire Wolds. *Naturalist* (*Hull*) **108**, 3–7.

Ratcliffe, D. A. (ed.) (1977). "A Nature Conservation Review" Vol. 1. Cambridge University Press, Cambridge.

Richards, O. W. (1926). Studies on the ecology of English heaths. III. Animal communities of the felling and burn successions of Oxshott Heath, Surrey. *J. Ecol.* **14**, 244–281.

Schlettwein, C. H. G. and Giliomee, J. H. (1987). Comparison of insect biomass and community structure between fynbos sites of different ages after fire with particular reference to ants, leafhoppers and grasshoppers. *Ann. Univ. Stell. Ser. A3 (Landbouw.)* **2**(2), 1–76.

Shirt, D. B. (ed.) (1987). "British Red Data Books: 2. Insects". Nature Conservancy Council, Peterborough.

Southwood, T. R. E., Brown, V. K. and Reader, P. M. (1979). The relationships of plant and insect diversities in succession. *Biol. J. Linn. Soc.* **12**, 327–348.

Speight, M. C. D. (1989). Saproxylic invertebrates and their conservation. *Council of Europe, Nat. Environ. Ser.* **42**, 1–82.

Thomas, C. D. (1985a). Specializations and polyphagy of *Plebejus argus* (Lepidoptera: Lycaenidae) in North Wales. *Ecol. Entomol.* **10**, 325–340.

Thomas, C. D. (1985b). The status and conservation of the butterfly *Plebejus argus* L. (Lepidoptera: Lycaenidae) in north west Britain. *Biol. Conserv.* **32**, 29–51.

Thomas, J. A., Thomas, C. D., Simcox, D. J. and Clarke, R. T. (1986). Ecology and declining status of the silver-spotted skipper butterfly (*Hesperia comma*) in Britain. *J. Appl. Ecol.* **23**, 365–380.

Usher, M. B. (1979). Natural communities of plants and animals in disused quarries. *J. Environ. Manage.* **8**, 223–236.

Usher, M. B. (1980). An assessment of conservation values within a large Site of Special Scientific Interest in North Yorkshire. *Field Studies* **5**, 323–348.

Usher, M. B. (1987). Creation of a new pond, Hopewell House, North Yorkshire. *In* "Conservation Monitoring and Management" (R. Matthews, ed.), pp. 12–18. Countryside Commission, Cheltenham.

Usher, M. B. and Gardner, S. M. (1988). Animal communities in the uplands: how is naturalness influenced by management? *In* "Ecological Change in the Uplands" (M. B. Usher and D. B. A. Thompson, eds), pp. 75–92. Blackwell, Oxford.

Usher, M. B. and Smart, L. M. (1988). Recolonization of burnt and cut heathland in the North York Moors by Arachnids. *Naturalist* (*Hull*) **113**, 109–111.

Warren, M. S. and Stephens, D. E. A. (1989). Habitat design and management for butterflies. *Entomologist* **108**, 123–134.

Warren, M. S., Thomas, C. D. and Thomas, J. A. (1984). The status of the heath fritillary butterfly *Mellicta athalia* Rott. in Britain. *Biol. Conserv.* **29**, 287–305.

Weiss, S. B. and Murphy, D. D. (undated). Thermal microenvironments and the restoration of butterfly habitat. Manuscript, 9 pp.

Wells, T. C. E. (1989). The re-creation of grassland habitats. *Entomologist* **108**, 97–108.

Wells, T., Bell, S. and Frost, A. (1982) "Creating Attractive Grasslands using Native Plant Species". Nature Conservancy Council, Shrewsbury.

Wells, T. C. E., Cox, R. and Frost, A. (1989). The establishment and management of wildflower meadows. *Focus Nat. Conserv.* **21**, 1–27.

APPENDIX I. Results of a postal survey of the 15 regions of the Nature Conservancy Council asking what aspects of national nature reserve management are primarily targeted at the creation or re-creation of habitats for arthropod conservation. For NCC regions the letters indicate England (E), Scotland (S) and Wales (W).

Taxa	National nature reserve (and region)	Management operations
Invertebrates generally	Cairngorms, Glen Tanar (S, north-east)	Enclosures to encourage natural regeneration of native pine woodland
Invertebrates generally	Creag Meagaidh, Morrone Birkwood (S, north-east)	Enclosures to encourage natural regeneration of native birch woodland, with juniper (*Juniperus communis*) at Morrone
Invertebrates generally	St. Cyrus (S, north-east)	Gorse (*Ulex europaeus*) scrub clearance
Invertebrates generally	Rhum (S, north-west)	Restoration of mixed woodland, periodic monitoring of arthropods
Invertebrates generally	Cairnsmore (S, south-west)	Manage hill pasture by swiping/grazing
Invertebrates generally	Thrislington Plantation (E, north-east)	Magnesian limestone grassland transplantation experiment
Invertebrates generally	Woodwalton Fen (E, East Midlands)	Rotational mowing of fen vegetation
Invertebrates generally	Barnack Hills and Holes, Barton Hills, Castor Hanglands and Saltfleetby-Theddlethorpe Dunes (E, East Midlands)	Grassland management
Invertebrates generally	Mottey Meadows (E, West Midlands)	Leaving uncut grass margins to hay fields
Invertebrates generally	Old Winchester Hill, Beacon Hill, Martin Down, Prescombe Downs, Aston Rowant, and Ashford Hill (E, south)	Invertebrate conservation a major aim of both grassland and scrub management
Invertebrates generally (Lepidoptera in particular)	Tentsmuir Point (S, south-east)	Management of dune pasture and dune slacks; grazing experiments with goats; cutting and herbiciding to control invasive scrub; eradication of conifers and *Hippophaë* (sea buckthorn)

Appendix I cont'd.

Taxa	National nature reserve (and region)	Management operations
Invertebrates generally (Lepidoptera and Odonata in particular)	Taynish (S, south-west)	Control invasive birch and willow scrub on mires; control invasive gorse scrub by cutting interconnecting glades; graze coastal grassland to improve floristic diversity; dam outflows of mires to create areas of open water
Invertebrates generally	Avon Gorge (E, south-west)	Clearance and maintenance of woodland glades; widening of rides
Invertebrates generally	Ebbor Gorge (E, south-west)	Clearance and maintenance of woodland glades; removal of scrub from limestone grassland areas
Invertebrates generally	Rodney Stoke (E, south-west)	Maintenance of old limestone grassland areas by scrub clearance and grazing; widening of rides in woodland
Invertebrates generally	Shapwick Heath (E, south-west)	Maintenance of meadows and woodland fringes by grazing and mowing of old droves
Invertebrates generally	Barrington Hill (E, south-west)	Maintenance of old meadows by grazing and mowing, leaving scrubby fringes with old hedgerows
Invertebrates generally	Braunton Burrows (E, south-west)	Removal of scrub from grassland areas and subsequent mowing; grazing experiment (Soay sheep) in progress
Invertebrates generally	The Lizard (E, south-west)	Grazing of cliff grassland areas by cattle, sheep and ponies; scrub clearance
Invertebrates generally	Yarner Wood (E, south-west)	Clearance and maintenance of woodland glades and widening of rides
Invertebrates generally	Axmouth-Lyme Regis Undercliffs (E, south-west)	Scrub clearance and subsequent mowing of old grassland areas
Invertebrates generally	Hambledon Hill (E, south-west)	Scrub clearance and grazing (cattle/sheep) on rich chalk grassland site; specific works to favour *Lysandra bellargus* (Adonis blue)

Appendix I cont'd.

Taxa	National nature reserve (and region)	Management operations
Invertebrates generally	Hog Cliff (E, south-west)	Scrub clearance and grazing (cattle/sheep) on rich chalk grassland site; specific works to favour *Eurodryas aurinia* (marsh fritillary). Coppicing of old hazel areas in woodland blocks
Invertebrates generally	Holt Heath (E, south-west)	Widening of rides in blocks of old oak woodland
Invertebrates generally	Studland Heath (E, south-west)	Clearance and maintenance of woodland glades
Invertebrates generally	All Dorset heathlanc reserves (E, south-west)	Removal of encroaching pines and birch from areas of open heath
Invertebrates generally	Highbury (E, West Midlands)	Coppicing
Invertebrates generally	Wyre Forest (E, West Midlands)	Coppicing
Invertebrates generally (Odonata at the ponds)	Beinn Eighe (S, north-west)	Management of glades, restoration of woodland and creation of ponds
Invertebrates generally (Odonata and Lepidoptera in particular)	Ariundle (S, north-west)	Management of glades and rides and creation of ponds
Invertebrates generally (Coleoptera and Diptera in particular)	Duncombe Park (E, north-east)	Management of old timber for saproxylic invertebrates including *Ischnomera cinerascens* and other species of Cerambycidae. Retention of over-mature trees and dead fallen branches and replanting of native tree species for the future
Invertebrates generally (particularly Odonata, Lepidoptera – *Coenonympha tullia* – and Araneae – *Dolomedes fimbriatus*)	Glasson Moss, Roudsea Moss, Thorne Moors (E north-west)	Creation and re-creation of lowland raised mire and open water habitats by damming old drains in peat, creating lagoons and clearing scrub, i.e. generally getting the water table nearer the surface and minimizing water loss

Appendix I cont'd.

Taxa	National nature reserve (and region)	Management operation
Invertebrates generally (Coleoptera in particular)	Moccas Park (E, West Midlands)	Management of old timber for saproxylic invertebrates particularly *Pyrrhidium sanguineum* (Cerambycidae), *Hypebeaus flavipes* (Melyridae) and *Ernoporus caucasicus* (Scolytidae). Includes retention of over-mature trees, stacking of large branches and boles in shade and replanting for the future
Dune invertebrates, especially *Argynnis aglaia* (dark green fritillary) and day-flying moths	Newborough Warren (W, north)	Sheep grazing of dune and slack communities to recover short swards previously maintained by high rabbit population densities
Freshwater invertebrates	Loch Leven (S, south-east)	Re-creation of lagoons and protection of sheltered areas to allow re-creation of reed (*Phragmites australis*); areas for wildfowl feeding
Freshwater invertebrates	Cairnsmore (S, south-west)	Clear and open hill drains; remove *Molinia* litter
Freshwater invertebrates	Ainsdale Sand Dunes (E, north-west)	Creation of permanent and seasonal pools in dune system
Freshwater invertebrates (Odonata in particular)	Coed Llety Walter (NCC and Woodland Trust) (W, north)	Repairs to dam to recover large water body with fringing woodland
Freshwater invertebrates (Odonata in particular)	Cors Erddreiniog (W, north)	Excavation of small ponds in peat body with depressed water table (*Molinia–Myrica* community)
Freshwater invertebrates (*Neuronia clathrata* in particular)	Chartley Moss (E, West Midlands)	Creation/retention of open water in pools/ditches
Brackish water invertebrates	North Solent (E, south)	Reverse trend for brackish lagoons to become freshwater lagoons
Odonata	Dunnet Links (S, north-west)	Enlargement of ponds
Odonata	Glen Strathfarrar (S, north-west)	Creation of ponds

Appendix I cont'd.

Taxa	National nature reserve (and region)	Management operations
Odonata	Rhum (S, north-west)	Two ponds dug in 1983, one in boggy area surrounded by pines (seven breeding species by 1989) and one less well sheltered (four or five breeding species); control of invasive scrub. Also possible introduction sites for common frog and/or toad
Odonata	Cairnsmore (S, south-west)	Create hunting/hawking runs for dragonflies
Odonata	Loch Lomond (S, south-west)	Creation of pool (*ca* 260 cm^2) in 1986, sheltered from the west but exposed to sunshine; attempt to control Canadian waterweed (*Elodea canadensis*); temporary fencing needed to keep out cattle
Odonata	Moine Mhor (S, south-west)	Damming of old drainage ditches to increase breeding sites
Odonata	Cors Bodeilio, Cors Geirch (W, north)	Pond excavation
Odonata	Dyfi (W, Dyfed, Powys)	Dam creation to re-create wet raised mire. The series of pools are ideal for breeding Odonata
Odonata	Cors-y-Llyn (W, Dyfed, Powys)	Excavation of a pond in a wet meadow and peat cutting
Odonata	Newham Fen (E, north-east)	Creation of additional deeper pools in the mire as refugia for Odonata
Odonata	Saltfleetby-Theddlethorpe Dunes (E, East Midlands)	Enlargement of series of bomb craters as breeding sites for dragonflies
Odonata	Bure Marshes (E, East Anglia)	Management of the reserve's dyke system
Odonata	Thursley Common (E, south-east)	Provision and management of a variety of aquatic habitats for a range of species of Odonata
Odonata	Avon Gorge (S, south-west)	Excavation and maintenance of old woodland pond

Appendix I cont'd.

Taxa	National nature reserve (and region)	Management operations
Odonata	Gordano Valley (E, south-west)	Excavation and maintenance of ditch system; clearance of encroaching scrub from reedbeds
Odonata	Ebbor Gorge (E, south-west)	Maintenance of small pond
Odonata	Shapwick Heath (E. south-west)	Excavation and maintenance of ditch system; upkeep of small ponds
Odonata	Braunton Burrows (E, south-west)	Excavation and maintenance of ponds
Odonata	The Lizard (E, south-west)	Excavation and maintenance of ponds/scrapes
Odonata	Dorset Heaths (5 NNRs) (E, south-west)	Creation and management of heathland pools principally for firefighting purposes
Odonata	North Solent (E, south)	Opening up 16 old freshwater ponds which have become silted up or shaded over; aim is to re-establish habitats for a suite of species, especially *Sympetrum sanguineum*
Odonata	Chartley Moss (E, West Midlands)	Ditch maintenance
Orthoptera	Castle Hill (E, south-east)	Grassland management for *Decticus verrucivorus* (Tettigoniidae)
Orthoptera	Stoborough Heath (E, south-west)	Reintroduction of grazing to favour *Decticus verrucivorus* (Tettigoniidae)
Hemiptera	Aston Rowant (E, south)	Re-creation of open, stony ground, especially to conserve *Hallodopus montadoni* and *Taphropeltus hamulatus*
Lepidoptera	Dunnet Links (S, north-west)	Management of glades and rides
Lepidoptera	Whitlaw Mosses (S, south-east)	Management of peripheral grassland
Lepidoptera	Morton Lochs (S, south-east)	Creation of clearings in secondary woodland; bracken (*Pteridium aquilinum*) control
Lepidoptera	Cairnsmore (S, south-west)	Spray bracken to create clearings; manage pasture to create herb-rich communities

Appendix I cont'd.

Taxa	National nature reserve (and region)	Management operations
Lepidoptera	Coed Gorswen (W, north)	Ride and glade creation
Lepidoptera	Walberswick (E, East Anglia)	Management of stands of reed (*Phragmites australis*) for Lepidoptera dependent on this as a food plant
Lepidoptera	Wyre Forest (E, West Midlands)	Coppice management for butterflies, particularly *Argynnis adippe* (high brown fritillary); encouragement of young birch (*Betula* spp.) for *Endromis versicolora* (Kentish glory)
Lepidoptera	Thursley Common (E, south-east)	Maintenance of a variety of age classes of invasive *Betula pendula*
Papilio machaon (swallowtail)	Bure Marshes, Hickling Broad (E, East Anglia)	Cutting sedge stands and invading scrub
Strymonidia pruni (black hairstreak)	Monks Wood (E, East Midlands)	Manage stands of blackthorn (*Prunus spinosa*); small amounts of annual cutting if necessary; careful timing of cuts so as not to destroy eggs
Strymonidia pruni (black hairstreak)	Castor Hanglands (E, East Midlands)	Blackthorn (*Prunus spinosa*) scrub to be managed primarily for *S. pruni* wherever it occurs
Plebejus argus (silver-studded blue)	Westleton Heath (E, East Anglia)	Management of *Calluna* stands
Agricia artaxerxes (northern brown argus)	Cragbank Wood (S, south-east)	Management of peripheral grassland
Lycaena dispar (large copper)	Woodwalten Fen (E, East Midlands)	Planting/seeding of great water dock (*Rumex hydrolapathum*) to retain dock-rich fen
Argynnis adippe (High brown fritillary), *Boloria euphrosyne* (pearl-bordered fritillary)	Gait Barrows (E, north-west)	Re-create coppice woodland; 400 m of ride widening and 0.33 ha of coppicing annually
Mellicta athalia (heath fritillary)	Blean Woods (E, south-east)	Re-creation of open, coppiced woodland to encourage common cow-wheat (*Melampyrum pratense*), the larval foodplant
Eurodryas aurinia (marsh fritillary)	Cors Geirch (W, north)	Grazing of fen meadow with Welsh mountain ponies

Appendix I cont'd.

Taxa	National nature reserve (and region)	Management operations
Eurodryas aurinia (marsh fritillary)	Dyfi (W, Dyfed, Powys)	Re-creation of herb-rich meadows with devil's-bit scabious (*Succisa pratensis*); management techniques include grazing (not very successful) and rotovation followed by re-seeding
Eurodryas aurinia (marsh fritillary)	Rhos Llawr Cwrt (W, Dyfed, Powys)	Restoration of meadow by cattle and pony grazing
Eurodryas aurinia (marsh fritillary)	Findlandrigg Wood (E, north-west)	Habitat management (mowing, winter fires) to re-create open habitats to promote devil's-bit scabious (*Succisa pratensis*)
Carterocephalus palaemon (chequered skipper)	Loch Sunart Woodlands (S, north-west)	Maintenance and enlargement of rides and glades by scrub and tree removal
Epione paralellaria (dark bordered beauty)	Newham Fen (E, north-east)	Maintenance of larval food plant (*Salix repens*) and fen carr habitat within constraints imposed by other management objectives
Siona lineata (black veined moth)	Wye (E, south-east)	Grassland management
Hymenoptera: Aculeata	Lower Derwent Valley (E, north-east)	Proposed creation of south-facing sandy "cliff" faces
Hymenoptera: Aculeata	Walberswick (E, East Anglia)	Re-excavation of sand pits to create sandy cliff faces
Hymenoptera: Aculeata	Thursley Common (E, south-east)	Creation and maintenance of sandy patches within *Calluna* heath. Creation of south-facing sandy cliffs (*ca* 1.5 m high) for sand wasps
Coleoptera	Newham Fen (E, north-east)	Creation of additional deeper pools in the mire as refugia for *Agabus uliginosus* (Dytiscidae)
Coleoptera	Thursley Common (E, south-east)	Management of heathland pools for *Graptodytes flavipes* (Dytiscidae)

Appendix I cont'd.

Taxa	National nature reserve (and region)	Management operations
Araneae	Thursley Common (E, south-east)	Diversification of the age structure of *Calluna* heath by mowing of discontinuous 1–2 m strips
Araneae: *Singa hamata*	Monks Wood (E, East Midlands)	Hand removal of developing scrub in rough grassland about every 5 years; if grass needs to be cut, only 25% of field in a year and no cutting in successive years

12

The Conservation of Insects on Arable Farmland

J. W. DOVER

I. Introduction ... 294
II. Background to the Development of Conservation Headlands ... 294
A. Gamebird Population Trends, Mortality and Pesticides ... 294
B. Philosophy underlying the Development of Conservation Headlands ... 295
III. Design and Agronomy of Conservation Headlands ... 296
A. Guidelines for the use of Conservation Headlands ... 296
B. Selective Weed Control ... 296
C. Preventing Weed Invasion from Boundary Vegetation ... 297
D. Screening Fungicides for Insecticidal Activity ... 297
E. Reduced Spray Drift ... 298
F. The Costs of using Conservation Headlands ... 298
G. Reduced Fertilizer Use in Conservation Headlands ... 298
IV. Effect of Conservation Headlands on Chick-food Insects and Chick Survival ... 299
A. Chick-food Insect Abundance ... 299
B. Gamebird Brood Size and Population Trends ... 299
C. Insect Conservation ... 301
V. Butterflies and Conservation Headlands ... 301
A. Butterfly Transects ... 301
B. Population Trends ... 304
C. Butterfly Host Plants in Conservation Headlands ... 305
D. Observations of Butterfly Behaviour in Field Margins ... 306
E. Movement Patterns of Satyrid Species on Farmland ... 309
VI. Conservation headlands as a Tool in Conservation ... 312
A. Benefits to Other Species ... 312
B. Conservation Headlands as Buffers against Spray Drift ... 313
C. Wildlife Corridors and Conservation Headlands ... 313
VII. Conclusions ... 314
References ... 315

I. INTRODUCTION

The conservation of insects on farmland is a relatively new subject. Concern for the effects of intensive agriculture on farmland wildlife has been recently, and effectively, coupled with concern at the scale of agricultural overproduction to produce a powerful argument in favour of the deintensification of farming (Section III.G). However, the financial resources allotted to nature conservation have largely been used to save some of the premier wildlife habitats which have been disappearing at an alarming rate since 1949 (Newbold, 1989), many as a result of the post-war intensification of agriculture. In consequence, insect conservation on farmland has developed from the pragmatic requirements of the farming community itself; not from a desire to preserve insects *per se*, but because they are central to the management of wild, but cropped species — such as game birds — or because insect predators and parasitoids have the potential to control pest species. However, the management practices developed for pragmatic reasons may have major benefits for other farmland wildlife, and may be adopted for altruistic ends.

This paper reports on The Cereals and Gamebirds Research Project of the Game Conservancy Trust. The principal aim of this project was to devise a practical management system whereby the decline in wild grey partridge (*Perdix perdix* L.) populations could be halted and reversed. The management system has become known as "conservation headlands". The development of the conservation headland concept, its effects on the insect food of the partridge, on partridge populations and on beneficial and aesthetic insect species (with particular reference to butterflies), and its potential for use in nature conservation are described.

II. BACKGROUND TO THE DEVELOPMENT OF CONSERVATION HEADLANDS

A. Gamebird Population Trends, Mortality and Pesticides

In Britain, the wild population of the partridge has declined by approximately 80% since the 1940s (Potts, 1986). Work by Blank *et al.* (1967) and Potts (1980) showed that the key factor in this decline was chick mortality, and that this was related to pesticide use. The use of synthetic pesticides since the late 1940s has escalated with a low percentage of cereals sprayed with herbicides in 1950 (13%) to 63% in 1960 (Woodford,

1964) and in excess of 90% by the early 1970s (Potts, 1986). The number of compounds registered for use as pesticides has also increased from just 11 in 1957 to over 80 in 1985 (Sotherton, 1989).

For the first 21 days of life the chicks of *P. perdix* require arthropod food — principally insects — whilst their gut systems develop the ability to process plant material (Green, 1984; Green *et al.*, 1987). Pesticides increase chick mortality by disrupting the food chain, reducing the availability of arthropod food (Southwood and Cross, 1969; Potts, 1986).

Not all insects in the cereal ecosystem are suitable or utilized by partridge chicks. There are three main groups of "preferred" chick-food items: Hemiptera — the heteropteran bugs, in particular *Culocoris* spp.; Coleoptera — the chrysomelid, curculonid and small diurnal carabid beetles; and larvae of the Tenthredinidae (sawflies especially *Dolerus* spp.) and Lepidoptera (Potts, 1986).

Insecticides, herbicides and fungicides are active in the reduction of insect abundance. Insecticides and some fungicides which have an insecticidal action (such as the compound pyrazophos — see Section III.D below) kill these non-target insects as well as the pest species. Herbicides have a major, but indirect effect by removing the host plants of these non-target species (Rands, 1986).

A reduction in food availability means that chicks have to forage over a larger area (Rands, 1986) and poor nutrition results in lower survival and increased susceptibility to predators (Potts, 1986).

B. Philosophy Underlying the Development of Conservation Headlands

It is clear that if chick mortality was mediated by poor food availability, then the only way to reduce this mortality would be to provide more food. The only way to provide more chick-food insects in the arable environment would be to reduce the impact of pesticides. However the wholesale abandonment of agricultural advances achieved over several decades is clearly not an acceptable option.

Green (1984) showed that the grey partridge had a preferred feeding area — the field margin (*sensu* Greaves and Marshall, 1987). Field margins tend to be lower yielding than the rest of the field because of soil compaction, shading and other competitive effects. The reduction in pesticide use on such an area would have the maximum benefit to the quarry species (being close to the nesting cover and preferred feeding area) and reduce yields less than in the centre of the field.

This essentially was the motivation behind the development of conservation headlands: the need to reduce the impact of modern farming systems on a wild, but cropped, species in a manner acceptable to the farming community with as little yield loss as possible. Moreover, the technique had to be reasonably simple to establish in practice.

III. DESIGN AND AGRONOMY OF CONSERVATION HEADLANDS

A. Guidelines for the Use of Conservation Headlands

Conservation headlands are the outer 6-m-wide edges of cereal fields which receive reduced and selective pesticide inputs. The concept is under continuous development (*cf.* Rands, 1985; Boatman and Sotherton, 1988) since the technique depends on the *management* of the outer 6 m of the cereal crop. Hence, the effects of new pesticides on the insect fauna of cereal fields need to be quantified as do novel uses and combinations of those already available. The 1989–1990 guidelines for establishing conservation headlands in cereal fields are given in Table I, and may be generally summarized as no wide-spectrum broadleaved herbicide and no insecticides after 15 March in the year of harvest (Boatman and Sotherton, 1988).

B. Selective Weed Control

Certain grass weed herbicides that have been examined for their selective action are recommended for use to reduce unwarranted yield loss and agronomic problems (Boatman, 1987; in press). Grass weeds do not

TABLE I. Summary of the 1989–1990 guidelines for the selective use of pesticides on conservation headlands.

	Autumn spraying	Spring spraying
Insecticides	Yes (avoiding drift into hedgerows)	No
Fungicides	Yes	Yes (except pyrazophos)
Growth regulators	Yes	Yes
Herbicides		
Grass weeds	Yes	Yes
	(but only those chemicals approved for use)	
Broadleaved weeds	No	No
	(except for selective control of certain species, e.g. cleavers)	

significantly contribute to the insect fauna that conservation headlands were developed to encourage.

Specific problem broadleaved weeds such as cleavers (*Galium aparine* L.) can cause unacceptable reductions in yield, together with increased harvesting and grain cleaning costs (Boatman and Sotherton, 1988). Selective compounds have been screened in field trials, resulting in dosage and timing recommendations which allow conservation headlands to be established where this weed occurs and for use in "fire-brigade" actions (Boatman *et al.*, 1988; Boatman, 1989; Bain and Boatman, 1989).

C. Preventing Weed Invasion from Boundary Vegetation

Several weed species such as cleavers and barren brome (*Bromus sterilis* L.) are thought to invade the crop from reservoirs in the boundary vegetation. In order to prevent the invasion of these species into conservation headlands, a 1-m-wide "sterile strip" is created by applying a persistent broad-spectrum herbicide between the field boundary vegetation and the crop (Boatman and Wilson, 1988). The herbicide is applied using a tractor-mounted, hooded, flat fan nozzle to minimize spray drift.

D. Screening Fungicides for Insecticidal Activity

It has been known for some time that certain fungicides also exert an insecticidal action (see references in Sotherton and Moreby, 1988). As the concept of conservation headlands is designed around the requirement not to kill "beneficial" or "non-target" organisms in the headland area, information on the insecticidal activity of fungicides against the non-pest fauna was essential. Of 27 fungicidal compounds screened for insecticidal activity, only one, pyrazophos, was shown to be toxic (Sotherton *et al.*, 1987; Sotherton and Rands, 1987; Sotherton and Moreby, 1988). The toxicity of this particular compound rivals that of some commercial insecticides (Sotherton *et al.*, 1988), hence pyrazophos is specifically not recommended for use on conservation headlands.

The use of insecticidal compounds, including pyrazophos, on headlands may also depress the natural pest control exerted by polyphagous and aphid specific predators on cereal aphids. Sotherton and Moreby (1988) found that aphid pest numbers exceeded the economic threshold (i.e. the point below which it costs more to spray insecticide than to suffer a slight yield loss) on winter wheat plots sprayed with the compound pyrazophos whereas the pests remained below the economic threshold on plots which did not receive this nominally fungicidal compound.

E. Reduced Spray Drift

The use of conservation headlands also reduces the potential for pesticide spray drift into boundary vegetation. This benefit accrues as a result of leaving off certain compounds, such as wide-spectrum broadleaved herbicides. The results (Dover and Cuthbertson, 1989) also indicate the extreme care which must be taken when spraying around field margins up to the crop edge if the drift of toxic compounds into the boundary vegetation is to be minimized.

F. The Costs of using Conservation Headlands

The yield loss resulting from the use of conservation headlands is approximately 5–10% of the normal headland yield (Boatman and Sotherton, 1988). There are additional, perhaps less obvious, costs such as increased expenditure on selective compounds, separate harvesting and storage of grain, seed cleaning, increased drying costs etc. However these costs are dependent on the specific cereal crop and on the market for which it was grown (e.g. animal feed or seed). These costs can be calculated and have been examined in detail by Boatman and Sotherton (1988) and Boatman (1989). Such information is essential for farm managers and conservation organizations who need to know the management, husbandry and fiscal costs as well as the environmental benefits of managing farmland for nature conservation.

G. Reduced Fertilizer Use in Conservation Headlands

Further development of the conservation headland concept is possible. Eutrophication is one of the major causes of loss of diversity in plant communities (Green, 1972). A reduction in the use of fertilizers may have important effects on the non-crop flora of cereal fields particularly rare arable weeds (Boatman and Wilson, 1989) and also in reduced drift of fertilizer into the boundary vegetation. Boatman and Wilson (1989) report some preliminary findings of reduced fertilizer use in conservation headlands. A combined reduction in pesticide and fertilizer inputs at the margins of cereal fields could have substantial wildlife benefits. Reduced inputs of nitrogenous fertilizer are likely to feature in the extensification scheme currently being developed by the European Community. This scheme differs from set-aside schemes in that farmers are encouraged to continue farming under extensification, but less intensively and with lower inputs, whereas land is taken out of production under set-aside with dubious wildlife benefits (Boatman and Wilson, 1989).

IV. EFFECT OF CONSERVATION HEADLANDS ON CHICK-FOOD INSECTS AND CHICK SURVIVAL

A. Chick-food Insect Abundance

In order to test the effect of reduced pesticide inputs on chick-food items, and also the resulting impact on grey partridge brood sizes, large-scale field trials were undertaken. For example in 1983, an experiment was carried out on an 11-km^2 arable farm in Hampshire; 65% of the farm was in cereal crops (521 ha). In this experiment (Rands, 1985), 37 fields were used. These were divided into three blocks with two treatment plots (fields with fully sprayed or conservation headlands) per block.

In the above example chick-food insects, sampled using sweep nets, were significantly more abundant in conservation headlands (Table II) (Rands, 1985). The following year, a larger number of headlands were sampled from three farms showing broadly similar results (Table II) (Sotherton *et al.*, 1985).

B. Gamebird Brood Sizes and Population Trends

Grey partridge brood sizes were assessed in the late summer over a number of years on farms with fully sprayed and conservation headlands, by determining the size and composition of coveys by field counts. The results for 1983–1986 in Hampshire and eastern England are given in Table III for the grey partridge and for the pheasant (*Phasianus colchicus* L.) (1984–1986), whose chicks probably also require insect food (Hill 1985; Rands and Sotherton, 1991). The results show that brood size for both of these species was significantly greater in fields which had conservation headlands rather than fully sprayed ones.

Using radio-tracking techniques, Rands (1986) also found that in fields which had margins with conservation headlands, grey partridge broods

TABLE II. The abundance of wild gamebird chick food insects in sprayed and conservation headlands during June on a Hampshire farm (1983) and on three study farms (1984).

	1983 Mean (± SE) prey items per 50 sweeps		1984 Mean (± SE) prey items per 20 sweeps	
Prey species	Sprayed headlands	Conservation headlands	Sprayed headlands	Conservation headlands
Heteroptera	53.9 ± 21.4	163.2 ± 44.8**	26.5 ± 5.8	49.2 ± 8.8*
Larval Tenthredinidae and Lepidoptera	3.5 ± 1.0	5.0 ± 1.0 NS	3.7 ± 1.2	4.4 ± 0.8**
Chrysomelidae, Curculonidae and small Carabidae	4.2 ± 0.5	12.3 ± 2.8*	5.4 ± 0.9	6.5 ± 1.3 NS
Total prey	62.2 ± 21.5	180.0 ± 45.2***	34.4 ± 6.2	60.4 ± 9.6*
No. headlands sampled	19	18	46	42

NS = $p > 0.05$, * $p < 0.05$, ** $p < 0.02$, *** $p < 0.001$.
From Sotherton *et al.* (1985).

TABLE III. Grey partridge and pheasant brood sizes on blocks of cereal fields with sprayed and conservation headlands in Hampshire and eastern England.

		Mean brood size (± 1 SE)			
		Grey partridge		Pheasant	
Study area	Year	Sprayed headlands	Conservation headlands	Sprayed headlands	Conservation headlands
	1983	4.7 ± 0.5	8.4 ± 0.4**		
	1984	7.4 ± 0.8	10.0 ± 0.6**	3.2 ± 0.5	6.9 ± 0.5***
Hampshire	1985	3.3 ± 1.1	6.2 ± 1.2*	3.0 ± 1.0	4.6 ± 0.6*
	1986	5.9 ± 1.6	6.2 ± 1.0 NS	2.0 ± 0.5	5.9 ± 0.7**
Eastern	1984	4.7 ± 0.3	7.8 ± 0.6***		
England	1985	2.7 ± 0.4	4.0 ± 0.7*	2.6 ± 0.3	3.7 ± 0.4**
	1986	4.8 ± 0.6	8.7 ± 1.5***	3.4 ± 0.6	3.5 ± 0.7 NS

NS = $p > 0.05$, * $p < 0.05$, ** $p < 0.02$, *** $p < 0.001$.
From Sotherton *et al.* (1989).

had smaller home ranges with a higher proportion of the range made up of the headland.

The spring pair densities of grey partridges at the Hampshire study farm showed a substantial increase over the 1984–1986 period compared with other nearby estates (Rands, 1991). By 1986, the spring breeding densities of the grey partridge at the study farm had increased from 4 pairs km^{-1} in 1983 to 12 pairs km^{-1}, which was close to the highest historical figures for the farm. The results elegantly demonstrated the potential of conservation headlands in reversing the pesticide-mediated decline in Grey partridge populations.

C. Insect Conservation

Experiments using conservation headlands (above) have demonstrated that insect conservation through the manipulation of pesticide use at the cereal field margin (the headland) is possible. It is clearly not feasible to carry out a study of the effect of conservation headlands on every insect species found in field margins. However the impact on chick-food items may be considered to be an indication of the substantial benefits to insect conservation which may be obtained using conservation headlands. However, one group of insects, of little or no importance in the diet of gamebird chicks, has been studied in connection with conservation headlands over a number of years — the butterflies.

V. BUTTERFLIES AND CONSERVATION HEADLANDS

A. Butterfly Transects

In order to compare the number of butterflies seen in field margins with fully sprayed and conservation headlands, Rands and Sotherton (1986) devised a modification of the butterfly transect technique (Pollard *et al.*, 1975). Fourteen field margins from the two headland regimes were paired up on the basis of the habitat-edge type (e.g. hedgerow, grass bank, wood edge etc.), orientation, approximate length and, wherever possible, cereal crop. The transect route was walked once a week throughout the summer and the numbers of each butterfly species observed in the field boundary and adjacent headland recorded for each of the 14 field margins.

The results of the transect comparisons for 1984–1988 (Dover, 1991; Dover *et al.*, 1990) Dover, unpublished data) are presented in Table IV. The principal results show that, overall, there tend to be more species of butterfly and more individual butterflies in field margins which have conservation headlands. Also, where the statistical comparison of individual species abundance between the two regimes was possible, in only 3.4% of cases were there significantly more butterflies in field margins

TABLE IV. The number of butterflies (per km) seen over fully sprayed and conservation headlands, Hampshire, 1984–1988.

	1984		1985		1986		1987		1988	
	C	S	C	S	C	S	C	S	C	S
Hesperiidae										
Ochlodes venata (B. & G.)	4.7	0.3***	4.8	1.7^{NS}	1.3	2.8^{NS}	2.5	0.5^{NT}	4.2	0.84*
Thymelicus lineola (Ochs.)	11.4	0.6***	2.9	0.4^{NT}	0.8	1.9^{NS}	—	—	0.4	0.0^{NT}
+ *Thymelicus sylvestris* (Poda)										
Lycaenidae										
Celastrina argiolus (L.)	8.1	3.9*	1.5	1.7^{NT}	—	—	—	—	—	—
Lycaena phlaeas (L.)	1.1	0.3^{NT}	—	—	—	—	—	—	—	—
Polyommatus icarus (Rott.)	5.0	0.3***	—	—	0.3	0.0^{NT}	—	—	—	—
Nymphalidae										
Aglais urticae (L.)	26.7	11.6***	61.4	30.4***	50.4	39.4*	28.5	5.9***	28.1	17.6*
Argynnis paphia (L.)	0.8	0.0^{NT}	—	—	—	—	—	—	—	—
Inachisio (L.)	10.6	5.0***	11.1	2.6***	1.0	0.3^{NT}	5.4	0.9**	2.7	0.4^{NT}
Ladoga camilla (L.)	0.3	0.0^{NT}	—	—	—	—	—	—	—	—
Polygonia c-album (L.)	—	—	0.5	0.0^{NT}	0.8	0.6^{NT}	0.4	0.0^{NT}	—	—
Vanessa atalanta (L.)	0.0	0.3^{NT}	0.0	0.9^{NT}	1.5	0.0^{NT}	1.7	0.0^{NT}	0.0	2.1^{NT}
Vanessa cardui (L.)	—	—	3.4	2.6^{NS}	5.2	2.2^{NT}	0.0	0.5^{NT}	8.1	1.3***

Pieridae										
Anthocharis cardamines (L.)	3.1	0.0***	0.5	0.4^{NT}	2.8	3.6^{NS}	1.7	0.9^{NT}	1.5	0.0^{NT}
Gonepteryx rhamni (L.)	10.0	1.7***	6.8	1.7**	0.8	0.6^{NT}	1.2	0.0^{NT}	0.8	0.0^{NT}
Pieris brassicae (L.)	15.6	10.5^{NS}	3.9	6.1^{NS}	178.5	92.9***	42.6	18.2***	46.2	36.8^{NS}
Pieris napi (L.)	37.3	5.8***	68.7	51.3*	66.9	14.9***	17.0	3.7***	10.0	8.8^{NS}
Pieris rapae (L.)	4.7	3.3^{NS}	1.0	1.3^{NT}	66.9	17.4***	2.9	0.5^{NT}	20.8	6.3***
Satyridae										
Aphantopus hyperantus (L.)	14.5	6.4***	8.2	2.6*	5.7	19.3***	18.2	16.4^{NS}	16.2	22.2^{NS}
Coenonympha pamphilus (L.)	3.1	0.0**	—	—	—	—	—	—	—	—
Lasiommata megera (L.)	2.5	1.1^{NS}	—	—	—	—	—	—	—	—
Maniola jurtina (L.)	33.7	12.7***	51.7	13.5***	52.7	38.6**	69.9	19.6***	61.7	31.8***
Melanargia galathea (L.)	12.5	0.0***	1.0	0.0^{NT}	0.5	0.3^{NT}	0.0	0.5^{NT}	1.2	0.0^{NT}
Pararge aegerta (L.)	2.8	0.6*	0.5	0.9^{NT}	—	—	0.0	2.3^{NT}	0.4	0.0^{NT}
Pyronia tithonus (L.)	25.9	16.0***	51.7	8.7***	36.2	24.6**	30.2	12.8***	30.1	21.3^{NS}
Total butterflies	234.4	80.4***	279.6	126.8***	472.3	259.4***	222.2	82.7***	232.4	149.4***
% in C and S	74.5	25.5	68.2	31.8	64.6	35.4	72.9	27.1	60.9	39.1
Transect length (m)	3594	3622	2068	2300	4005	3626	2418	2193	2595	2390

C = conservation headlands; S = fully sprayed headlands. NT = no test; NS = $p > 0.05$, * $p < 0.05$, ** $p < 0.01$, *** $p < 0.001$.
From Sotherton *et al.* (1989).

with fully sprayed headlands compared with 96.6% of cases showing more butterflies in field margins with conservation headlands.

There was a reduction in the number of species of butterfly seen during transects over the 5 years of the study. This was most likely to be due to a reduction in the number of edge habitat types from four in 1984 to two in 1988 coupled with poor summers during 1985–1988 (Dover, 1991, Dover *et al.*, 1990).

B. Population Trends

In order to examine the impact of using conservation headlands on butterfly populations at the study farm, a reference population trend was required — in this case the south and south-east (SSE) regional data for the Institute of Terrestrial Ecology's National Butterfly Monitoring Scheme (NBMS) (Pollard *et al.*, 1986). Data were obtained for this region for the years 1984–1988. In order to compare the population trends at the Hampshire study farm with those from the NBMS, a composite index value was obtained for the cereal fields by using the mean of the numbers of butterflies seen per km in both headland regimes. For each of six butterfly species (three satyrids and three pierids) the population index for the initial year (1984) was set at 100, and deviations from this initial value were plotted in subsequent years. The NBMS data were treated in the same manner, with the initial 1984 value being set at 100. It should be noted that data from the Game Conservancy's butterfly transects were from different transect routes each year, because of crop rotations, whilst the NBMS data was from the same transect route each year. Some caution must therefore be exercised in the comparison of the two data sets (Fig. 1, 1984–1987 from Dover *et al.*, 1990, unpublished data 1988).

As the three pierid species have open population structures and are highly mobile species (especially *Pieris rapae* L. and *Pieris brassicae* L.) it was not expected that it would be possible to distinguish departures from the regional trends, as any increases in population size due to conservation headlands would be quickly diluted by dispersal. The data presented in Fig. 1 broadly reflect this expectation. Although the amplitude of the population trends is large for *P. rapae* and *P. brassicae*, the general trends are quite compatible. The differences in amplitude probably arise from the NBMS data being collected largely from nature reserves whilst the experimental data was collected from an arable farm. The more sedentary species with closed population structures were however expected to show differences in trends if conservation headlands were having some beneficial effect. This seems to be the case, with marked declines from

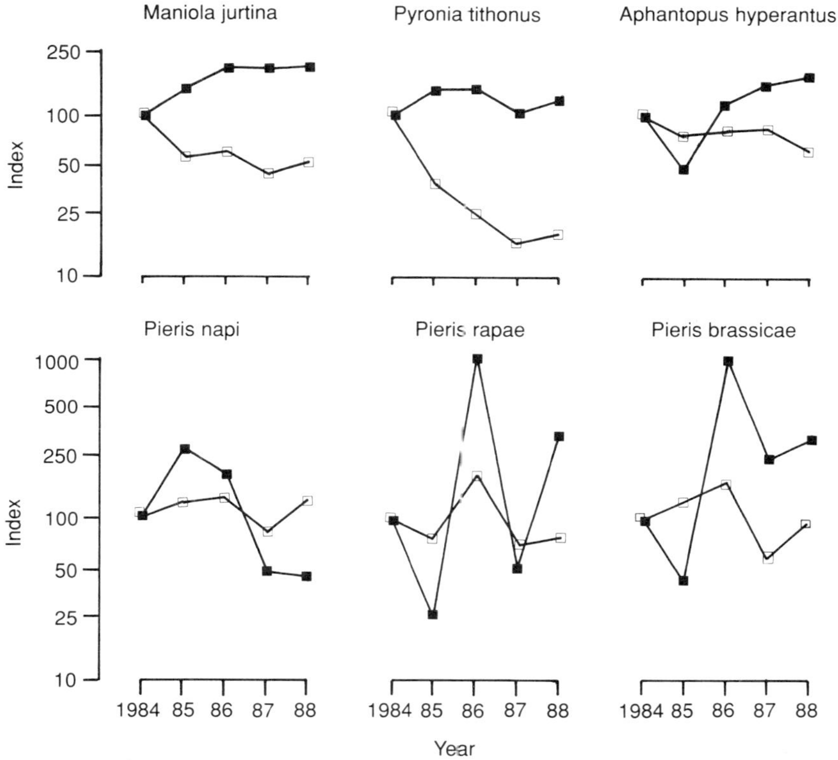

Fig. 1. The relative index of abundance for six butterfly species at a farm with conservation headlands (closed squares) and the south–south-east region of the National Butterfly Monitoring Scheme (open squares) (1984–1988). Index of abundance starting from an arbitrary value of 100 in 1984. Note log scale. From Dover *et al.* (1990).

the initial 1984 value for the regional data, whilst the Hampshire study farm has generally shown increases over this initial level.

C. Butterfly Host Plants in Conservation Headlands

The host plants of some of the common pierid species, charlock (*Sinapis arvensis* L.) and rape (*Brassica napus* L.), are present in conservation headlands (Dover *et al.*, 1990) and are exploited as larval foodplants. Some of the increased numbers of such species (*P. napi*, *P. rapae* and *Anthocharis cardamines* (L.)) in field margins with conservation headlands may arise partly because of this (Dover *et al.*, 1990). However, the host plants of the satyrid, hesperiid, nymphalid and lycaenid butterflies are not found in conservation headlands. Indeed, the grasses used by satyrid and hesperiid butterflies are selectively controlled using herbicides. It is fortunate that their foodplants are eliminated by the initial spraying of headlands, because all these species of satyrids and hesperids overwinter on or at the base of their host grasses. Any laid on grasses in the conservation headland would have been destroyed in late summer or autumn following harvest and ploughing.

D. Observations of Butterfly Behaviour in Field Margins

Observations were made of butterfly behaviour in order to examine the reasons underlying the increased abundance of butterflies in field margins with conservation headlands. Study areas were established both in field margins with conservation and fully sprayed headlands adjacent to short hedges (approx. 2 m high). Butterflies were followed through these field margins and their behaviour, categorized as flight, feeding, resting (basking with wings open and settled with wings closed), interacting (with other butterflies), ovipositing and walking over vegetation, was dictated into a small cassette tape recorder. In addition, the location of the behaviour was recorded as was any species of plant visited for nectar. These data were subsequently timed and transcribed (Dover, 1989a) and behaviour profiles constructed (Dover, unpublished data).

For some species such as *P. rapae*, there are strong differences in behaviour between the two regimes. In field margins with fully sprayed headlands (*P. rapae* males in 1986) most of the recorded activity (95.9%) was flight with very little time spent feeding (0.7%), resting (0.5%) or interacting (3.0%). The majority of time (65%) being spent in association with the hedgerow (Table V; Dover, unpublished data). In the field margins which had conservation headlands, the majority of activity was recorded in association with the headland (89.4%) with a reduced proportion of time spent in flight (42.6%), increased feeding (32.3%), increased resting (20.8%) and also increased time spent interacting (4.4%). This change in activity patterns is presumably a response to increased nectar sources present in conservation headlands as a result of omitting wide-spectrum broadleaved herbicides. However, the degree of exploitation of nectar resources in conservation headlands may vary with the season. For example, *P. napi* males only fed in the headland and only used one species of plant (*S. arvensis*) in field margins which had conservation headlands in the spring of 1986 although nectar was available in

TABLE V. The behaviour of *Pieris rapae* and *Pyronia tithonus* males in cereal field margins with conservation and fully sprayed headlands 1986. Mean time in seconds (± 1 SE). All behaviour = the sum of flight, feeding, resting and interacting.

n	Headland regime	All behaviour		Flight		Feeding		Resting		Interacting	
Pieris rapae											
36	Fully sprayed										
	Hedge	39.5 ± 5.52	**	37.7 ± 5.22	**	0.4 ± 0.34		0.3 ± 0.24		1.1 ± 0.38	NS
	Head	21.2 ± 4.75		20.5 ± 4.51		0.0		0.0		0.7 ± 0.47	
66	Conservation		+++		+++						+++
	Hedge	16.0 ± 5.21	***	8.5 ± 1.92	***	3.2 ± 2.33	***	4.2 ± 3.76	**	0.2 ± 0.09	
	Head	135.5 ± 29.37		56.1 ± 10.84		45.7 ± 12.27		27.3 ± 12.44		6.4 ± 1.33	
Pyronia tithonus											
21	Fully sprayed										
	Hedge	374.4 ± 66.57	***	139.2 ± 22.69	***	101.6 ± 36.70	***	122.8 ± 35.74	NS	10.8 ± 5.79	
	Head	109.9 ± 69.82		12.0 ± 3.23		0.0		97.1 ± 67.90		0.6 ± 0.34	
22	Conservation		++		+++		+++		NS		+++
	Hedge	186.8 ± 56.29	NS	90.0 ± 26.90	***	28.3 ± 15.48	NS	67.7 ± 34.72	NS	0.8 ± 0.51	
	Head	138.1 ± 51.87		45.8 ± 17.75		11.8 ± 8.44		78.9 ± 29.19		1.5 ± 0.88	

n = No. of butterflies. Hedge = activity associated with the hedge, hedgebank and sterile strip. Head = activity associated with the headland. Comparison of the proportion of time spent in the two subareas of the field margin (hedge vs head) within a spray regime: ** $p < 0.01$, *** $p < 0.001$. Comparison of the proportion of time spent in the two subareas of the field margin between spray regimes (conservation vs fully sprayed): $^{++}$ $p < 0.01$, NS = $p < 0.05$.

the hedgerow. By the time of the emergence of the summer generation of these butterflies, a wider range of nectar plants was available in both the hedgerow and headland, and feeding was evident from both areas of the field margin (Fig. 2; Dover, 1989b).

The behavioural response of *P. rapae* to conservation headlands may be regarded as the "best case", and there is likely to be a continuum in the level of exploitation of resources in conservation headlands by different species. For example, although *P. tithonus* males fed in the conservation headland studied in 1986, the proportion of time spent feeding was 21% in fully sprayed field margins and 12% in field margins with conservation headlands (Table V; Dover, unpublished data). Studies in 1985 and 1986 (Dover, unpublished data) showed little or no feeding by the males of this species in conservation headlands. This was also the case for the females of this species, showing exploitation of headland nectar in 1986 but not in 1987 (Fig. 3; Dover, 1989b). One of the common names of *P. tithonus* is the hedge brown, and it is well known for its association with shrubs (e.g. Thomas, 1986). This species may be less opportunistic in its exploitation of nectar sources, having a preference for hedgerow nectar. However the data of Table V clearly show of that more of these butterflies are found in association with field margins which have conservation headlands than those with fully sprayed ones. It is possible that, when hedgerow nectar is scarce, unsuitable or unavailable, then headland nectar would be used as an additional or alternative nectar source.

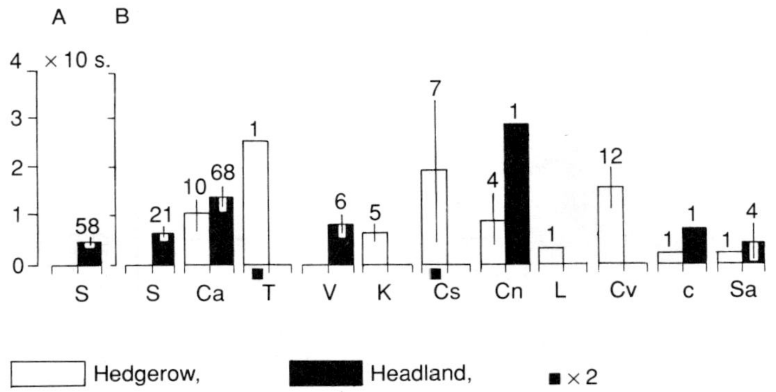

Fig. 2. Flower visit frequency (numbers), duration and location by male *Pieris napi* in field margins with conservation headlands. A, spring brood; B, summer brood. S, *Sinapis arvensis*; Ca, *Cirsium arvense* (L.); T, *Taraxacum officinale* Weber; V, *Viola arvensis* Murray; K, *Knautia arvensis* (L.); Cs, *Centaurea scabiosa* L.; Cn, *Carduus nutans* L.; L, *Lapsana communis* L.; Cv., *Clematis vitalba* L.; C, *Cirsium vulgarae* (Savi); Sa, *Silene alba* (Miller). From Dover (1989b).

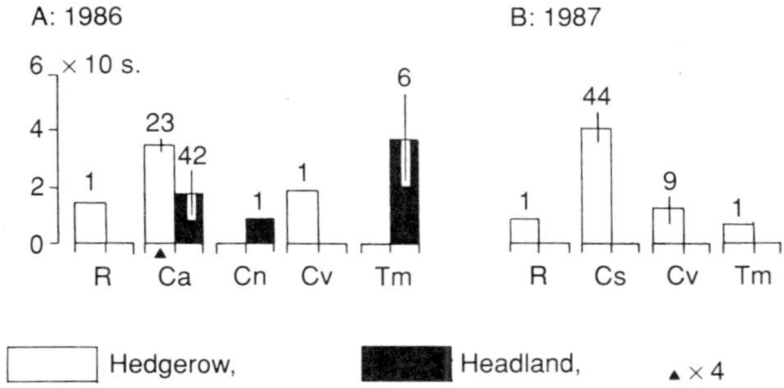

Fig. 3. Flower visit frequency (numbers), duration and location by female *Pyronia tithonus* in field margins with conservation headlands in 1986 and 1987. R, *Rubus fruticosus* L.; Tm, *Tripleurospermum maritimum* (L.); see Fig. 2 for further abbreviations. From Dover (1989b).

The increased abundance of butterflies in conservation headlands probably arises principally from two factors: changes in butterfly distribution to exploit the nectar resources in conservation headlands and, for some bivoltine species, temporary increases in numbers as a result of more larval food plants being available in the headlands. The increases in population size suggested by Fig. 1 for some univoltine satyrid species could possibly result from the fact that more nectar was available to females, which might increase their fecundity and longevity. This could also be true of other species (including the pieridae) whose vagility and open population structures would make such observations difficult (Section V.B).

E. Movement Patterns of Satyrid Species on Farmland

Some authors (Stern and Smith, 1960; Watt *et al.*, 1974; Murphy, 1983; Murphy *et al.*, 1983; Wiklund and Karlsson, 1984) have suggested that nectar availability has a profound effect on the fecundity and longevity of butterflies. For nectar sources to benefit butterfly species which have closed population structures, they must be located within the colony's boundaries. One of the principal questions which arises from this is "how far do butterflies move and what proportion of the population exhibit such mobility?". To try to answer this question, mark–release–recapture (MRR) experiments were made in 1987 in a 234-ha block of farmland. Butterflies were marked principally in two woods, two game cover crops (Anon., 1979), a green lane, and in the field margins radiating out from the woods. Butterflies were also marked in the field margins surrounding a narrow copse in a 66.5-ha block of cereal fields to the south of the larger block. The aim was to see if butterflies, particularly the satyrid species, moved between the large permanent and semi-permanent landscape elements of arable farmland.

Only 325 *P. tithonus* were captured in 1987; of the 170 recaptures of this species only 4.7% showed movement away from the previous capture site (note that the small arable block was considered to be a single habitat in this context). However, rather more *M. jurtina* were captured (1222); of the 328 recaptures, 16% showed movement out of, into, or between the permanent or semi-permanent habitat blocks; some of these movements were in excess of 1 km in length (Fig. 4) (Dover, Clarke and Rew, unpublished data).

During this study, it also became evident that the game cover crops used as marking stations made good butterfly habitats on farmland (Fig. 4; Dover, 1988; Dover and Rew, unpublished data). The usefulness of crops consisting of Jerusalem artichoke (*Helianthus tuberosus* L.) and canary grass (*Phalaris tuberosa* L.) as butterfly habitats may have resulted from their more permanent nature (>10 years; Anon., 1979), sheltered aspect, low pesticide and fertilizer inputs, and patchy growth allowing the invasion of grasses and broadleaved plants.

More detailed studies of butterfly movement patterns were made in 1988 in order to determine how compact the populations of "closed population" satyrid species might be, and to examine the extent and frequency of movement within the population boundaries; for it was evident that the low level of "long distance" movement detected in 1987 might allow the recolonization or reinforcement of extinct or damaged populations, but was clearly insufficient to allow localized (discrete) nectar sources to be exploited to any great extent. *A. hyperantus*, *P. tithonus* (Fig. 5) and *M. jurtina* all exhibited compact populations, with unsheltered grass banks and tracks being unsuitable habitat (Dover, unpublished data).

The actual area of permanent habitat available to these species was 1.22 ha or 3.1% of the study area, but if the grass banks and track are excluded, the butterflies habitat area was 0.67 ha or just 1% of the total field area. The results show that, whether through behavioural or physical phenomena, the populations of these species were restricted to a well-defined part of the study site and were predominantly "closed". This has important implications for, if nectar is really acting as a limiting resource on population size and growth, and if it is deficient in intensively farmed land, then any additional nectar sources must be located within population

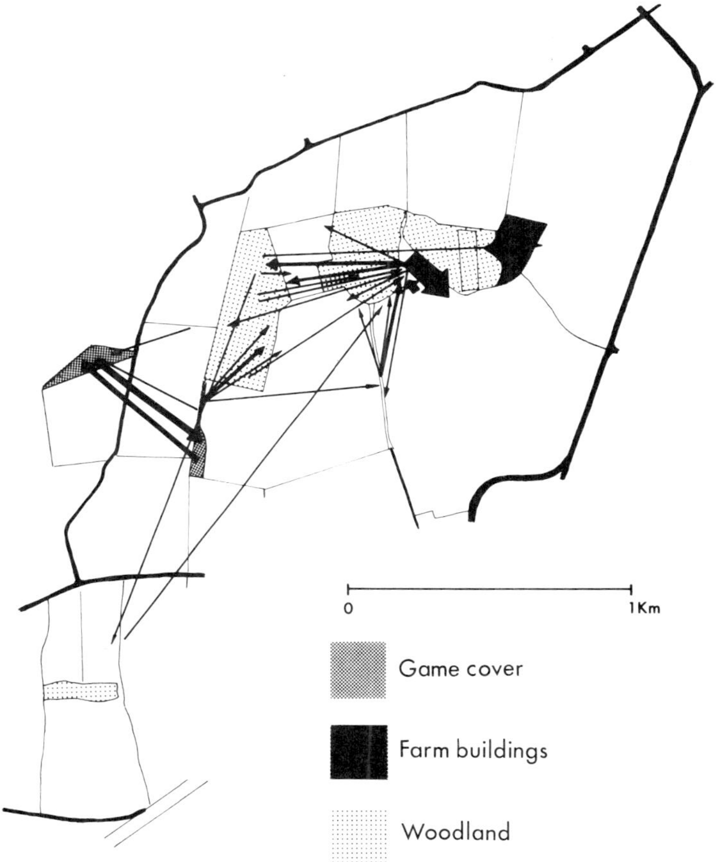

Fig. 4. Movement patterns of *Maniola jurtina* butterflies in 300 ha of farmland, Hampshire 1987. Thinnest arrowed line: 1 butterfly; thickest: 16. Thin, unarrowed, lines: field boundaries; thick: roads.

boundaries if they are to be valuable. The *P. tithonus* population illustrated in Fig. 5 does in fact have one hedgebank which is extremely rich in nectar. However, this bank is unique in its nectar richness in the whole (11 km^2) of the farm; other populations might require far more nectar than is available.

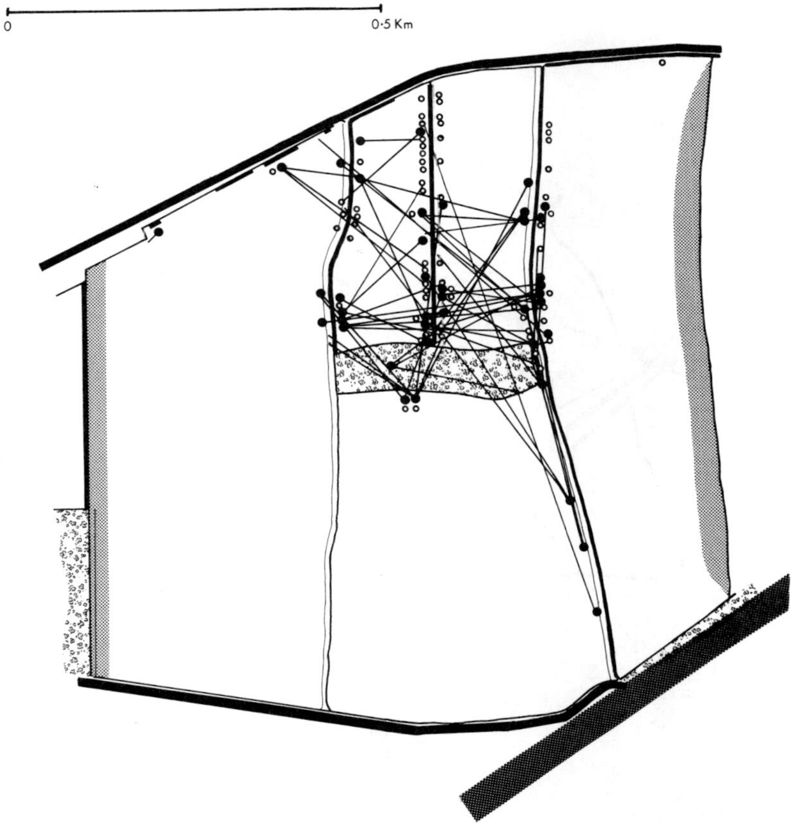

Fig. 5. Movement patterns of *Pyronia tithonus* butterflies in 66.5 ha of arable farmland, Hampshire 1988. Open circles: recapture sites along the same field edge; closed circles: recapture sites between field margins. Lines join origin and recapture sites. White stippling: railway; fine black stippling: areas not sampled; blotchy stippling: woods. Thick lines: roads; medium lines: hedges; thin lines: grass banks; thinnest lines: farm tracks.

VI. CONSERVATION HEADLANDS AS A TOOL IN CONSERVATION

A. Benefits to Other Species

Conservation headlands benefit a wide range of species including partridges, pheasants and butterflies, rare arable weeds (Wilson, 1988) and small mammals (Tew, 1988). It is likely that other beneficial, non-target and aesthetically desirable species will benefit from the extended use of conservation headlands on farmland. For example the Syrphidae, whose larvae are aphid predators, require nectar as adults for egg maturation, and Cowgill (1989) has demonstrated flower visit preferences in arable field margins.

Polyphagous predators, especially members of the Carabidae and Staphylinidae, are well known for their potential in the control of cereal aphids (Sunderland and Vickerman, 1980). These species are frequently found in association with field boundaries which they use as overwintering habitats, dispersing from them into the crop in the spring (Sotherton, 1984, 1985; Coombes and Sotherton, 1986). Some of the small Carabidae are also eaten by gamebird chicks, and have been shown to be more abundant in conservation headlands. Conservation headlands may increase the predatory potential of polyphagous predators by providing alternative prey in years when cereal aphid outbreaks do not occur. They may also increase the abundance of these predators by reducing the impact of pesticides on field boundary vegetation, which makes up the overwintering cover, and on exposure to insecticidal compounds (see also Section III.D).

B. Conservation Headlands as Buffers against Spray Drift

Boatman *et al.* (1989) point out that new and replacement hedges are commonly being planted in farmland, often to link semi-natural habitats or as "wildlife corridors". Recent work has been carried out on overwintering refugia for predatory arthropods (Thomas and Wratten, 1988; Chiverton, 1989; Thomas, 1989). These are earth ridges which are sown with grasses. When located across the centre of large fields, they effectively reduce the field size, facilitating a more rapid colonization of the crop by predators. The use of conservation headlands next to these structures may buffer the developing hedges, hedgebanks and "ridges" against some pesticide drift. The development of low nitrogen inputs to conservation headlands (Section IIF) increases the usefulness of conservation headlands. The reduction of fertilizer drift into hedgebanks also reduces the likelihood of hedgebanks developing a pernicious annual flora dominated by cleavers and barren brome.

C. Wildlife Corridors and Conservation Headlands

Although the idea that hedgerows are wildlife corridors that link isolated habitats has received wide publicity, the data to support this view are extremely limited and difficult to verify (Green, 1985). Pollard *et al.* (1974) concluded that hedgerows were not essential as breeding sites or as corridors for birds. However, hedges and their associated hedgebanks provide host plants and nectar sources for many of the common butterfly species. Unquantified observations suggest that butterfly flights along hedgerows are impeded by gaps when the wind is strong. Even in moderate wind, little activity is evident along hedgerows except on the lee side. Further, there is little butterfly activity in places where the habitat is either an unsheltered grass bank or track (Fig. 5). The concept of wildlife corridors is attractive; Dover (1990) found that pierid butterflies followed the contours of hedgerows and copse edges when in flight.

The distribution of butterflies in cereal fields was also strongly associated with the field margin. Ninety-eight per cent of butterflies observed during transects in a spring barley field in 1989 were found by walking around the perimeter of the crop, a circuit of the crop 6 m in from the crop edge and along a track separating the crop from the hedgerow along one of the field edges (total length = 2.4 km). This compared with just 2% of the butterflies which were found by walking the 3 km of tram-lines which ran up and down the field (Dover, 1990).

A hedgerow is a relatively narrow corridor, usually being rather less than the ideal 3.3 m wide recommended by the Game Conservancy (Anon., 1986). However, the creation of conservation headlands either side of a hedgerow could effectively widen the corridor by a further 12 m, providing extra nectar and host plants for many insects during the spring and summer (Boatman *et al.*, 1989).

VII. CONCLUSIONS

Although conservation headlands were developed to encourage game conservation, the result has been to create a powerful tool for reducing the impact of intensive agriculture on the remaining farmland wildlife habitats. In addition, conservation headlands may provide extra nectar and host plant resources for those species which live within or at the margin of those habitat elements. At the time of writing, there are some 250 km of conservation headland, grant-aided, in the Breckland Environmentally Sensitive Area. At least a further 480 km of conservation headland exist in the rest of Britain, paid for by farmers. If most farms had conservation headlands in at least some of their cereal fields, the net benefit to insect conservation of farmland would be enormous.

Acknowledgements

Thanks are due to my colleagues in the CGRP for their support and stimulating discussions. In particular, I would like to thank Dr N. W. Sotherton. I would also like to thank the directors, management and farm workers of the Marydown Estate for their continuing cooperation, in particular Mr H. R. Oliver-Bellasis. I also thank Mrs K. Gobbett, Dr S. A. Clarke, Miss L. Rew, Mr J. Knight and Mr O. Cheeseman for their help in the collection of butterfly field data. The butterfly work was supported by the Nature Conservancy Council, the World Wide Fund for Nature and a Natural Environment Research Council Special Topics in Agriculture and the Environment Award (G2/044). The core funding for the development of conservation headlands was from the British cereal farmers who subscribed to the Cereals and Gamebirds Research Project of the Game Conservancy Trust. Thanks are due to the following journals for use of their figures: *The Entomologist, Entomologists' Gazette, Ecological Entomology* and *The British Crop Protection Council*.

REFERENCES

Anonymous (1979). "Game and Shooting Crops". Game Conservancy Advisory Booklet No. 2 Game Conservancy, Fordingbridge.

Anonymous (1986). "The Grey Partridge". Game Conservancy Advisory Booklet No. 4. Game Conservancy, Fordingbridge.

Bain, A. B. and Boatman, N. D. (1989). The potential of quinmerac (BAS 518H) for the selective control of cleavers (*Galium aparine*) in field boundaries. "Proceedings of the 1989 Brighton Crop Protection Conference — Weeds". pp. 1057–1063.

Blank, T. H., Southwood, T. R. E. and Cross, D. J. (1967). The ecology of the partridge. I. outline of population processes with particular reference to chick mortality and nest density. *J. Anim. Ecol.* **36**, 549–556.

Boatman, N. D. (1987). Selective grass weed control in cereal headlands to encourage game and wildlife. "Proceedings of the 1987 Brighton Crop Protection Conference — Weeds", Vol. 1, pp. 277–284.

Boatman, N. D. (1989). Conservation headlands: agronomic consequences and costs to the farmer. *Game Conserv. Rev. 1988* **20**, 50–53.

Boatman, N. D. (1989). Selective weed control in field margins. "Proceedings of the 1989 Brighton Crop Protection Conference — Weeds", pp. 785–795.

Boatman, N. D. and Sotherton, N. W. (1988). The agronomic consequences and costs of managing field margins for game and wildlife conservation. *Aspects of Appl. Biol.* **17**, 47–56.

Boatman, N. D. and Wilson, P. J. (1988) Field edge management for game and wildlife conservation. *Aspects Appl. Biol.* **16**, 53–61.

Boatman, N. D. and Wilson, P. J. (1989). Cereal extensification: a vital tool for conservation. *Game Conserv. Rev. 1988* **20**, 63–65.

Boatman, N. D., Freeman, K. J. and Green, M. C. E. (1988). The effects of timing and dose on the control of *Galium aparine* (cleavers) and other broadleaved weeds by fluroxypyr in cereal headlands. *Aspects Appl. Biol.* **18**, 117–128.

Boatman, N. D., Dover, J. W., Wilson, P. J., Thomas, M. B. and Cowgill, S. E. (1989). Modification of farming practice at field margins to encourage wildlife. *In* "Biological Habitat Reconstruction" (G. P. Buckley, ed.), pp. 299–311. Belhaven, London.

Chiverton, P. A. (1989). Creation of within-field overwintering sites for natural enemies of cereal aphids. "Proceedings of the 1989 Brighton Crop Protection Conference — Weeds", pp. 1093–1097.
Coombes, D. S. and Sotherton, N. W. (1986). The disperal and distribution of polyphagous predatory Coleoptera in cereals. *Ann. Appl. Biol.* **108**, 461–474.
Cowgill, S. (1989). The role of non-crop habitats on hoverfly (Diptera: Syrphidae) foraging on arable land. "Proceedings of the 1989 Brighton Crop Protection Conference — Weeds", pp. 1103–1109.
Dover, J. W. (1988). Butterflies and game cover. *Game Conserv. Rev. 1987* **19**, 84–87.
Dover, J. W. (1989a). A method for recording and transcribing observations of butterfly behaviour. *Entomol. Gaz.* **40**, 95–100.
Dover, J. W. (1989b). The use of flowers by butterflies foraging in cereal field margins. *Entomol. Gaz.* **40**, 283–291.
Dover, J. W. (1990). Butterflies and wildlife corridors. *Game Conserv. Rev. 1989* **21**, 62–64.
Dover, J. W. (1991). Butterflies and Conservation Headlands. *In* "Future of Butterflies in Europe: Strategies for Survival" (J. G. van der Made, ed.). Pudoc, Wageningen, in press.
Dover, J. W. and Cuthbertson, P. (1989). The reduction of spray drift into field boundaries: the importance of conservation headlands. *Game Conserv. Rev. 1988* **20**, 54–56.
Dover, J. W., Sotherton, N. W. and Gobbett, K. (1990). Reduced pesticide inputs on cereal field margins: the effects on butterfly abundance. *Ecol. Entomol.* **15**, 17–24.
Greaves, M. P. and Marshall, E. J. P. (1987). Field margins: definitions and statistics. *In* "Field Margins — BCPC Monograph No. 35" (J. M. Way and P. G. Greig-Smith, eds), pp. 3–10. BCPC, London.
Green, B. H. (1972). The relevance of seral eutrophication and plant competition to the management of successional communities. *Biol. Conserv.* **4**, 378–384.
Green, B. J. (1985). "Countryside Conservation: the Protection and Management of Amenity Ecosystems", 2nd edn. George Allen & Unwin, London.
Green, R. E. (1984). The feeding ecology and survival of partridge chicks (*Alectoris rufa* and *Perdix perdix*) on arable farmland in East Anglia. *J. Appl. Ecol.* **21**, 817–830.
Green, R. H., Rands, M. R. W. and Moreby, S. J. (1987). Species differences in diet and the development of seed digestion in partridge chicks *Perdix perdix* and *Alectoris rufa*. *Ibis* **129**, 511–514.
Hill, D. A. (1985). The feeding ecology and survival of pheasant chicks on arable farmland. *J. Appl. Ecol.* **22**, 645–654.
Murphy, D. D. (1983). Nectar sources as constraints on the distribution of egg masses by the checkerspot butterfly, *Euphydryas chalcedona* (Lepidoptera: Nymphalidae). *Environ. Entomol.* **12**, 463–466.
Murphy, D. D., Launer, A. E. and Ehrlich, P. R. (1983). The role of adult feeding in egg production and population dynamics of the checkerspot butterfly, *Euphydrays editha*. *Oecologia* **56**, 257–263.
Newbold, C. (1989). Semi-natural habitats or habitat re-creation: conflict or partnership? *In* "Biological Habitat Reconstruction", (G. P. Buckley, ed.) pp. 9–17. Belhaven Press, London.
Pollard, E., Hooper, M. D. and Moore, N. W. (1974). *Hedges.* Collins, London.
Pollard, E., Elias, D. O., Skelton, M. J. and Thomas, J. A. (1975). A method of assessing the abundance of butterflies in Monks Wood National Nature Reserve in 1973. *Entomol. Gaz.* **26**, 79–88.
Pollard, E., Hall, M. L. and Bibby, T. J. (1986). "Research and Survey in Nature Conservation. No. 2. Monitoring the Abundance of Butterflies 1976–1985". Nature Conservancy Council, Peterborough.

Potts, G. R. (1980). The effects of modern agriculture, nest predation and game management on the population ecology of partridges, *Perdix perdix* and *Alectoris rufa*. *Adv. Ecol. Res.* **11**, 2–79.

Potts, G. R. (1986). "The Partridge: Pesticides, Predation and Conservation". Collins, London.

Rands, M. R. W. (1985). Pesticide use on cereals and the survival of grey partridge chicks: a field experiment. *J. Appl. Ecol.* **22**, 49–54.

Rands, M. R. W. (1991). The effects of pesticides on grey partridge chick survival and breeding density, *J. Appl. Ecol.* in press.

Rands, M. R. W. and Sotherton, N. W. (1986). Pesticide use on cereal crops and changes in the abundance of butterflies on arable farmland. *Biol. Conserv.* **36**, 71–82.

Rands, M. R. W. and Sotherton, N. W. (1991). The impact of the selective use of pesticides at the edges of cereal crops on wild gamebird stocks in Britain. *In*: "Congress of the XVIII Union of Game Biologists", Krakow, August 1987 in press.

Sotherton, N. W. (1984). The distribution and abundance of predatory arthropods over-wintering on farmland. *Ann. Appl. Biol.* **105**, 423–429.

Sotherton, N. W. (1985). The distribution and abundance of predatory Coleoptera over-wintering in field boundaries. *Ann. Appl. Biol.* **106**, 17–21.

Sotherton, N. W. (1989). Farming methods to reduce the exposure of non-target arthropods to pesticides. *In* "Pesticides and Non-target Invertebrates" (P. C. Jepson, ed.). Intercept Books, Wimborne.

Sotherton, N. W. and Moreby, S. J. (1988). The effects of foliar fungicides on beneficial arthropods in wheat fields. *Entomophaga* **33**, 87–99.

Sotherton, N. W. and Rands, M. R. W. (1987). Predicting, measuring and minimising the effects of pesticides on farmland wildlife on intensively managed arable land in Britain. *In* "Pesticide Science and Biotechnology; Proceedings of the Sixth IUPAC Congress of Pesticide Chemistry" (R. Greenhalgh and T. R. Roberts, eds), pp. 433–437. Blackwell Scientific, Oxford.

Sotherton, N. W., Rands, M. R. W. and Moreby, S. J. (1985). Comparison of herbicide treated and untreated headlands on the survival of game and wildlife. "1985 Brighton Crop Protection Conference – Weeds", Vol. 3, pp. 991–998.

Sotherton, N. W., Moreby, S. J. and Langley, M. G. (1987). The effects of the foliar fungicide pyrazophos on beneficial arthropods in barley fields. *Ann. Appl. Biol.* **111**, 75–87.

Sotherton, N. W., Jepson, P. C. and Pullen, A. J. (1988). Criteria for the design, execution and analysis of terrestrial non-target invertebrate field tests. *In* "Field Methods for the Study of Environmental Effects of Pesticides. BCPC Monograph No. 40" (M. P. Greaves, B. P. Smith and P. W. Greig-Smith, eds), pp. 183–190.

Sotherton, N. W., Boatman, N. D. and Rands, M. R. W. (1989). The "Conservation Headland" experiment in cereal ecosystems. *Entomologist* **108**, 135–143.

Southwood, T. R. E. and Cross, D. J. (1969). The ecology of the partridge III Breeding success and the abundance of insects in natural habitats. *J. Anim. Ecol.* **38**, 497–509.

Stern, V. M. and Smith, R. F. (1960). Factors affecting egg production and oviposition in populations of *Colias philodice eurytheme* Boisduval (Lepidoptera: Pieridae). *Hilgardia* **29**, 441–454.

Sunderland, K. D. and Vickerman, G. P. (1980). Aphid feeding by some polyphagous predators in relation to aphid density in cereal fields. *J. Appl. Ecol.* **17**, 389–396.

Tew, T. (1988). The effects of conservation headlands on small mammals. *Game Conserv. Rev. 1987* **19**, 88–90.

Thomas, J. A. (1986). *RSNC Guide to the Butterflies of the British Isles*. Country Life, Twickenham.

Thomas, M. B. (1989). Creation of island habitats to enhance populations of beneficial insects. "Proceedings of the 1989 Brighton Crop Protection Conference – Weeds", pp. 1097–1093.

Thomas, M. B. and Wratten, S. D. (1988). Manipulating the arable crop environment to enhance the activity of predatory insects. *Aspects Appl. Biol.* **17**, 57–66.

Watt, H. B., Hoch, P. C. and Mills, S. (1974). Nectar resource use by *Colias* butterflies: chemical and visual aspects. *Oecologia* **14**, 353–374.

Wiklund, C. and Karlsson, B. (1984). Egg size variation in satyrid butterflies: adaptive vs historical, "Bauplan" and mechanistic explanations. *Oikos* **43**, 391–400.

Wilson, P. J. (1988). The conservation of rare and vanishing arable weeds. *Game Conserv. Rev. 1987* **19**, 80–83.

Woodford, E. K. (1964). Weed control in arable crops. "Proceedings of the 7th British Weed Control Conference", Vol. 3, pp. 944–962.

13

The Utilization and Value of Non-domesticated Insects

M. G. MORRIS, N. M. COLLINS, R. I. VANE-WRIGHT AND J. WAAGE

I.	Introduction	319
II.	Services from Insects	321
	A. Chemical Pest Control: Insects as Monitors of Environmental Effects	321
	B. Natural Control of Pests	322
	C. Genetic Resources for Biological Control	323
	D. Pollination	325
	E. Decomposition	328
III.	Goods from Insects	330
	A. Livestock	330
	B. Deadstock	332
	C. Insect Products	335
	D. Food	336
IV.	Planning for Resource Management	338
	A. Maintaining Insect Diversity	338
	B. Using Insects to Identify Key Areas for Conservation	341
	C. Maintaining Key Areas through the Benefits from Utilization	342
V.	Conclusions	343
	References	344

I. INTRODUCTION

In this chapter we discuss the practical and commercial use which humans can make of insects in the wild state. Only a few insects, in no way representative of the abundance of insect species present on earth, have been domesticated. One view of the practicality of the conservation of biodiversity is to make greater use, and to recognize the potential, of the range of organisms useful to humans. This approach has already contributed to the conservation of plants, many of which have actual or potential uses as alternative crops (National Research Council, 1979), sources of fuel, fixers of nitrogen, and especially as drugs or medicines (Farnsworth and Soejarto, 1985). It is important to persuade sceptics that insects have a similar value because of increasing doubt that more conventional approaches to conservation can by themselves slow down the rate of species' extinction, particularly in the tropics. The World Conservation Strategy (IUCN/UNEP/WWF, 1980) recognized the importance of combining conservation with development, and more recent work has refined and developed this approach (Anon., 1987); a new version of the World Conservation Strategy is in preparation.

Entomology has had two important historical strands: intellectual curiosity in the diversity of insect life, and the practical need to control insect pests, in the context both of human health and the protection of agricultural stock or crops. In neither aspect has conservation been important until recently. The study of insect diversity, driven by a variety of beliefs and philosophical approaches, has been seen as an end in itself. Although amateurs have contributed greatly to this aspect of entomology, the view was developed that this intellectual study of insect diversity should also be supported by public funding of universities, national museums of natural history and biological research institutes.

At about the same time that the need to conserve the insect resource was perceived as urgent, funding for research based solely on intellectual curiosity began to decline. Emphasis on market principles (the practical value of research), and recognition that the numbers of insect species were much greater than previously thought, contributed to this decline.

It is against this background that the conservation of insect diversity must be viewed. The emphasis on usable results from research indicates that the practical value of insects must be recognized as a growing factor in conservation. Yet, as even this field is under-funded, conservationists must not overestimate practical value, nor neglect other, often less tangible, aspects. Intellectual curiosity will continue to be a major driving force in insect conservation.

Although this chapter emphasizes the services which insects provide, and the practical use which is made of non-domesticated insects and insect products, it is recognized that insects are used extensively in scientific research and contribute to aesthetic enjoyment, although in a minor way. It is clear that calculation of the "value" of insects in these contexts is difficult, if not impossible.

II. SERVICES FROM INSECTS

A. Chemical Pest Control: Insects as Monitors of Environmental Effects

Only about 200 species of insects are considered to be serious pests of agriculture, but this small assemblage of species has had a substantial impact on the human condition. Naturally, effort has focused on the elimination of these species, but this has generated in its wake some interesting issues for insect conservation.

Since the introduction of organochlorine compounds such as DDT for insect pest control in the 1950s, synthetic chemical insecticides have dominated insect control methods. Like many breakthroughs, chemical pesticides have proved to be a mixed blessing. The evolution of these chemical products over the past four decades has reflected growing problems of biochemical resistance in pest species, accumulation in food chains (e.g. Hunt and Bischoff, 1960; Ratcliffe, 1980) and human toxicity (Gear, 1983). As a result, compounds used today tend to be safer to humans and animals and less persistent in the environment than earlier products. Special problems still remain, such as the dangers to bat populations (already very reduced) from persistent wood preservatives (Cooke and Mitchell-Jones, 1988). Destruction of a pest's natural enemies often results from the use of non-selective pesticides, and can have the net effect of exacerbating the pest problem in the long term (Waage, 1989).

Most of the pesticides used around the world are still broad spectrum in their action, killing a range of insect species, pest and non-pest. Concern over their effects on non-target organisms continues, particularly where such chemicals are used over large areas of natural habitats, as in the current aerial spray campaigns against desert locust (*Schistocerca gregaria*) and tsetse flies (*Glossina* species) in Africa, and against temperate forest pests such as gypsy moth (*Lymantria dispar*) in North America and pine beauty (*Panolis flammea*) in northern Britain.

In monitoring undesirable side effects of insecticides on the environment, non-pest insects are an obvious tool; they are susceptible to insecticides, they occur in a diversity of habitats, they are easily sampled, and they are near the bottom of the important food chains which often end in rare mammalian, avian and reptilian predators. In testing new pesticides, agrochemical companies frequently assess the diversity of non-target insect species in croplands using conventional ecological indices (Brown, 1989). By contrast, programmes to assess the side effects of spray campaigns on more natural ecosystems (such as those mentioned above) will often focus on a few key species in particular habitats (Grant, 1989). As pesticide use comes under increasing scrutiny, the value and use of non-domesticated insects as environmental indicators will grow.

Although pesticide use has been broadly associated with the decline of some indigenous insect species, such as butterflies, there is little specific evidence and most insect conservationists perceive pesticides as a minor threat compared with habitat destruction (Thomas, 1991). Agroecosystems, which are the most heavily sprayed, are generally monocultures low in insect diversity compared with natural ecosystems. Of the species likely to be eliminated by pesticide use there, very few would be missed by any but the most ardent entomologist. An important exception is those insects which are themselves natural enemies of insect pests.

B. Natural Control of Pests

Virtually all insect pests are hosts to a complex of predators and parasitoids, many of which are very prey or host specific. A greater dependence on the natural pest control which these insects give, and a complementary decrease in dependence on chemical pesticides, form a key element of modern, integrated pest management (IPM). Diverse complexes of natural enemies can be threatened by pesticide use, particularly if this is widespread, and in the chequerboard of modern agricultural lands in Europe, for example, there may often be few unsprayed places to hide (Greig-Smith, 1988; Burns, 1989).

A long-term project to assess the effect of pesticides on the farmland environment, the 7-year Boxworth Project, is revealing persistent effects of pesticides on natural enemy numbers (Anon., 1987). Areas with full insurance pesticide use, where chemicals are applied on a routine, scheduled basis independent of pest numbers, have been compared with areas where spray regimes are based on measurement of pest threshholds and other methods to reduce the application of pesticides. Compared to these, the full insurance fields showed a decline in numbers of generalist predatory invertebrates, such as spiders and Carabidae, over the years. This was accompanied by a reduction in predation, which was measured separately. Less mobile carabids suffered the greatest long-term suppression, while numbers of highly dispersive predators, such as empid flies, changed little in the full insurance fields. However, some species of predators actually appeared to increase (e.g. the carabid *Trechus quadristriatus*), possibly because their biology isolates them from contact with pesticides and reduces numbers of other predators, thereby reducing competition. Among non-target organisms, Collembola seem to be most affected, with their numbers much lower on full insurance fields than fields where pesticide use was more limited. This may be significant as they are important alternative prey for Carabidae.

The conservation of insect natural enemies indigenous to the crop environment is one of the several technologies of biological pest control. Their numbers in crops are reduced not only by pesticides but by the simplified environment of the crop itself. Monocultures of single crops often lack the flowers, shelter and alternative insect food which many natural enemies require. The contribution of vegetational diversity to natural enemy numbers and the suppression of pest outbreaks is well documented (Altieri and Letourneau, 1984).

Conservation of natural enemies has been an effective method in a range of agricultural systems. Species of *Formica* ants have long been protected in parts of continental Europe because of their role as general predators, particularly in forests (Wells *et al.*, 1983). The effectiveness of a wide range of polyphagous predators known to be enemies of cereal aphids (Edwards *et al.*, 1979) can be improved by appropriate habitat management of field boundaries in combination with selective use of pesticides (Poehling, 1989). This results in a weedy refuge for insect predators and their alternative prey, which are benefits for natural control of pests and also for gamebirds feeding on insects. Gamebird interests have long benefited from modifications of chemical control in cereal fields, especially the management of field boundaries and, since set aside, the establishment of "conservation headlands" (Hudson and Rands, 1988). Together with predators of pests, other insect species such as butterflies have also benefited (Rands and Sotherton, 1986). More attention needs to be given to long-term effects and the role of field margins and conservation headlands in the ecology of populations, particularly metapopulations, of such species. Other ways of constructing refugia include strip spraying and strip cropping, both of which aim to leave part of the crop untouched, as a source of later immigration of natural enemies into treated areas.

Despite these and other difficulties, encouraging natural enemies through cultural and other means is increasingly being seen as one element in reducing the use of chemical pesticides. There are many spin offs to such a policy, particularly in the developed world.

C. Genetic Resources for Biological Control

A very different form of biological pest control generates an even more direct argument for the conservation of non-domesticated insects. Introduction, or "classical biological control", seeks to redress the ecological imbalance created when a crop-feeding insect or a weed is introduced into a new region of the world without its specific insect natural enemies, and consequently becomes a serious pest. In an introduction programme, these specific natural enemies are generally sought in the pest's area of origin, and the most promising are studied. After appropriate safety screening and quarantine, they are released in the country where the pest is a problem, in the hope of returning pest populations to a low, stable level over time. According to the CAB International Institute of Biological Control's database of introductions against insect pests (BIOCAT), about 600 insect natural enemy species have been used in over 4000 separate introduction programmes against 500 pest species over the past century. About 30% of these introductions have led to the establishment of the natural enemy, and a smaller fraction to substantial success, resulting in the long-term reduction of pest populations (Waage and Greathead, 1988, and unpublished results).

Weed pests tell a similar story, although here plant-feeding insects are the control agents and safety screening therefore needs to be even more intense. Over 100 specific weed-feeding insects have been introduced in about 600 separate introductions against 100 weed species, with a success rate of about 30% (Julien, 1987; Waage and Greathead, 1988, and unpublished results). Notable successes in the past have been the control of prickly pear cactus (*Opuntia* spp.), water fern (*Salvinia molesta*) and St John's Wort (*Hypericum perforatum*).

In the area of origin of an exotic insect or weed pest, the species may be rare, even unknown, and its specific natural enemies likely to be more so. The pest or plant may also come from a natural habitat quite different from the natural or agricultural habitat which it has invaded, and one which may be threatened. A recent magazine advertisement by a North American conservation organization illustrated several useful species of plants and animals of the Amazonian region threatened by destruction of their habitats there. One was *Cyrtobagous salvineae*, the weevil specific to water fern, which has successfully controlled this invasive weed in Australia, Papua New Guinea and Sri Lanka (Julien, 1987). The argument behind this example is clear – endangered natural habitats harbour not only desirable species but potential pests like water fern, and the loss of such habitats may eliminate the reservoir of specific biological control agents which could be used to control the pest in its exotic range.

Floating in the Amazon besides water fern are other innocuous plants which are serious exotic pests elsewhere, including water hyacinth (*Eichornia crassipes*). Indeed, tropical America is, for unknown reasons, the primary source area for most tropical exotic weeds, and hence for their insect control agents. Preservation of neotropical flora and fauna is therefore of great relevance to weed control world-wide.

However, the problem of preserving the natural habitat of potential weed control agents is not restricted to threatened tropical systems. Europe is the region of origin of many of the weed and insect pest problems of North America, Australia, New Zealand and temperate South America. In these regions, relatively innocuous European wildflowers such as ragwort (*Senecio jacobaea*) and thistles (*Carduus* and *Cirsium* spp.), and shrubs like bramble (*Rubus fruticosus* agg.), gorse (*Ulex europaeus*) and heather (*Calluna vulgaris*) have become serious invasive weeds, threatening both agriculture and conservation areas. Many of these weeds have been successfully controlled by the introduction of specific European insect species, but today the source areas for some of these insects are rapidly disappearing. The intensification of agriculture and the increase in application of herbicides to farmland and road verges is eliminating dense populations of plants such as knapweeds (*Centaurea* spp.) and spurges (*Euphorbia* spp.) in Eastern Europe and with them source areas for the specialized insects which could be used for their control in North America (D. Schroeder, personal communication).

The world's natural enemies of pests and weeds, and the habitats in which they live, provide a reservoir for the future biological control of new exotic pests spread around the world by increasing transportation. The invasion of Europe in recent years by such tropical exotic pests as the whitefly (*Bemisia tabaci*), the thrips (*Frankliniella* sp.) and the leafminer (*Liriomyza trifolii*) illustrates this phenomenon, which is mirrored by even more dramatic exotic invasions in the tropics. As a form of pest control, biological control is ecologically safe when done properly. Modern methods of selection of insect natural enemies for specificity to the pest help to reduce to an absolute minimum any risk to the fauna or flora of the country of introduction. Furthermore, for difficult problems, like invasive exotic weeds in conservation areas, biological control is the only feasible method of treatment. All others, such as mechanical and chemical control, are likely to cause as much damage to the endangered indigenous flora and fauna as the weed itself!

It goes without saying that avoidance of pest problems by preventing introduction of unwanted alien species is the cheapest and most ecologically sound option (see Howarth and Ramsay, Chapter 4). However, the reality is that pest outbreaks will remain enormously damaging into the foreseeable future. The protection of insect natural enemies of both indigenous and exotic pests, and their habitats, is therefore an important component in the conservation of non-domesticated insects.

D. Pollination

Although many of the major orders of insects are known to have been in existence by the Permian, much of the radiation of the class occurred from the late Cretaceous onwards and is associated with the evolution of angiosperm plants. Lepidoptera and aculeate Hymenoptera, in particular, do not appear in the fossil record until angiosperms were well established. Insects and plants present a wide variety of more or less mutualistic relationships, and their coevolution has been well studied in many cases, and is suspected in many more. One of the most important of these relationships is pollination. Although this term itself suggests benefit only to the plant partner, many devices have evolved to give advantages to the insect, notably nectar as a food source, while some plants have mechanisms to utilize insects with a minimum of reciprocal benefit. Few animals other than insects are important pollinators in temperate regions, but in the tropics birds, bats and some other mammals are also effective pollinators.

Honey bees are among the best known insect pollinators, being morphologically adapted for pollen collection, but, being domesticated, are outside our survey. Many of the evolutionarily more complex pollinating mechanisms involve plants which are not utilized by humans, and in these cases no services can be considered as being rendered by the insect pollinators. Many crop plants depend on insects for their pollination (Free, 1970), but there are also anemophilous and self-pollinated species. In crops such as apple, pollination by both insects and wind may occur (Free, 1964). Important pollinators occur in the Coleoptera, Lepidoptera, Diptera and particularly in the Hymenoptera; some species of Hemiptera and Thysanoptera are also known to be pollinators. It is not always easy to distinguish regular, important pollinators from occasional ones or mere flower visitors.

In some cases, populations of non-domesticated insect pollinators have been increased by rearing and release, and by introductions following establishment of the crop in new areas, The small weevil *Elaeidobius kamerunicus* has been extensively imported and cultured in south-east Asia for pollination of oil palm (*Elaeis guineensis*) (Greathead, 1983). The leafcutter bee (*Megachile rotundata*) (Stephen, 1961) and alkali bee (*Nomia melanderi*) (Stephen, 1965) have been employed as alternatives to honey bees for alfalfa pollination in North America. Both species have been encouraged with artificial nests and breeding sites. Natural pollinators are frequently affected adversely by modern intensive agriculture, giving additional reasons for reducing use of chemicals (Section IIB). Under specialized conditions, insect pollinators may be dispensed with altogether. Artificial pollination of tomato by simple mechanical means is a standard practice when the crop is grown under glass.

An important aspect of complex pollinator–plant relationships is the way in which the morphology of the flower is matched to that of the pollinator. Tongue length of the insect is related to corolla length in a range of different species. A famous example is *Xanthopan morgani* ssp. *praedicta*, associated with the orchid *Angraecum sesquipedale* in Madagascar, which was "predicted" to exist by Darwin on the grounds that the plant's flowers required a specialized pollinator with an exceptionally long tongue: a similar association has recently been studied in detail by Nilsson *et al.* (1985). Intraspecific matching of pollinators occurs in red clover (*Trifolium pratense*), in which cultivars with corollas of different lengths tend to be pollinated by different species of bumble bees (Bombidae), which vary in their length of tongue. This relationship is one which breaks down on occasion, as some short-tongued bumble bees have learned to bite into the base of the corolla, thus obtaining nectar without pollinating the flowers. In studies of the pollination of red clover, which is highly self-incompatible, the importance of the whole ecology of bumble bees has been found to be important. For instance, yield of clover seed was more than twice as high in fields near woodland, in which bumble bees nested, than in comparable fields in open country (Hawkins, 1965).

Specialized plant–insect pollination systems can be extremely complex. Two well-known examples are yucca and yucca moths (*Tegeticula* spp., Incurvariidae) and figs and fig wasps (Hymenoptera, Chalcidoidea, Agaontidae). Pollination of yucca by *Tegeticula* is obligatory for both plant and moth; no other insect can pollinate the plant, and larvae of the moth can develop only in fertilized ovaries of an appropriate species of yucca; unfertilized ones are aborted. Fig wasps have complex morphological and life history adaptations (Wiebes, 1979). Male wasps have reduced eyes and antennae and are apterous; they never leave the fig receptacles in which they develop. In the edible fig (*Ficus carica*), receptacles of three different kinds are produced at different times of the year. Neuter flowers are produced, in which the fig wasps, *Blastophaga psenes* in this case, breed. A complex cycle, in which the female wasps invade or breed in the different receptacles, is evidence of a very close relationship between plant and pollinator, one which appears to be evolutionarily ancient (Murray, 1985).

A feature of yucca moths, and especially fig wasps, is the diversity of species which have coevolved with different species of their hosts. There is considerable variation in the details of the pollination process and life history of the pollinating insects, although the basic pattern remains the same.

Despite the interest of these specialized relationships, particularly in endeavouring to understand the process of coevolution, most insect-pollinated crops are serviced by non-specialist species. In temperate crops especially, there is considerable variation in the most effective species of pollinators in both space and time. A diversity of solitary bees pollinate fruit trees in areas where honey bees are scarce or absent. The pollinating insects of field crops may vary considerably from year to year. Probably many more species of insects are important in pollination than is currently realized. As with the insects providing other services, it must be assumed that potential pollinators must be conserved as well as those whose value as pollinators is well established.

E. Decomposition

All ecosystems may be divided into the plant subsystem, the herbivore subsystem and the decomposition subsystem. The plant subsystem fixes energy and matter through photosynthesis and absorption of nutrients through roots. Net gains (net primary production, NPP) may be distributed (a) in the plant subsystem as increased biomass of perennial growth (where the ecosystem is in a subclimax, building phase); (b) through consumption by animals in the herbivore subsystem; or (c) by being transferred to the decomposition subsystem as plant litter or detritus (Swift *et al.*, 1979). In natural, mature ecosystems, the bulk of NPP passes through the decomposition subsystem, where it is oxidized and broken down into simple and stable carbon and nitrogen compounds.

The decomposition subsystem is characterized by an extraordinary array of soil-living fungi, bacteria, protozoa and invertebrates, whose relative importance and proportions show broad regional differences. In brief, the microflora (bacteria and fungi) and microfauna (nematodes and protozoa) are dominant in high latitudes, but the larger mesofauna (including Collembola and termites) and macrofauna (including earthworms, millipedes, woodlice, beetles and molluscs) come to play a much greater role in the hotter, wetter tropical ecosystems. Certain insects come to dominate decomposition processes, notably the termites (Isoptera). In tropical rain forests and savannas, termites alone consume up to 16% and 24% of litter production, respectively, and, through their burrowing and nesting habits, they expose even more organic material to the action of bacteria, fungi and other decomposers (Collins, 1981, 1988).

Modern industrialized farming has tended to treat the soil as no more than a structured form of blotting paper, whose only role is to absorb, and release on demand, the various chemicals that are applied. This is taken to its extremes in the horticultural practice of hydroponics, where plant roots are bathed in an "ideal" solution of nutrients in water running through plastic tubes. Productivity may be high, but these developments ignore the basic ecological principle that in a balanced and stable ecosystem production is equalled by decomposition.

The resulting imbalance is characterized in one of two ways: either the decomposer community, bereft of its normal input of detritus, turns its attention to crops and becomes a "pest", or the productivity rises above the capacity of the decomposer fauna, and organic wastes accumulate. The former condition is well illustrated by termites. Throughout the tropics, field crops such as maize and sugar cane, and forestry plantations (particularly exotics), are attacked by those species of termites which can survive agricultural operations. Even in grazing lands, termites that normally feed on grass litter will turn to standing grass when overgrazing reduces litter production (Collins, 1982). In most ecosystems, even agricultural ones, termites are actually very beneficial in aerating and turning the soil and recycling organic matter. Unfortunately, agricultural management is not sophisticated enough to maintain the optimal termite populations.

In mid-latitudes and temperate ecosystems, where water availability and temperature limit the rate of oxidation of organic matter, waste accumulates as artificially raised productivity outstrips the decomposer community. Examples are the accumulation of cattle dung in The Netherlands, with consequent release of much unwanted methane and nitrogen into the atmosphere, and the problem of excess wheat and barley stubble in the United Kingdom, with again unwelcome side effects from straw burning in late summer. A special case is the problem of placental mammal (sheep and particularly cattle) dung in Australia, where the native decomposers, notably dung beetles, are able to cope only with marsupial dung (Waterhouse, 1974). It has been estimated that 2.4 million ha of pasturage are unavailable in any one year because of the presence of cowpats (Norris, 1974).

Insects and other invertebrates can be employed to solve some of these problems, with direct economic benefits as a result, but the technology is rather unsophisticated as yet. Dung beetles have been successfully imported from Africa and elsewhere to deal with the Australian problem (Bornemizza, 1976) — an interesting parallel to biological control of pests by introduced insect enemies. Other entrepreneurial activities include the use of fly maggots to break down fish and other waste to supply, paradoxically enough, the coarse fishing industry with bait. Earthworms (though not insects!) are increasingly used to convert sewage and other organic waste into a high-quality fertilizer, and worm biomass that can be used to feed poultry and other livestock. The live worms that are used to set up the cultures are said to be worth over US $2.00 per pound.

These few examples serve to demonstrate that insects and other invertebrates play a very important role in natural decomposition processes. In man-made habitats there is potential for harnessing their capabilities to help reduce the waste that accumulates when productivity is artificially raised.

III. GOODS FROM INSECTS

A. Livestock

There is a long tradition of trade in entomological livestock in Europe and North America. Amateur entomologists in particular buy eggs, larvae or pupae of indigenous Lepidoptera to rear them through for their collections or simply for the pleasure of releasing them into the wild. Entomological suppliers also provide material to schools and universities for teaching and to research institutes for experimental work. Exotic groups that are frequently traded as live material include stick insects (Phasmida) and praying mantids (Mantodea), as well as non-insect invertebrates such as spiders, scorpions, molluscs and myriapods, and, of course, the ever-popular Lepidoptera.

This trade has not been evaluated in financial terms, but it is reasonable to assume that turnover is low, probably less than £1 000 000 per year. However, a great boost has resulted from the development of the "butterfly house" industry.

Butterfly houses, also known as butterfly centres, farms or "worlds", have become a popular attraction, growing from one or two displays in the United Kingdom in the late 1970s to about 60–70 world-wide in 1989. Britain is the main centre, with 38 commercial houses currently in operation, but numbers are increasing in mainland Europe, Canada, the USA and Australia. The display usually consists of a walk-through glasshouse, in which free-flying tropical butterflies are exhibited amongst hothouse vegetation.

Analysis of the UK industry based on 1986 data revealed that over 300 species of butterflies and moths had been displayed, but that a mere two dozen common species formed the core of most displays (Collins, 1987). The better ones may show up to 70 species at any one time, and about 150 over the summer season. About 500 000 individual butterflies were used in 1986, one-third being bred on site and two-thirds brought in from dealers in Britain and abroad. The main tropical suppliers were in the Philippines and Malaysia, but India, Taiwan and the USA were also important (Table I). More recently, dealers in Costa Rica, Peru and

TABLE I. The number of Lepidoptera species exported from 25 producer countries to British butterfly houses in 1986 (from Collins, 1987).

Argentina	10	Madagascar	1
Australia	5	Malta	1
Brazil	7	West (peninsular) Malaysia	86
Canada	1	Peru	3
Costa Rica	22	Philippines	61
Federal Republic of Germany	2	Poland	1
Ecuador	4	Singapore	1
Spain	2	Sri Lanka	22
France	24	Taiwan	35
India	40	Trinidad	3
Indonesia	1	USA	32
Japan	21	South Africa	2
Kenya	1		

Indonesia have been active. There is no evidence of use of threatened species, in contrast to the deadstock trade, and virtually all the material is captive bred.

In 1986, the gross gate takings of the British industry, based on a total of 44 establishments, were estimated rather conservatively at £5 000 000, with about £600 000 expenditure on live butterfly pupae. Although the number of British butterfly houses has reduced to 38, mainly as a result of the imposition of the Zoo Licensing Act, 1981, world-wide growth means that gross gate takings would certainly now exceed £10 000 000 per year. The investment in capital is often substantial, ranging from a minimum of about £20 000 for glasshouses and equipment to £200 000 for a major display and breeding centre, such as the one in Penang, Malaysia.

In so far as the butterfly house industry has put a market value on 300 or more species of insect, there is a broad benefit to the conservation/development partnership. Rural people in the countries of origin derive some financial benefit and presumably come to place a certain value on the wild insects and habitats from which they draw their breeding stock. More substantial returns to conservation are hard to evaluate, but one case stands out. Shipstern Wildlife Reserve has a 9600-ha private forest reserve in north-east Belize. Sustainable use of the reserve is a major objective for the two entrepreneurs who own the reserve, one of whom is a keen entomologist. They help to achieve this by supplying butterflies from breeding operations in the reserve to their own butterfly houses in the UK and mainland Europe. As supplies grow they will also be marketed to other establishments. The reserve is currently being developed, but when profits increase they will be used to improve its infrastructure as well as the visitor centre, research facility and the butterfly breeding operation, thus benefiting conservation and improving the local economy.

B. Deadstock

There are two elements to the trade in dead insects; these can be designated the "decorative" and "specialist" trades. The decorative trade is discussed under insect products (Section III.C, below), but inevitably there is some overlap with the specialist trade. Many more insects are utilized as deadstock than living material. The turnover world-wide is unknown, but certainly runs into tens of millions of dollars. The Taiwan deadstock trade alone was worth \$20–30 million in the late 1970s (Inskipp and Wells, 1979), but it has certainly declined considerably since then. Although high-quality specimens need careful handling, the deadstock trade is not dependent on fast transportation and careful logistics, as is the live trade. Butterflies for the decorative trade can be handled more robustly than those destined for specialists, and in Taiwan, at least formerly, were processed in a production line. Tens, if not hundreds of millions of butterflies were distributed from the Taiwanese factories, which, as well as using specimens caught locally, imported material from Madagascar, Latin America and south-east Asia for processing and redistribution. As a result, the display cases of butterflies for sale in the world's major cities and airports are much the same whether the purchaser is in Taipei, Caracas, Singapore or Heathrow. Large numbers of cheap butterflies are used in the decorative trade, which is often characterized as "high volume, low value and low quality". This trade accounts for the bulk of the world turnover.

However, the butterflies supplied by the specialist trade include high-value, high-quality material for hobbyists, serious collectors and even research workers. Many specimens are not in this category, and a hobbyist wishing to collect Papilionidae, for example, can accumulate about 80% of the world's species without stepping further than the nearest post office (Table II): 61% of the 573 species listed by Collins and Morris (1985) are available for under £5 each (many for less than 60p), and by shopping around and sharing purchases in bulk with others, many species can be obtained very cheaply. A further 18% of species come in at £5–£100, but only the wealthy enthusiast can afford the 1–2% available at over £100 per specimen. The highest recorded price for one butterfly was the equivalent of US \$1500 for a male *Ornithoptera "allottei"* (almost certainly a natural hybrid) sold at a Paris auction in 1966, although specimens of other birdwings have been advertised for up to US $7000, and high prices have undoubtedly been paid on the black market for protected birdwings and rare parnassians.

A feature which particularly distinguishes the specialist trade from both the live and decorative markets is that it concentrates on the rare, valuable and often threatened species. Four of the species most conspicuously in this category are now listed on Appendix I of the Convention on International Trade in Endangered Species of Wild Fauna and Flora (CITES), and are thereby banned from international commercial trade. Queen Alexandra's birdwing, *Ornithoptera alexandrae*, arguably the world's most endangered species and totally protected by law in Papua New Guinea (its only country of occurrence) since 1966, was fetching over US $2000 for a fine specimen, presumably from an old collection! The Jamaican Homerus swallowtail, *Papilio homerus*, was advertised for up to US $2800 in the USA in 1984, before the ban. The other two CITES Appendix I species, the Corsican swallowtail, *Papilio hospiton*, and the Luzon peacock, *Papilio chikae*, were heavily traded. $25 was perhaps an average price for a specimen of *P. hospiton*, and a specimen was recently (and illegally) offered at US $35. *P. chikae* illustrates a common feature of the specialist trade: when a previously unknown, or poorly understood, species first comes on the market it fetches very high prices, but if it continues to be available, prices fall rapidly; in 1988 the

TABLE II. The availability of the world's Papilionidae as deadstock on the open market 1980–1985; based on minimum prices advertised and reflecting the price of males, females being generally more expensive (after Collins and Morris, 1985).

	Number of available species			Number of unavailable species	Total world spp.
	Less than £5	£5–100	More than £100		
Baroniinae	1	–	–	–	1
Parnassiinae					
Parnassiini	12	13	2	12	39
Zerynthinii	6	3	–	5	14
Papilioninae					
Leptocircini	105	17	–	31	153
Troidini	85	29	3	27	144
Papionionini	143	38	1	40	222
Papilionidae					
Total	352	100	6	115	573

species was offered (again illegally) at US $35. Trade in several other valuable butterflies is now monitored by CITES (Appendix II species). These include the genera *Bhutanitis* (pairs can reach US $200+, depending on species), *Teinopalpus* (US $75 for a good pair of *T. imperialis*), and the apollo (*Parnassius apollo*). The last is one of the most widely distributed of all Papilionidae, and most specimens are not very costly (*ca.* $2–20); it is listed in CITES partly because of the very large numbers of specimens traded, partly because so many local populations are at risk, and partly, it must be admitted, for reasons which remain obscure.

Although it has been thought necessary to ban or monitor the trade in these few species, it must be emphasized that the trade in deadstock has been known to have a serious adverse effect on insect (primarily butterfly) populations only where the extent of the species' habitat has already been severely reduced (for instance by forest clearance in the case of *O. alexandrae*). In industrialized countries there is justifiable concern about the activities of commercial collectors who irresponsibly catch large numbers of species that are already threatened, usually by changes in land use. The protection of the law must be extended to these species, but this should not be in the form of undiscriminating bans on the collection of all, or even many, species. Legislation which has the effect of stifling interest in insects by banning most forms of collecting attacks the wrong targets. Long lists of "protected" insects are counterproductive: the Bern Convention Appendix II species strike the right kind of balance. Protection should be positive, not negative; recovery plans for the restoration of species' populations, and of course their habitats, should be an integral part of conservation laws. Again, the Bern Convention contains several of the positive features needed in effective legislation.

In most tropical countries, the benefits from the butterfly trade, in both its economic and educational aspects, outweigh any small negative impacts. The close relationship between utilization and conservation of the butterfly resource needs to be emphasized. It is unrealistic to expect the rural poor of the Third World to conserve a resource which is of no perceived direct or indirect benefit to them.

Before independence was established in Papua New Guinea, the country's legislators were concerned that trade in butterflies tended to benefit foreign entrepreneurs rather than the indigenous population. Commercial collecting was therefore restricted to native Papua New Guineans, and as a further enlightened step, the Insect Farming and Trading Agency was set up to service the small band of butterfly farmers in the country. The operation of the Agency has been described by the US National Academy of Sciences (1983) and Hutton (1985), and the establishment of a butterfly farm by Parsons (undated). The form of farming almost exclusively practised in Papua New Guinea is technically ranching. Foodplants are established (habitat enrichment) in a butterfly garden, fenced against pigs, and these plants, such as *Aristolochia* vines, attract butterflies from the forest. The resulting larvae are reared without protection from predators and parasites until just before pupation. In essence, the Agency is a means of ensuring that the profits from the deadstock trade (Papua New Guinea does not permit the export of live insects) go mostly to the producers, and not to middlemen. The Agency is being used as a model for similar schemes to benefit rural poor and conservation of the resource elsewhere, for instance in Irian Jaya. In other countries, butterfly farming often operates in combination with the live trade; however, some operators dislike producing dead butterflies for the trade.

Butterfly farming has been regarded as of conservation benefit because it does not deplete wild stocks. However, it is unlikely that butterflies are threatened by commercial collecting under normal circumstances, and in fact some of the material supplied by the Papua New Guinea Agency is wild-caught. Rule-of-thumb conservation practices, for example in utilizing the popular *Delias* species, are sometimes applied by those collecting wild-caught butterflies commercially. Farming does produce the high-quality specimens valued by most specialists.

Lepidoptera are by no means the only insects utilized in the deadstock trade. In fact, informal study of dealers' catalogues suggests that the proportion of Coleoptera and orthopteroid insects in trade is increasing. Tropical Cerambycidae, Buprestidae and Dynastinae (Scarabaeidae) are some of the popular groups, whilst trade in beetles such as *Carabus s.l.* in Europe is considerable. It is unlikely that many of these insects, except Phasmida, are farmed to any extent.

C. Insect Products

Rather few products, other than food, are produced by insects. Lac, cochineal and beeswax are examples, but are mainly given by insects in domestication, particularly in the case of beeswax. However, wild bees' nests do yield wax, as well as honey, mainly to indigenous societies. Lac and cochineal are produced by species of Coccoidea, as are other less well-known gums and waxes. Silk is also mainly produced by domesticated species; indeed, the silkworm, *Bombyx mori*, is not known in the wild state. Various species of saturniine moth produce different silks, mainly used locally, but sometimes exported. Examples are Shantung, Tasar, Muga and Eri silks.

A link with the deadstock trade (see Section II.B) is provided by the "butterfly pictures" manufactured in Taiwan, for example. The ingenious use of cut butterfly wings to make stereotyped pictures arouses conflicting opinions, but the pictures are said to be popular with some tourists. Insects, often embedded in plastic, are used in a wide range of decorative or "fancy" goods. In these products insects are used only as embellishments, so that the product is not of specific or functional use.

Many other insect products have little more than an anecdotal or curiosity value. Blowfly larvae are used on quite an intensive scale as angling bait in western countries. A more surprising, and now superseded use of such larvae was to scavenge in soldiers' wounds, so keeping them relatively free from gangrene. Still in a surgical vein, it is said that the mandibles of certain ants have been used to suture cuts, in the absence of more conventional methods. Some insects produce drugs, the most notorious being "Spanish fly" (cantharidine, extracted from the meloid beetle *Lytta vesicatoria* and other species), a dangerous aphrodisiac which has, or had, genuine pharmaceutical uses. Finally, the tannin in cynipid galls on oak, most notably those of *Andricus kollari* (oak marble gall), has been used in the preparation of leather and the manufacture of ink. However, as galls are only induced, not produced, by the gall "causer", this may not be a valid example of an insect product.

D. Food

The use of insects as human food, in European-based societies, is beset with cultural and social taboos (Holt, 1988). Despite being an omnivore of quite remarkable range, western man has never taken to eating insects on any scale. Entomophagy has been for cranks, or seen to have only curiosity value.

Leaving aside any cultural connotations, the nutritional value of typical and well-known insects, such as the locust, *Locusta migratoria*, the honey bee, *Apismellifera* and the pupae of the housefly, *Musca domestica* is high (Table III). Many insects contain a good deal of protein, much of which is available (Dreyer and Wehmeyer, 1982). Although the amino acid composition of insect protein is not ideal for mammal nutrition, this is easily balanced by the addition of suitable plant protein (DeFoliart, 1989). The calorific value of insects is usually high, and they often contain useful minerals and vitamins (Taylor, 1975).

In cultures other than those of the western world, insects are often a significant part of the diet (Bodenheimer, 1950; Taylor, 1975; DeFoliart, 1989). In Africa many species are utilized, but predominantly vari-

TABLE III. The nutritional value of some conventional foods and of insects (percentages by weight; data from Taylor, 1975).

	Protein	Fat	Carbohydrate	Water	Minerals
Beef	17.4–19.4	15.8–25.1	0	56.7–63.9	0.8–0.9
Chicken	20.6–23.4	1.9–4.7	0	73.7	1.0
Milk	3.5	3.7	4.9	87.2	0.7
Locusta	15.3–46.1	2.4–9.6	6.8–7.5	10.5–70.6	0.8–5.0
Musca (dry pupae)	63.1	15.5	–	3.9	5.3
Apis (larvae)	15.4	3.71	0.4	70.2	2.2

ous Orthoptera (grasshoppers and tettigonids, especially locusts and *Schistocerca* in northern and central areas) and large moth caterpillars (especially Saturniidae such as the "Mopani worm", *Gonimbrasia belina*; Silow, 1976; Malaisse and Parent, 1980). Known as finkubala in Zambia, these insects are prepared by removing the bitter-tasting guts and are then boiled and fried, or stewed or dried in the sun (Brandon, 1987). Most insects require less preparation than this.

In south-east Asia a greater variety of insect life is consumed. The giant water bug *Lethocerus indicus* is a particular delicacy, only the males being used, as it is said that the characteristic flavour derives from their sex pheromone glands. The natives of New Guinea eat insects such as cicadas and the grubs of sago beetles. It is said that when the butterfly trade was developing, some tribesmen had to be persuaded not to eat the colonial larvae of the swallowtail *Papilio laglaizei*, because they were more valuable as a resource for the deadstock trade. The use made by Australian aborigines of "wichety grubs" and "baked bogong moths" is perhaps a trifle more familiar (Taylor, 1975). Koreans make good use of grasshoppers and the pupae of *Bombyx mori*, which are a by-product of silk manufacture. The Japanese have similar tastes, and it is recorded that, following an operation in 1987, the late Emperor Hirohito ate little other than rice with sugar and cooked wasps (a mixture of larvae, pupae and adults of *Vespula lewisi*; Mitsuhashi, 1988).

Use of insects as food is also widespread in South and Central America (DeFoliart, 1989). In Mexico particularly, a very wide variety of insects is consumed. Pre-Columbian North Americans evidently also made good use of insects for food, notably Orthoptera (Madsen, 1989).

There are some problems in increasing the use of insects as human food. Compared with other animal food, insects are small, making difficulties in collecting and processing. There are some risks from parasites, poisoning, allergens and disease, though these are likely to decrease with greater familiarity with the food resource. The unpredictability of wild sources of insect food is also a problem. Particularly in the Third World there are good reasons why the use of insects for human food may increase. The nutritional value of the food is good. Economic necessity may induce many people to make greater use of insects.

"Green" issues may be more important in western Europe, where the conversion rate to conventional animal protein is inferior in terms of conversion/unit area. Medicinal use, and survival in wilderness areas, are rather more specialized aspects. However, the novelty of entomophagy in western culture may give it a considerable boost in the next few years. Perhaps less debatable is the apparent potential for the use of "farmed" insects as feedstocks for poultry, pigs, fish and even astronauts (DeFoliart, 1989)!

IV. PLANNING FOR RESOURCE MANAGEMENT

A. Maintaining Insect Diversity

This section is an overview of the maintenance of the diversity of insects, particularly in what is termed the "wider countryside" in the UK and other parts of the developed world, and in the "natural environment" of developing countries. In any planning of conservation these issues are complementary to conservation in protected areas. Particular effort is put into the selection, acquisition and management of protected areas because wildlife conservation is either a primary or secondary objective of maintaining these areas. In the wider countryside, conservation has to compete with a wide range of other forms of land use. Consequently, it is likely to be at a disadvantage when specialized management for specific conservation objectives is required. However, there is no doubt that protected areas, however large, numerous and well managed, on their own are not a complete answer to conservation problems. Even in the most intensively farmed landscapes some wildlife, particularly insects, survives. A major aim of conservationists is to enhance, perpetuate and improve this survival.

The irony has long been recogized that the developed temperate countries (the north) have most of the world's conservationists and relatively low biodiversity, whilst the tropics (or south) are much richer in species and biodiversity generally yet have few workers in the conservationist field. The geographical analysis of a group such as the Papilionidae shows the difference in biodiversity distribution clearly (Collins and Morris, 1985). Broadly, the difference in abundance of biologists is a result of the relative affluence of the two areas. There is in consequence of these two major differences a need to adopt separate approaches to the conservation of insect diversity in the two kinds of area, whilst recognizing aspects of conservation that they have in common.

In the tropics, the main problem is to prevent, or more realistically slow down, the present loss of natural habitats, particularly the richest ones, of which rain forest is one of the most important. There have recently been some hopeful signs of slight improvement in what is often regarded as an intractable, even hopeless, cause. Amelioration of Third World debt through "forgiveness", rescheduling, or exchange for conservation of protected areas is now well advanced. Linking of development with conservation is something more than just a pious hope, at least in some states (Anon., 1987). Despite these signs, however, the overall position remains grim. Growth of human population is rampant, particularly in India, south-east Asia and in South and Latin America. Poverty, internal strife and militarism, and overexploitation of resources are endemic in many parts of the world, and unlikely to improve in the short term. Transmigration of human population, at present small (though not insignificant in areas such as Irian Jaya; Petocz, 1984), is likely to be a major problem in the coming decades as a result of global climate change, especially sea level rise and increased desertification (Tickell, 1989). Climatic change itself may require radical reassessment of priorities for the conservation of biodiversity in the medium term. Against this stark picture three important issues stand out.

1. The utilization of insect biodiversity, as both services and goods, needs to be encouraged. At international, national and regional levels such efforts are unlikely to be very significant, but they may be crucial at local and village level. The educative effect of promoting a regard for the value of the insect resource is often forgotten when its low commercial value is calculated.
2. The global network of protected areas must be improved in number, quality and degree of protection and management; this is discussed below.
3. The scientific effort put into describing the insect resource in all its aspects must be improved. Throughout the twentieth century, estimates of the numbers of insect species on earth have been rising (Stork, 1988). Although there are clearly no easy answers to the paucity of scientific manpower in the tropics, emphasis on insect biodiversity as a practical resource gives some hope for improvement. There are many contraindications, however (see Section I). Development of a greater interest in insects would be a step in the right

direction. Thus, it is important that the laudable aim of conserving insect diversity should not inhibit unduly the scientific, personal and even commercial collecting of insects. The conservation, utilization and scientific interest of a self-renewing resource such as insect biodiversity form the basis of a resource management strategy (Morris, 1986).

Although habitat destruction is also the main problem in conserving the insect resource of the developed temperate countries, many of the details are different. Poverty, famine, war and population growth are not the problems they are in the developing world. On the other hand, a long history of exploitation, development and intensification of agriculture has greatly reduced the extent of natural and semi-natural communities, particularly in highly populated countries such as the UK. Since the beginning of the Second World War, agricultural production has greatly increased, with consequent reduction in the abundance of many insect species. During the same period, however, there has been growing interest and activity in wildlife conservation and, more recently, in more general environmental concern and green issues. Support for conservation has risen markedly through membership of such bodies as the National Trust, Royal Society for the Protection of Birds, World Wide Fund for Nature, county wildlife and nature conservation trusts, and many others. The relative species poverty of the British (and European) insect fauna, long history of interest in entomology (amateur and professional) and considerable body of scientific information on all aspects of conservation are factors which distinguish Britain and the rest of Europe, together with other developed countries, from Third World ones. Despite these characteristics, there are many aspects of conservation in the developed world which are highly unsatisfactory. Conservation interests compete on unequal terms with those of development and commerce. The very popularization of conservation can lead to misunderstanding of scientific aims and rejection of what is seen as elitism.

The maintenance of insect biodiversity in the wider countryside during the last quarter of this century and into the next is based on a number of important considerations. First, the realization that the intensification of agriculture is finite and probably past its peak is vital. Although environmentally sensitive areas (ESAs), set aside, and more recently extensification have yet to be assessed, the likelihood is that they will tend to enhance insect biodiversity. Second, the processes of recording and evaluating the insect fauna (e.g. Heath *et al.*, 1984; Shirt, 1986) are well advanced, even if progress is slower than is desirable. The initiative of the Nature Conservancy Council in commissioning reviews of major orders is to be particularly welcomed. Third, conservation management, of both species–populations and protected areas, is well developed. Butterflies (Thomas, 1984) are a very good, but by no means the only group of insects to be studied for conservation purposes. The conservation of a resource such as calcareous grassland has a considerable entomological input (Hillier *et al.*, 1990). When these aspects are considered together, they support the assertion of Bradshaw (1977) that restoration of disturbed sites, re-establishment of lost species and the creation of new areas of interest on degraded and abandoned sites are the conservation challenges for the future. Considerable progress has been made in the re-establishment of species in both its theoretical (Morris and Thomas, 1989) and practical aspects (Thomas, 1989). Studies of natural recolonization of re-created habitat types (e.g. Morris, 1990a,b,c) and secondary succession (Brown and Southwood, 1987) are important in this context.

Conservation in the wider countryside is often contrasted with site safeguard; the two are, of course, complementary. The selection, and to some extent the acquisition, of protected sites in Britain is generally satisfactory (Ratcliffe, 1977), though less so in some other parts of Europe. However, improvements in defining the objectives of management of protected sites and in management itself are certainly possible. Most key areas in Britain are plagioclimax, and rotational management, together with a suite approach to a number of similar reserves, is often appropriate, particularly for insects (Morris, 1990). In the view of some conservationists, emphasis on re-establishment and re-creation is misplaced because of its supposed devaluing of the near uniqueness of natural and semi-natural sites. This misconception must be countered if the best is to be achieved from conservation in the wider countryside as well as in protected areas.

B. Using Insects to Identify Key Areas for Conservation

Because key, or protected, areas can be conceived as islands in a sea of less important land, the equilibrium theory of biogeography (MacArthur and Wilson, 1967) has been fruitful in stimulating studies aimed at identifying the biogeographical basis of reserve selection (Usher, 1986). Theoretical aspects of geometry, size, number of fragments and colonization/extinction rates have been combined with the use of real biological criteria. Although entomological criteria have not been extensively employed in this field, some use has been made of insect surveys in the selection of reserves. Richness of Heteroptera, and especially Auchenorrhyncha, was a supporting criterion in the choice of calcareous grassland sites in the British Nature Conservation Review (Ratcliffe, 1977). Unfortunately there has been little agreement on which groups of insects to select for the purpose. Disney (1986) advocated Diptera and parasitic Hymenoptera; Speight (1986) proposed Carabidae, Syrphidae and Symphyta. Popular groups such as butterflies are probably too limited in numbers of species to be satisfactory in reserve selection in northern Europe, though they are valuable in assessing the effects of management (Thomas, 1991). In aquatic habitats, water beetles (Eyre *et al.*, 1986) and Trichoptera (Green, 1989) have been used. In tropical regions well-known insect groups such as the Papilionidae (Collins and Morris, 1985) may have considerable value for the recognition of areas of high or special diversity, especially when subjected to new methods of critical faunas or network analysis (Vane-Wright *et al.*, 1991; Margules, 1989).

The suitability of insect groups in assessment work has been judged mainly on ecological grounds. The use of utilisation data offers an alternative approach. At present, ruderal communities and the early stages of secondary succession are probably undervalued in comparison with plagioclimax communities, such as lowland heath, wetlands and calcareous grasslands. Yet many of the potential weed species mentioned earlier under Section IIC Biological Control are characteristic of such successional communities.

C. Maintaining Key Areas through the Benefits from Utilization

In conservation philosophy the concepts of "wilderness" and "non-intervention areas" have a firmly established place. On the other hand, there has been a pronounced shift from an "elitist" approach to reserve establishment to a more popularist "reserves are for people" rationale. At the present time, the pendulum appears to be swinging back again, with fears being expressed for the long-term security of the Pennine Way or the proposed South Downs National Park (to give two UK examples) in the face of rapidly increasing pressure from visitors.

Utilization benefits are an important component in the debate on these issues, more particularly in the Third World tropics. Although the numbers, range and administration of protected areas often leave much to be desired, the physical extent of most of the best examples far outstrips what western Europe can offer, though this is less true of North America. The selection of these areas can often still be made on comprehensive ecological criteria, in which a full range of altitudinal and spatial variation in habitat type can be conserved — at least on paper. What is often not so clearly publicized is the degree of exploitation which accompanies the declaration of these areas as protected, whether that exploitation be of dwellers in or near the reserve or of its wildlife. It is clear that few nations can afford the luxury of having large (up to 1 000 000 ha) reserves which are unused. The utilization of wildlife in all acceptable forms must be an integral part of the management and sustainment of these areas. As yet, the economic and practical details of such utilization are very sketchy. Utilization of non-domesticated insects is an important component in their development.

V. CONCLUSIONS

Some of the earth's greatest wealth is contained in natural ecosystems and the biological communities that inhabit them. Whilst insects are rarely, if ever, the main focus of resource exploitation, they underpin the stability of these communities by their genetic diversity and their complex web of ecological interactions. The direct or commercial value of insects lies in a wide range of goods and services, many of which are overlooked by the standard cost-benefit analyses of primary industries. This paper draws attention to these benefits, and seeks ways of providing economic support for the conservation of insect-rich habitats from the profits of exploitation of natural communities, either in cash or kind.

The services provided by insects include natural suppression of agricultural and medical pests and weeds, pollination of agricultural and silvicultural crops, and decomposition of polluting and hazardous wastes. It is generally only when these services break down that they become appreciated. The diversity of insects is itself a service, acting as it does as a genetic resource for biological control programmes, and as a pool of biochemical products that is only beginning to be assessed for its usefulness.

Insect goods are widely traded. The range of species and products traded for nutritional, aesthetic and scientific reasons probably outstrips all other forms of life, although the total economic turnover is modest. In some developing countries insects may account for 10% or even more of the protein consumed by rural people.

Despite these actual and potential benefits, the resources are not being conserved adequately. Insects, along with many other lesser known organisms, are being swept aside by development in tropical and temperate climates alike. The papers presented in this volume testify to this fact. The sites of richest diversity must be identified and managed as genetic resources for the future. All branches of entomology can contribute, and indeed should contribute, to such an analysis. The developed nations, who benefit most from the world's biological diversity, can implement the necessary research and facilitate non-destructive management of natural ecosystems through economic inducements. It is time to realize that insects, in common with other life forms, are neither inexhaustible nor indestructible.

REFERENCES

Altieri, M. A. and Letourneau, D. K. (1984). Vegetational diversity and insect pest outbreaks. *CRC Crit. Rev. Plant Sci.* **2**, 131–169.

Anonymous (World Commission on Environment and Development). (1987). "Our Common Future". Oxford University Press, Oxford.

Bodenheimer, F. S. (1950). "Insects as Human Food". W. Junk, The Hague.

Bornemizza, G. F. (1976). The Australian dung beetle project 1965–1975. *Aust. Meat Res. Comm. Rev.* **30**, 1–30.

Bradshaw, A. D. (1977). Conservation problems in the future. *Proc. R. Soc. Lond. B* **197**, 77–96.

Brandon, H. (1987). The snack that crawls. *Int. Wildl.* (March–April), 16–21.

Brown, R. A. (1989), Pesticides and non-target terrestrial invertebrates: an industrial approach. *In* "Pesticides and Non-target Invertebrates" (P. C. Jepson, ed.), pp. 19–42. Intercept Ltd, Wimborne, Dorset.

Brown, V. K. and Southwood, T. R. E. (1987). Secondary succession: patterns and strategies. *In* "Colonisation, Succession and Stability" (A. J. Gray, M. J. Crawley and P. J. Edwards, eds), pp. 315–337. British Ecological Society Symposium No. 26. Blackwell Scientific Publications, Oxford.

Burns, A. J. (1989). Long term effects of pesticides on natural enemies of cereal crop pests. *In* "Pesticides and Non-target Invertebrates" (P. Jepson, ed.), pp. 177–193. Intercept Ltd, Wimborne, Dorset.

Collins, N. M. (1981). The rôle of termites in the decomposition of wood and leaf litter in the Southern Guinea savanna of Nigeria. *Oecologia* **51**, 389–399.

Collins, N. M. (1982). The interaction and impact of domestic stock and termites in a kenyan rangeland. *In* "The Biology of Social Insects" (M. D. Breed, C. D. Michener and H. E. Evans, eds) pp. 80–84. Westview Press, Boulder, Colorado.

Collins, N. M. (1987). "Butterfly Houses in Britain: the Conservation Implications". IUCN, Cambridge, UK, and Gland, Switzerland.

Collins, N. M. (1988). Termites. *In* "Malaysia" (Earl of Cranbrook, ed.), pp. 196–211. Key Environment Series. Pergamon, Oxford.

Collins, N. M. and Morris, M. G. (1985). "Threatened Swallowtail Butterflies of the World: the IUCN Red Data Book". IUCN, Gland.

Cooke, A. S. and Mitchell-Jones, A. J. (1988). Dangers to bats. *Batchat* **10**, 1–5.

DeFoliart, G. R. (1989). The human use of insects as food and as animal feed. *Bull. Entomol. Soc. Am.* Spring, 22–35.

Disney, R. H. L. (1986). Assessments using invertebrates: posing the problem. *In* "Wildlife Conservation Evaluation" (M. B. Usher, ed.), pp. 271–293. Chapman and Hall, London.

Dreyer, J. J. and Wehmeyer, A. S. (1982). On the nutritive value of Mopanie Worms. *S. Af. J. Sci.* **78**, 33–37.

Edwards, C. A., Sunderland, K. D. and George, K. S. (1979). Studies on polyphagous predators of cereal aphids. *J. Appl. Ecol.* **16**, 811–823.

Eyre, M. D., Ball, S. G. and Foster, G. N. (1986). An initial classification of the habitats of aquatic Coleoptera in North-east England. *J. Appl. Ecol.* **23**, 841–852.
Farnsworth, N. R. and Soejarto, D. D. (1985). Potential consequence of plant extinction in the United States on the current and future availability of prescriptive drugs. *Econ. Bot.* **39**, 231–240.
Free, J. B. (1964). Comparison of the importance of insect and wind pollination of apple trees. *Nature* **201**, 726–727.
Free, J. B. (1970). "Insect Pollination of Crops". Academic Press, London.
Gear, A. (1983). "The Organic Food Guide". Essex.
Grant, I. F. 1989. Monitoring insecticide side-effects in larger-scale treatment programmes: tsetse spraying in Africa. *In* "Pesticides and non-target Invertebrates" (P. Jepson, ed.), pp. 43–69. Intercept Ltd, Wimborne, Dorset.
Greathead, D. J. (1983). The multi-million dollar weevil that pollinates oil palms. *Antenna* **7**, 105–107.
Green, P. C. (1989). The use of Trichoptera as indicators of conservation value: Hertfordshire gravel pits. *J. Environ. Manage.* **29**, 95–104.
Greig-Smith, P. W. (Ed.) (1988). "The Boxworth Project. Pest and Disease Control Systems. Their Ecological and Economic Effects in Intensive Cereal Production". 1988 Report. ADAS, Tolworth Laboratory, 138 pp.
Hawkins, R. P. (1965). Factors affecting the yield of seed produced by different varieties of red clover. *J. Agric. Sci. (Cambridge)*, **65**, 245–253.
Heath, J., Pollard, E. and Thomas, J. A. (1984). "Atlas of Butterflies in Britain and Ireland". Viking, Harmondsworth.
Hillier, S. H., Wells, D. A. and Walton, D. W. H. (eds) (1990). "Calcareous Grasslands: Ecology and Management". Bluntisham Books, Bluntisham, Cambs.
Holt, V. M. (1988). "Why Not Eat Insects?" (reprint with new introduction). British Museum (Natural History), London.
Hudson, P. J. and Rands, M. R. W. (1988). "The Ecology and Management of Gamebirds". Blackwell Scientific Publications, Oxford.
Hunt, E. G. and Bischoff, A. I. (1960). Inimical effects on wildlife of periodic DDD applications to Clear Lake. *Calif. Fish Game* **46**, 91–106.
Hutton, A. F. (1985). Butterfly farming in Papua New Guinea. *Oryx* **19**, 158–162.
Inskipp, T. and Wells, S. M. (1979). "International Trade in Wildlife". Earthscan, London.
IUCN/UNEP/WWF (1980). "World Conservation Strategy", IUCN, Gland, Switzerland.
Julien, M. H. (1987). "Biological Control of Weeds", 2nd edn. CAB Int., Farnham Royal, Slough.
MacArthur, R. H. and Wilson, E. O. (1967). "The Theory of Island Biogeography". Princeton University Press, Princeton.
Madsen, D. B. (1989). A grasshopper in every pot. *Nat. Hist.* **1989**(7), 22–25.
Malaisse, F. and Parent, G. (1980). Les chenilles comestibles du Shaba méridonal (Zaïre). *Nat. Belg.* **61**, 2–24.
Margules, C. R. (1989). Introduction to some Australian developments in conservation evaluation. *Biol. Conserv.* **50**, 1–11.
Mitsuhashi, J. (1988). Rice with cooked wasps: an Emperor Hirohito's favorite dish. *Food Insect Newsl. (Madison)* **1**(2), 2.
Moore, N. W. (1987). "The Bird of Time". Cambridge University Press, Cambridge.
Morris, M. G. (1986). The scientific basis of insect conservation. *Proc. 3rd Eur. Congr. Entomol.* **3**, 357–367.
Morris, M. G. (1990). The effects of management on the invertebrate community of calcareous grassland. *In* "Calcareous Grasslands: Ecology and Management" (S. H. Hillier, D. A. Wells and D. W. H. Walton, eds). Bluntisham Books, Bluntisham, Cambs.

Morris, M. G. (1990a). The Hemiptera of two sown calcareous grasslands. I. Colonisation and early succession. *J. Appl. Ecol.* **27**, 367–378.
Morris, M. G. (1990b). The Hemiptera of two sown calcareous grasslands. II. Differences between treatments. *J. Appl. Ecol.* **27**, 379–393.
Morris, M. G. (1990c). The Hemiptera of two sown calcareous grasslands. III. Comparisons with the Auchenorrhyncha faunas of other grasslands. *J. Appl. Ecol.*, **27**, 396–409.
Morris, M. G. and Thomas, J. A. (1989). Re-establishment of insect populations, with special reference to butterflies. *In* "The Moths and Butterflies of Great Britain and Ireland" (A. M. Emmet and J. Heath, eds), Vol. 7 (1), pp. 22–36. Harley Books, Colchester.
Murray, M. G. (1985) Figs (*Ficus* spp.) and fig wasps (Chalcidoidea, Agaonidae): hypotheses for an ancient symbiosis. *Biol. J. Linn. Soc.* **26**, 69–81.
National Research Council (1979). "Tropical Legumes: Resources for the Future". National Academy Press, Washington.
Nilsson, L. A., Jonsson, L., Rason, L. and Randrianjohany, E. (1985). Monophily and pollination mechanisms in *Angraecum arachnites* Schltr. (Orchidaceae) in a guild of long-tongued hawk-moths (Sphingidae) in Madagascar. *Biol. J. Linn. Soc.* **26**, 1–19.
Norris, K. R. (1974). General biology. "The Insects of Australia, Suppl. 1974", Vol. 5, pp. 21–24. CSIRO, Melbourne University Press.
Parsons, M. (undated). "Insect Farming and Trading Agency Farming Manual". Division of Wildlife, Bulolo, PNG.
Petocz, R. J. (1984). "Conservation and Development in Irian Jaya. A Strategy for Rational Resource Utilisation". WWF/IUCN for Directorate-General of Forest Protection and Nature Conservation, Bogor, Indonesia.
Poehling, H.-M. (1989). Selective application strategies for insecticides in agricultural crops. *In* "Pesticides and Non-target Invertebrates" (P. C. Jepson, ed.), pp. 151–175. Intercept Ltd., Wimborne, Dorset.
Rands, M. R. W. and Sotherton, N. W. (1986). Pesticide use on cereal crops and changes in the abundance of butterflies on arable farmland. *Biol. Conserv.* **36**, 71–83.
Ratcliffe, D. A. (ed.) (1977). "A Nature Conservation Review" 2 Vols. Cambridge University Press, Cambridge.
Ratcliffe, D. [A.] (1980). "The Peregrine Falcon". T. and A. D. Poyser, Calton.
Shirt, D. (ed.) (1986) "British Red Data Books. 2. Insects". Nature Conservancy Council, Peterborough.
Silow, C. A. (1976). Edible and other insects of mid-western Zambia. Studies in ethno-entomology II *Occasional Papers of the Institutionen för Allmän och Janforande Etnografi via Uppsala Universitet* (5), xvi + 223 pp.
Speight, M. C. D. (1986). Criteria for the selection of insects to be used as bio-indicators in nature conservation research. *Proc. 3rd Eur. Congr. Entomol.* **3**, 485–488.
Stephen, W. P. (1961). Artificial nesting sites for the propagation of the leaf-cutter bee, *Megachile* (*Eutricharaea*) *rotundata*, for Alfalfa pollination. *J. Econ. Entomol.* **54**, 989–993.
Stephen, W. P. (1965). Artificial beds for Alkali Bee propagation. *Agric. Exp. Stn, Oregon State Univ. Corvallis, Bull.* 598.
Stork, N. E. (1988). Insect diversity: facts, fiction and speculation. *Biol. J. Linn. Soc.* **35**, 321–337.
Swift, M. J., Heal, O. W. and Anderson, J. M. (1979). "Decomposition in Terrestrial Ecosystems. Studies in Ecology". Vol. 5. Blackwell Scientific Publications, Oxford.
Taylor, R. L. (1975) "Butterflies in my Stomach". Woodbridge Press, Santa Barbara.
Thomas, J. A. (1984). The conservation of butterflies in temperate countries: past efforts

and lessons for the future. *In* "The Biology of Butterflies" (R. I. Vane-Wright and P. R. Ackery, eds), pp. 333–353. Royal Entomological Society, London.

Thomas, J. A. (1989). Ecological lessons for the re-introduction of Lepidoptera. *Entomologist* **108**, 56–68.

Thomas, J. A. (1991). Rare species conservation: case studies of European butterflies. *In* "The Scientific Management of Temperate Communities for Conservation" (I. F. Spellerberg, M. G. Morris and F. B. Goldsmith, eds). pp. 141–197. 29th Symposium of the British Ecological Society. Blackwell Scientific Publications, Oxford.

Tickell, C. (1989). Environmental refugees. The human impact of global climate change. *NERC News* July 1989, 14–20.

US National Academy of Sciences (1983). "Butterfly Farming in Papua New Guinea". National Academy Press, Washington.

Usher, M. B. (ed.) (1986). "Wildlife Conservation Evaluation". Chapman & Hall, London.

Vane-Wright, R. I., Humphries, C. J. and Williams, P. R. (1991). What to protect? – systematics and the agony of choice. *Biol. Conserv.* (in press).

Waage, J. K. (1989). The population ecology of pest-pesticide-natural enemy interactions. *In* "Pesticides and Non-target Invertebrates" (P. Jepson, ed.), pp. 81–93. Intercept Ltd, Wimborne, Dorset.

Waage, J. K. and Greathead, D. J. (1988). Biological control: challenges and opportunities. *Phil. Trans. R. Soc. Lond. B* **318**, 111–128.

Waterhouse, D. F. (1974). The biological control of dung. *Sci. Am.* **230**, 100–109.

Wells, S. M., Pyle, R. M. and Collins, N. M. (1983). "The IUCN Invertebrate Red Data Book". IUCN, Cambridge, UK, and Gland, Switzerland.

Wiebes, J. T. (1979). Co-evolution of figs and their insect pollinators. *Annu. Rev. Ecol. System.* **10**, 1–12.

14

Conservation of Neotropical Environments: Insects as Indicators

KEITH S. BROWN JR

I. Introduction ... 350
A. Habitat and Biodiversity Conservation in the Neotropics ... 350
B. Conservation Preserves: Attitudes and Actions ... 352
C. Urban, Rural and Natural Habitats ... 352
D. Conservation: Theoretical and Practical Questions ... 354
II. Adapting Insects to the Tasks of Defining Options, Ranking of Priorities, Making Surveys and Monitoring Systems ... 357
A. Why Conserve Insects, Anyway? ... 357
B. The Taxonomic Tangle ... 358
C. Emerging Patterns of Utility: Comparison of Insects with Birds and Mammals, and Combination with These and with Selected Plants ... 359
D. Contribution of Insect Work to Ecological and Conservation Theory and Practice in the Neotropics ... 360
III. Tools for the Tasks: Description, Discrimination, Explanation ... 360
A. Geological Information ... 360
B. Existing Biological Information ... 362
C. Theories and their Demise ... 362
D. Endemism and Diversity ... 363
E. Indicator Groups ... 363
IV. Case Histories and Some Legends ... 366
A. Endemism and Diversity in Neotropical Forests: the Distant Past ... 366
B. Monitoring the Short-term Effects of Human Intervention: the Biological Dynamics of Forest Fragments Project in Manaus ... 372
C. Long-Term Effects of Human Intervention: the Brazilian Atlantic Forests ... 378
D. Identification of Areas with Exceptional Biological Importance: Palaeo-environments ... 380

E. The Spectre of Extinction: Facts and Myths 381
F. Management of Anthropic Mosaics for Sustainable Yield: Extractive Reserves 382
V. A Practical Guide to Conservation Action using Insects 387
A. How to Choose a Group for Use in Conservation 387
B. How to Predict Areas of Potential Importance 391
C. How to Undertake Surveys of Butterflies (or Other Insects) 391
D. How to Transform Survey Data into Useful Parameters 393
E. How to Relate to the Public Sector and the Local Community 393
F. Insects and the Conservation Transition 396
VI. Epilogue: Important Considerations for the Agenda, 1990–2010 397
A. The Primacy of Education and Grass-roots Movements: Community and Conservation 397
B. Biosphere Reserves as Ideal Insect Habitat 399
C. Economic, Cultural and Scientific Imperialism 399
D. The Importance of Optimism 400
References 401

I. INTRODUCTION

A. Habitat and Biodiversity Conservation in the Neotropics

Conservation of insects forms a major sector of general conservation of terrestrial habitats, genetic diversity and species interactions, since well over half the genes, biomass and energy transfers in terrestrial ecosystems probably involve insects. About a third of the world's insect species occur in the vast expanses of the neotropics (about 17 500 000 km^2 not including south-temperate regions), strongly concentrated in forest and woodland biomes. In the best known large insect group (butterflies), 45% of all neotropical species, with large overlap, can be found in each of three large "megadiversity" countries (Peru, Colombia and Brazil), with 30% in a single part of the last country (the Brazilian Shield), nearly half of these endemic to this ancient system (Table I) (see McNeely *et al.*, 1989).

The topography of the neotropics is very complex on a fine scale, even in the lowland Amazon Basin, giving a high degree of regional heterogeneity (Brown, 1978, 1982). This fact multiplies the niches available for insect species large and small; a single productive tree may host many hundreds of species from over 20 orders, the majority still undescribed (Erwin, 1988). The ecological diversification is further multiplied by stochastic historical factors, on time scales both short (months to years:

TABLE I. Number of species (total found or described, endemic (End.) and threatened (Thr.)) in various regions and groups. R = country ranking (world-wide scale). Many data were received from Tim Werner of the Conservation International Terrestrial Vertebrate Database (Washington, DC), but many figures are still lacking or indeterminable.

	World		Neotropics		Colombia				Peru				Brazil			
Group	Total	Thr.	Total	Thr.	Total	End.	Thr.	R	Total	End.	Thr.	R	Total	End.	Thr.	R
Angiosperms	300 000	18 694			45 000			2	25 000			4	55 000			1
Palms	2 700		1 102	278	249	116	21		261	48	21		355	219	23	3
Orchids	23 500		8 266		3 000			1	2 000	850	300	4	2 500			2
Terrestrial vertebrates	23 890	1 868	9 612		2 867	326	77	3	2 610	312	55		3 030	700	310	1
Mammals	4 170	555	1 325	101	359	24	25	7	361	37	30	6	428	65	50	3
Primates	232		80		27	3		6	30	2	2	5	60	18	25	1
Rodents	1 750		651		93				107	25	10		122	39	7	1?
Birds	9 040	1 029	3 945		1 721	37	28	1	1 704	90		2	1 622	170	108	3
Parrots			143		51	5		4	49	2	15	5	71	25	14	2
Reptiles	6 458	149	2 481		383	103	34	6	297	94		8	467	17	9	4
Amphibians	4 122	48	1 864+		407	162		2	251	91		7	516	32	9	1
Freshwater fishes									840				1 500+			1
Insects	850 000	(described)	300 000?										100 000?		32?	1?
Butterflies	18 000		7 000		3 100			2	3 055			3	3 130[a]		23	1
Swallowtails	461	70	133	19	64	0	2	5	57	0	3	7	68[a]	13	9	4

[a] Of these, 2124 occur in the expanded Atlantic Forest region (including Uruguay and parts of Paraguay and Argentina, and the Araucaria, Cerrado and Caatinga biomes; see Fig. 2), with 951 endemic to this region; in the swallowtails, this region has 43 species, 18 endemic with 5 threatened. See details in Table IV.

unpredictable cold snaps, dry spells, floods, slides, windstorms and treefalls) and long (centuries to millenia: large tectonic movements and climatic fluctuations); in their combined action, these can lead to exalted environmental microheterogeneity with steeply and continuously rising species/area curves for insects, such that a few-hectare site with the right soils, latitude and topography can contain up to half the total species known from an entire country or region of millions of square kilometres (Brown, 1984; Lamas, 1985). The species in these systems live in small, dynamic and highly pressured microhabitat "islands" with intense turnover, quite prone to local extirpation but not to regional extinction. By extension, local communities are often unstable and fragile, while regional systems are highly resistant (*sensu* Pimm, 1986). What do such characteristics mean for the conservation of these species, communities and systems?

B. Conservation Preserves: Attitudes and Actions

The neotropical realm is peppered with large and small reserves and parks (Fig. 1) with varying levels of effectiveness and enforcement. These cover less than 15% of the total area, and not all are supported by local populations — though increasingly they are actually being proposed, secured and policed by local people who recognize their importance for future resources. Many of these reserves have been recommended and set up over the past 20 years, with appreciable participation of standardized insect surveys (endemism and diversity, especially of butterflies) (Fig. 1). While few conservation agencies may proclaim insects as important elements to be preserved, many now recognize that they are fundamental to system structure and function, giving the foundations upon which the "flagship species", often large fuzzy primates and vertebrate predators, must rest and subsist. The present is an ideal time to use insects to promote broad ecological and conservation understanding, sympathy and effective action in the neotropics. How can this moment be best exploited?

C. Urban, Rural and Natural Habitats

A landscape mosaic, such as predominates in many parts of the neotropics, will include different kinds of anthropic (man-made) systems with various levels of intensity of occupation and maintenance, interdigitated with primitive habitats of various sizes, shapes and degrees of disturbance. Insects are not uniformly distributed in such matrices, and tend to be strongly depressed in more artificial or chemical-controlled sectors. While

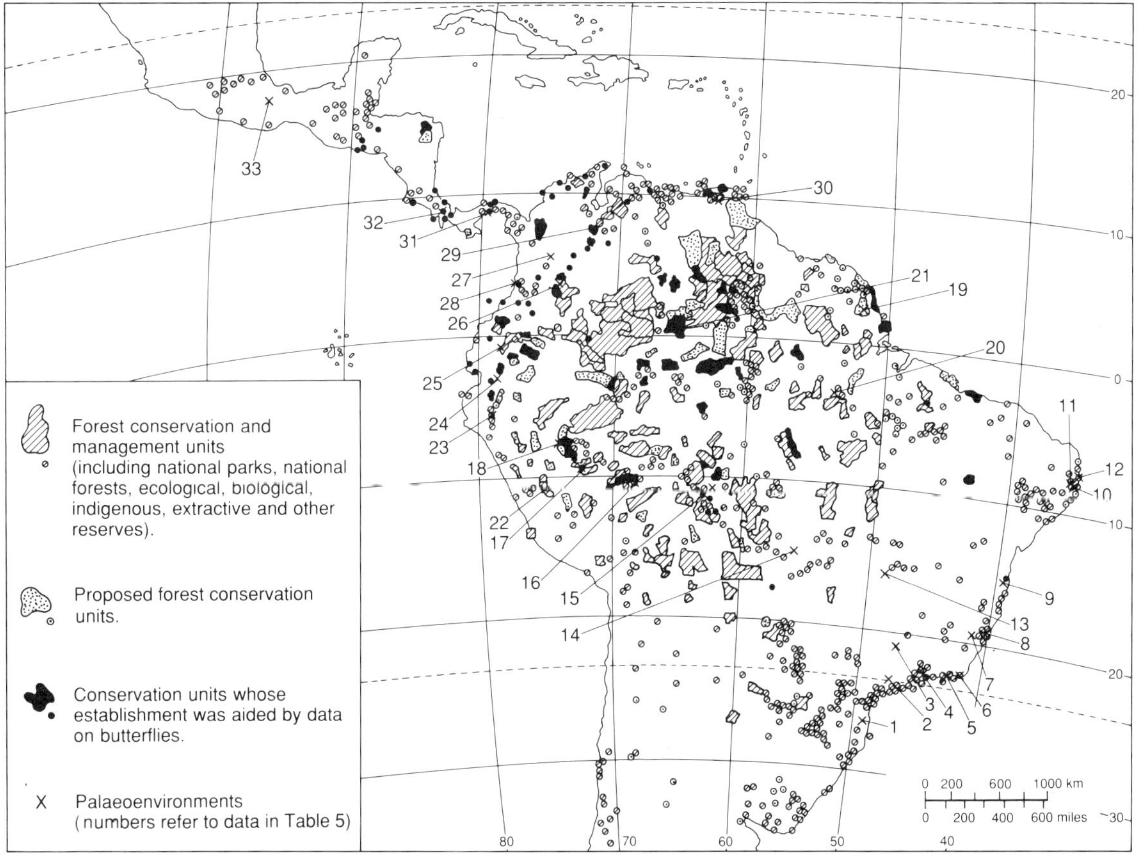

Fig. 1. Conservation units and palaeo-environments in tropical America, with indication of the contribution from insect data (updated from Brown, 1979).

natural disturbance of a tropical habitat, especially if unpredictable and not especially strong, can give an increase in species diversity and energy flow in the community, abuse with synthetic chemical fertilizers, insecticides or particulate pollution usually leads to strong reduction of diversity (especially in soil organisms and arthropods in general), breakdown of community structure and services, and irretrievable loss of the genetic heritage — frequently on a regional scale, today seen over vast areas of debilitated agroecosystems in industrialized nations (Chapter 13).

Maximum population density (of a few resident "pest" species) may occur in some anthropic habitats. Maximum diversity is usually found at ecotones of mildly disturbed natural vegetation; the latter as well as undisturbed patches include species found in no other habitats, while the purely anthropic sectors rarely shelter endemic, restricted or characteristic species in the neotropics. With most such species occurring outside official conservation units, and as much as 95% of some original neotropical biomes already converted or strongly disturbed, some people fear that many species (or at least locally adapted gene complexes) may have already been lost or reduced beyond recovery, with most of the rest soon to follow in their path (Myers, 1988). How resistant are neotropical insect species to transformation of the landscape by humans? What kind of landscape mosaics might provide maximum diversity and minimum species loss?

D. Conservation: Theoretical and Practical Questions

The fundamental questions which close each of the above sections bridge the gap between conservation biology and politics, data and projection, experiment and prediction, idealism and pragmatism. While scientific conservation must be based on extensive hard data, careful statistical analysis and solidly founded models for prediction, practical conservation deals with priorities and efficiency in allocation of funds and favours. In the neotropics, the basic questions of conservation — "What", "Why?", "Where?" and "How?" — are still strongly intertwined with "When?" and dependent on "Who?". Most agencies have accepted the modern response to "What?" as "Genetic diversity and ecological interactions" or their summation as "Evolutionary potential", while recent economic studies of natural resource usage alternatives have given ample answers to "Why?" (Lugo, 1988; Oldfield, 1989; Chapter 12). "Where?" and "How?" are still strongly fluid, empirical, dependent upon local and national politics and often little influenced by scientific data. Conservation biologists have made great progress in incorporating hard data and analysis into priority determination, however, and sociologists and anthropologists have collaborated with ecologists and economists to open up new vistas on methodology. Some more detailed questions about conservation theory and tactics, with suggested responses supported by data and results presented in the rest of this chapter and book, follow:

1. Some Scientific Questions for Conservation, and Possible Responses

Q1. How is biodiversity divided, and where is it concentrated?

R1. Mostly in terrestrial phytophagous or scavenging arthropods, especially Coleoptera, Lepidoptera, Hymenoptera, Diptera, Heteroptera, Orthoptera and Acari, and their various foodplants, in microheterogeneous tropical systems without severe fluctuations or constraints in the environment.

Q2. How much natural turnover exists in biodiversity?

R2. Slow but real, perhaps 1% per 10 000 years on a regional scale without human intervention or climatic change (Wilson, 1988), 10 times this in naturally or (low-level) anthropically disturbed systems, and thousands of times faster in strongly modified landscapes.

Q3. What are the effects of different types of natural and anthropic disturbance, and of mutual selective pressures between insects and their hosts, on biodiversity?

R3. Low-level disturbance multiplies niches and successional stages, increasing local diversity with little effect on the region. Stronger disturbance may decrease both local and regional diversity, or transfer it to lower taxonomic levels (some higher taxa disappear). Long-term disturbance (climatic changes) can do either, though no evidence exists for extensive species multiplication in insects. Habitat conversion or treatment with toxins (substances which reduce the efficiency of life in accumulating, conserving and controlling the flow and eventual dissipation of energy and matter) is usually accompanied by massive reduction in diversity. Coevolutionary diversification is mostly diffuse and very hard to test or demonstrate; it can even lead to a reduction in diversity (Futuyma and Slatkin, 1983; Spencer, 1988; Brown *et al.*, 1991).

Q4. How much diversity is truly divergent in terms of information content — chemistry, function, important community roles?

R4. Most species-level gene complexes are highly and informatively differentiated, at least in expression and function, and important, unique heritable elements are identifiable at subspecies, varietal and population levels also; thus, even local extinctions represent a significant loss of information and biosynthetic capacity, and should be regarded as undesirable.

Q5. How much biodiversity can survive cohabitation with man and his activities?

R5. Probably most survives at present in neotropical landscape mosaics, but massive conversion, burning and topsoil destruction could make large inroads in the future. Many insects seem to be quite resilient to environmental changes, and adaptable to new resources and challenges.

Q6. How much biodiversity has already succumbed to man's activities?

R6. In the neotropics, very little as yet, but acceleration is possible.

2. *Some Practical Problems in Conservation, and Suggested Solutions, incorporating the Above Questions and Responses (see McNeely* et al., *1989)*

P1. Where can the greatest biodiversity be most effectively preserved per unit of expense, area, justification and structuring?

S1. Where it is *found*, empirically, using field surveys with standardized methods and indicator groups.

P2. What indicator groups are most useful and reliable?

S2. Relatively large, ecologically and taxonomically diverse but individually habitat-faithful, well-studied, non-furtive, diurnal, easily identified organisms, with empirical studies often involving the primate/bird/amphibian/butterfly/palm/orchid/vine/fern octet; diversity and ease of survey, sampling and identification favour the second, fourth, sixth and last of these; other groups may be equally useful in many areas.

P3. What about theoretical approaches, such as Pleistocene refugia?

S3. Use as a last resort, and with great care, only when there is no way at all to look at the real situation.

P4. How can normal and integrated natural systems, with usual levels of diversity and turnover, be maintained?

S4. Maintain entire systems intact in large areas with minimal management. Trust nature in these cases, but manage small areas intensively. Maintain continuity, protect energy and matter sources; follow a general policy of knowledgeable benign neglect whenever possible.

P5. How can people be persuaded that conservation of biodiversity is important?

S5. Show them that it is in their own interest, including economically and in the short run; or better yet, help them to discover this fact by themselves and then tell *you* about it, and why you must help them accomplish this goal.

P6. How can anthropic and preserved natural systems best be integrated to guarantee the continuity of biodiversity?

S6. Take fullest advantage of all existing small remnants, guarantee corridors between them, and preserve some larger areas cutting across endemic and transitional regions, to include many gradients of heterogeneity and diversity, safe areas for poor competitors, and multiple replicates of local communities in a variety of stages of isolation, succession and natural decomposition and recomposition.

In the following sections, I will consider the roles of insects in the undertaking of the primary tasks in conservation of neotropical habitats, the tools available for these tasks, and some specific case histories, followed by a practical guide for conservation action using neotropical insects, and a vision of some principal conservation concerns over the coming (and crucial) two decades.

II. ADAPTING INSECTS TO THE TASKS OF DEFINING OPTIONS, RANKING OF PRIORITIES, MAKING SURVEYS AND MONITORING SYSTEMS

A. Why Conserve Insects, Anyway?

One could conceivably present a good case to encourage the extinction of a fair number of insect species that are disease vectors, plant pests or nuisance factors. This case would be no more defensible, however, than that for the eradication of large vertebrates that are disease reservoirs (zoonoses), plant devastators, river dammers or chicken predators. In fact, all of these species are important and must be preserved, for they fit together into balanced natural systems where the only really dangerous species is man. Most insects, even those regarded as noxious by man, serve important and even key functions in communities and provide cybernetic structure and services to these systems that are irreplaceable and necessary to their proper functioning in productivity, pollutant absorption, air and water renewal and purification, disease control, matter and energy conservation, and resource mobilization (Chapter 12). Some "do no more" than to prevent other insects from taking over the system; other regulate the conversion of specific types of plant biomass into animals, the intensity of carbon and mineral recycling, the renewal of disease-ridden or senescent natural populations, and the activity of the soil microflora. The rapid collapse and subsequent malfunction and wastage of systems overdosed with insect poisons, and the simplification and channelling of energy and mass flows in other communities even minimally treated with these toxins, testify to the key roles of insects in conservation and function of diversified tropical systems. Furthermore, once the insect community that comprises and controls the larger system has been devastated, its recomposition can be slow or unlikely, with a corresponding irretrievable loss of biological information.

B. The Taxonomic Tangle

In the neotropics, very few insect groups have over half of their species described. Those which do sometimes show too many names, since biosystematic study of living populations and their variation is almost nonexistent. The ecology of most groups is either unknown or (worse yet) only presumed from fantasized analogies with other groups that are presumably close, sympatric, or with similar morphology and function. Even so, three groups (Lepidoptera, Odonata and ants) have IUCN Specialist Groups to watch over their possibly threatened (or rarely collected) species. Local, regional or national species counts are impossible for all except a few limited groups which enjoy (or suffer from) extensive interest by amateurs or agronomists, either for their beauty or voraciousness. Use of insects as indicators in conservation is even more restricted, though rapidly expanding in many parts of the neotropics as knowledge grows and urgency closes in. Of the very few new systematists being trained (also noted in Chapter 3), only a small part use field work as a taxonomic tool to understanding the dynamics of diversity. Computer-constructed phylogenetic hypotheses, often based on limited adult external morphological characters questionably chosen and polarized, or complex high-technology schemes of macromolecular affinity, are presently far more publishable than natural history accounts and species or foodplant lists. This vicious taxonomic circle continues to slow down the potential use of insects and their ecological relationships in conservation work, or encourages their misuse, with much data (and material) collected never being employed for further analyses or illustrations. To supercede this sad state of affairs, many new ecologically-oriented systematists must be formed and informed — still feasible in Brazil, but increasingly difficult in more industrialized countries. Informatics, interlibrary fax systems, clear monographs and universal species cards, local surveys requiring identifications of hundreds of species, and active museums all contribute to help reduce this tangle, but like a kudzu vine it always grows in another direction. Until a dozen or more good-sized groups of insects with wide geographical and ecological distribution and differentiation are readily available for planning, surveying and monitoring conservation units, these potentially very useful animals will still be greatly under-utilized in efforts at preserving natural diversity.

C. Emerging Patterns of Utility: Comparison of Insects with Birds and Mammals, and Combination with These and with Selected Plants

Murphy and Wilcox (1986) and Wilcox *et al.* (1986) have discussed the usefulness of butterflies in conservation planning, by comparing their information content with that of bird and mammal communities in certain areas of western United States. They concluded, in these insect-poor communities, that butterflies are at least as useful as the vertebrates, less sensitive to smaller areas, and more apt for recognition of habitats and plant communities. Much greater usefulness can be expected in the insect-flooded and plant-rich tropical communities, where vertebrate groups can be scarce, furtive, and very slow and laborious to survey. Thus, over a half- or one-day bait/visual (non-collecting) survey for one or more indicator insect groups in a humid tropical forest, woodland or diversified savanna one can obtain data to:

1. Support or eliminate it as a viable option for biodiversity preservation.
2. Help to rank its importance in relation to other areas similarly surveyed.
3. Compare its characteristics quantitatively with those of the same and other areas at different times or under different regimes of disturbance or management.
4. Follow the population dynamics of key species that may indicate activity or phenology of associated host plants, predators, decomposers and competitors.
5. Locate and carefully monitor primitive, rare, threatened, declining or exploding, invading or pest species on the site, among other important and critical uses.

Such quick but not dirty surveys have been tested in a variety of neotropical habitats and have rarely failed to produce, in a single day, useful or necessary information related to local conservation efforts. When combined with surveys of ecologically unrelated animal and plant species, such as certain bird and flower groups, the insects provide a solid basis for many of the critical tasks related to the choice, implantation and management of preserved natural systems.

D. Contribution of Insect Work to Ecological and Conservation Theory and Practice in the Neotropics

Figure 1 has shown the neotropical conservation units created in the past 20 years, whose location, justification and establishment incorporated insect surveys of the sort just mentioned. While not based on insects alone, these processes recognized their usefulness in discovery of optimum sites for the preservation of biodiversity, of which they constitute such a large proportion in the tropics. This tried and tested programme of insect use in conservation planning is still being applied in many neotropical countries. In the following sections, I will describe how this approach can be made most useful in these efforts, over the coming two decades.

III. TOOLS FOR THE TASKS: DESCRIPTION, DISCRIMINATION, EXPLANATION (TABLE II)

A. Geological Information

In the neotropics, plants and phytophagous insects reach their highest diversity in areas with appreciably variable topography, rich soils and somewhat unpredictable climate (Brown, 1979, 1982). These elements are relatively independent of the biological sectors in the regional systems, and are often described and mapped by geomorphologists, pedologists, climatologists or geographers, without necessarily referring to biotic data. Properly understood and interpreted, these geoscientific data can be of great utility in preliminary delineation of potentially optimum conservation units (Table II). When three further physical factors are added — variation in the factors with time, vegetation structure affecting light and humidity and hydraulic systems, the last two like topography easily determined by remote sensing and clearly related to different levels of biodiversity — one has an effective battery of pre-planners for mapping probable ecosystems and seeking potential areas for later survey. In certain cases, mineral deposits are also well correlated with different communities and may be further used to set up plans for biological investigation.

TABLE II. Some tools for the tasks of site evaluation and monitoring for conservation of insects and their habitats in the neotropics.

Stage	Tool or information	May be used to	Sources
Geoscientific description and explanation	Climate and its seasonal variation	Predict vegetation and soils systems	Rainfall maps, climagrams
	Topography and drainage systems	Predict climate, soils, heterogeneity, systems	Radar images, contour maps
	Soil types, fertility, mosaics	Deduce palaeolimate and endemism, predict productivity and diversity	Geological maps, soil surveys, sampling
	Vegetation structure and its seasonal variation	Predict microclimates, heterogeneity, fauna	Remote sensing, vegetation maps
	Microclimate (light, humidity, resources)	Predict flora/fauna, heterogeneity, diversity and endemism	Local stations, field measurements
Biological description	Recent and older collections of diverse material	Preliminary ideas of endemism and rarity	Museums, private collections, literature
	Natural history accounts, ecological work in region	Estimations of disturbance, abundance, diversity, systems dynamics	Literature, correspondence, interviews
	Anthropic modification of system and its effect	Estimates of resistance or fragility, viability of conservation	Literature, conversations, observations, experiment
Biological discrimination	Phylogenetic analyses of species and groups	Arrive at hypotheses for relations and monophyly	Collections, characters, literature
	Biosystematic work with natural populations	Estimates of relations in real world	Field and laboratory experiments
	Standardized surveys and transects, yearly cycles	Develop concepts of endemism, diversity, rarity	Field work, transects in different seasons
	Quantitative indices for endemism and diversity	Transform concepts into comparable numbers	Mathematics, theory, literature
	Indicator groups (see Table III)	Make evaluation easier, more comparable	Community and system correlations
Explanation	Theories and models for processes, mechanisms, history and results of ecology and evolution	Extend prediction and evaluation to still unsurveyed sites, establish priorities	Literature, imagination, courage and a touch of bias

B. Existing Biological Information

Older collections of plants and insects, or records of vertebrates (including from tourists and hunters) can give a preliminary idea of the type of system present in a site (Table II). The most abundant data are often found with amateurs or naturalists who register, collect and keep non-random samples of orchids, bromeliads, palms, succulents, fruit trees, dragonflies, beetles, butterflies, moths, bees, ants, orthopterans, amphibians, reptiles, parrots, hummingbirds, emberizids (icterids and fringillids) and other songbirds, jays, rodents, marsupials, ungulates, carnivores and primates, alive or dead, from a variety of localities. This positive information, while often incomplete and obsolete, often displaced in time, space or along the disturbance continuum, and occasionally even erroneous, can be very useful in initial characterization of sites showing high biodiversity, endemism or rare species.

C. Theories and Their Demise

Many biogeographical and ecological theories have been advanced and used to permit preliminary identification of likely places of unusual biotic interest: continental drift, evolution rates and tendencies, palæoclimate, phylogenetic hypotheses, Pleistocene refuges, palaeobotany, palaeolimnology, speciation processes and prehistoric landscape modification by large animals including humans, have all made substantial contributions to conservation planning in the neotropics. Each case of correlation of the predictions with verified high endemism or species diversity (most localities show one or the other!) has been considered a further validation of the importance of this theoretical approach. The value of all these models cannot be denied, since most were originally derived from the characters they are later used to predict. Their efficiency in later prediction, however, is usually only as good as their original empirical correlation, so that they are at best only deviations from the direct use of the geological and biological information mentioned above. Their occasional failures, often as spectacular as their much-announced successes and explained away with complex "*ad hoc*" arguments, suggest that they should be used with care and always backed up by gathering real data in the field. I have always used these theoretical models to plan trips to new areas and to prepare for what might be encountered; I believe that they may increase the probability of finding really interesting areas, but they have also left me many times in unexpectedly sterile or monotonous habitats. Too many imponderables still exist in the formation, dynamics and maintenance of diversified neotropical communities, to permit a conservation programme to be based principally on simplified models and generalized theories.

D. Endemism and Diversity

Once an insect group is well known over a broad area, patterns of local or regional endemism and comparative species diversity can be seen and used to rank areas in terms of conservation potential. Because many insect species are very old and widespread (Coope, 1978; Brown, 1982, 1987), endemic patterns may best be sought at infraspecific levels, whose definitions should be strictly and provenly geographic, genetic and reasonably invariant over nuclear regions, based on a dense sampling net and, if possible, experiments. In principle, all local populations have some endemic genes, so general endemism should be sought in wider regional gene pools and related to fundamental adaptation to corresponding regional biotic or abiotic factors, including isolation and vegetation type. Endemism indices with a correction for hybridization (Brown, 1979, 1982) can be quite independent of sampling effort (above a minimum) and useful in identifying zones of diversity due to mixing rather than to species richness. All depend on careful phylogenetic analysis to identify monophyletic units in the group examined.

Diversity measures vary greatly in sophistication and are usually dependent on sampling effort. Given near-asymptotic values (species/time or species/area curves), which rarely are reached, simple numbers of species may be adequate for comparisons among areas. At lower levels, corrections may become necessary and can introduce bias into the diversity values, only partly mitigated by the use of information–theory transformations to evenness (H'). The application of diversity data to area comparisons should be made with caution and after attentive reading of Pielou's considerations (1975).

E. Indicator Groups

"Now all animals are created equal, but some are more equal than others" (George Orwell, *Animal Farm*, 1945). "All species are permissible as indicators, but not all are convenient, helpful, or edifying" (adapted from St Paul, I Cor. 10: 23). Table III lists some important attributes of insect groups that might be used as indicators, mostly related to ease of finding and evaluation and amount of information content. The proper

TABLE III. Classification of various groups of insects as indicators for biogeographical and ecological analysis and monitoring.

Desirable quality for an indicator group in ecology and biogeography	Collembola	Odonata (dragonflies, damselflies)	Isoptera (termites)	Hemiptera: Coreidae, Pentatomidae, Cygaeidae, Tingidae, Myridae	Homoptera: Membracidae, Cercopidae	Coleoptera: Carabidae, Cicindellidae, Elateridae, Cerambycidae, Chrysomelidae, Curculionidae
Taxonomically and ecologically highly diversified (many species in each locale or system)	++	++	++	++	++	++
Species have high ecological fidelity	++	++	++	++	++	++
Relatively sedentary	++	++	++	+	+	++
Species narrowly endemic, or if widespread, well differentiated (local or regional)	+	+	+	+	+	++
Taxonomically well known, easy to identify	+	++	++	+	+	++
Well studied (genetics, behaviour, biochemistry, ecology, biogeography)	+	+	++	+	+	++
Abundant, non-furtive, easy to find in field	+	++	+	+	+	+
Damped fluctuations (always present)	++	+	++	+	0	0
Easy to obtain large random samples of species and variation	+	+	+	0	0	+
Functionally important in ecosystem	++	+	++	+	+	++
Response to disturbance predictable, rapid, sensitive, analysable and linear	++	++	++	+	+	+
Associates closely with and indicates other species and specific resources	++	+	+	++	++	++
Total value as indicator (in 24)	19	18	20	14	13	19

Insect groups								
			Lepidoptera					(Aposematic)
Diptera: Asilidae, Tabanidae	Hymenoptera: Formicidae	Hym; Apoidea, Vespidae, Sphecidae	Sphingidae, Saturnoidea	Arctiidae	Papilionidae, Pieridae	Morphinae, Satyrinae (s.l.)	Bait-attracted Nymphalinae	Heliconiini, Ithomiinae
++	++	++	++	++	++	++	++	++
+	++	++	+	+	+	++	+	++
+	++	+	+	+	+	+	+	++
+	++	+	+	+	+	++	+	++
++	++	++	++	++	++	+	++	++
+	++	++	++	++	++	+	+	++
++	+	+	+	+	++	++	+	++
+	+	+	0	+	0	0	+	+
+	++	+	0	+	0	0	+	++
+	++	++	+	+	+	+	+	+
++	+	+	+	+	++	++	+	+
+	++	++	+	+	++	++	+	++
16	21	18	13	15	16	16	14	21

choice of a group of insect species as indicator for conservation of endemism and diversity is a fundamental tool for the tasks of conservation, potentially of inestimable value or enormous capacity to mislead (Landres *et al.*, 1988). An ideal group will greatly facilitate all aspects of a conservation programme, from initial identification of interesting sites through comparison, convincing the public and the authorities, and continuous monitoring and management, and provide a solid base for predicting where good places should occur elsewhere and then testing these predictions. Few groups have proven so useful, but diverse insect groups are probably safer to use than individual vertebrate species (Landres *et al.*, 1988; Table III). The following section describes a number of applications and results of the use of neotropical butterflies as indicators of a variety of environmental parameters important in conservation.

IV. CASE HISTORIES AND SOME LEGENDS

A. Endemism and Diversity in Neotropical Forests: the Distant Past

After suggestions made by Haffer (1969) and Vanzolini (1970, 1973) that Pleistocene climatic changes might have been important, through their effects on vegetation continuity, in the multiplication of species diversity in the Amazon forests, an analysis was undertaken of genetic differentiation patterns in three groups (over 165 species) of aposematic butterflies common in the neotropical forests. From biological experiments and phylogenetic analysis, it was concluded that the *species* showed a broad partitioning into only four endemic regions, corresponding to major biogeographical subdivisions of the neotropics, and related to ancient geological processes and large barriers (Fig. 2A). A quantitative formula was then developed for calculating endemism values per 30′ × 30′ quadrant in regional sets of recognized subspecies, shown to be produced by a limited number of adaptive genes (Sheppard *et al.*, 1985), serving as indicators of the general geographical differentiation in each species. A "native" subspecies to the nearest regional endemic centre (recognized by projection maps of all subspecies ranges) on the list was given a value of +1, a subspecies distributed across several centres received a 0 value, and mixed populations (characters from several different endemic subspecies) were assigned −1 overall. Thus, a quadrant in which half the resident species showed mixing of different regional subspecies characters gave a corrected endemism value of 0. This analysis was extended, through more that 1500 quadrant lists, to the limits of the neotropical forests. Isoline maps then showed a clear pattern of 43 regional centres of endemism, separated by "subspecies hybridization zones" where calculated values were negative for all nearby endemic centres (Fig. 2A). This pattern was grossly similar (though the centres had different shapes) for the three groups separately (Fig. 2B), suggesting a historical origin for the geographical differentiation, modified by the different ecological preferences of each group and their primary resources (vegetation structure, larval foodplants and preferred adult flowers). This historical factor was supported by an excellent correlation of the quantitative endemic centres — empirical biological phenomena in the present time — with a geoscientific model (paleoclimate, soils, geomorphology, vegetation structure) for probability of forest persistence during cold, dry periods in the past (especially the Würm-Wisconsin glaciation of 13 000–20 000 years BP), derived in part from study of Brazilian regions where forest persists today under unfavourable climatic conditions (Roraima and Goias) (Fig. 3).

In contrast to suggestions based on analysis of limited numbers of vertebrates from few localities, high species diversity in these butterflies was a very local phenomenon, negatively correlated with subspecies endemism; it tended to appear in strongly heterogeneous areas near the edges of the endemic centres (corrected endemism values near 0), and showed strong positive correlation with environmental transitions and unpredictable mild disturbance (Fig. 4) — that is, it was clearly a local ecological phenomenon, not a regional historical one. No "centres of diversity" could be found by isoline analysis, but many limited "hot spots" were found around the perimeters of endemic centres related to either intensive field work by naturalists, or (typically) the above-mentioned ecological factors. Regional diversity tended to appear along the "spine" of Central America, the eastern Andean slopes and the southern Amazon, corresponding to favourable climate and soils.

Models of "Pleistocene refugia", usually (and circularly) derived from biological data (endemism or diversity) rather than geological criteria, have been much used in theoretical analysis of conservation priorities in the neotropics, and were responsible for the establishment of many parks and reserves in the last 20 years (Fig. 1). While this *term* may be misused, the *use* is completely justified, since the endemism and diversity are real empirical factors which should be conserved — independent of their

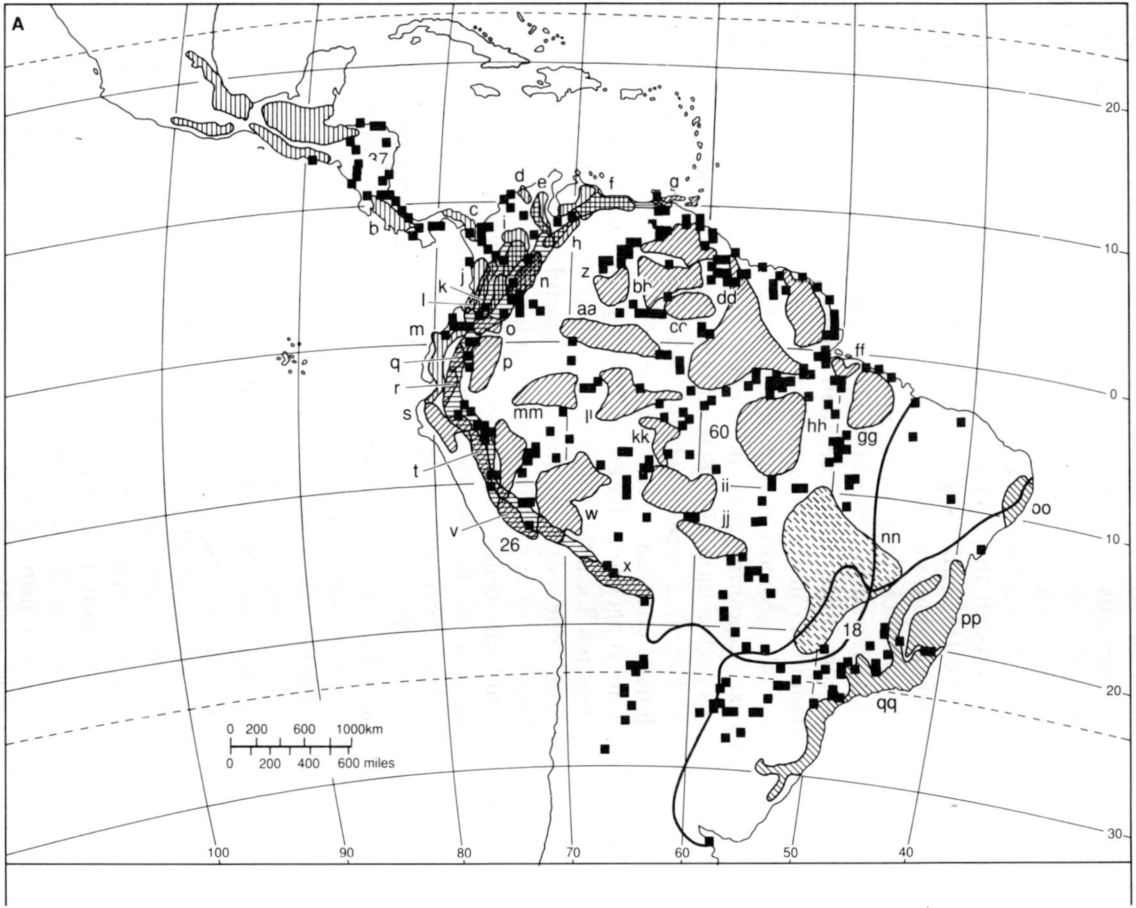
A
b
c
d
e
f
g
h
i
j
k
l
m
n
o
p
q
r
s
t
v
w
x
z
aa
bb
cc
dd
ff
gg
hh
ii
jj
kk
ll
mm
nn
oo
pp
qq
18
26
60
0 200 600 1000km
0 200 400 600 miles
100
90
80
70
60
50
40
30
20
10
0

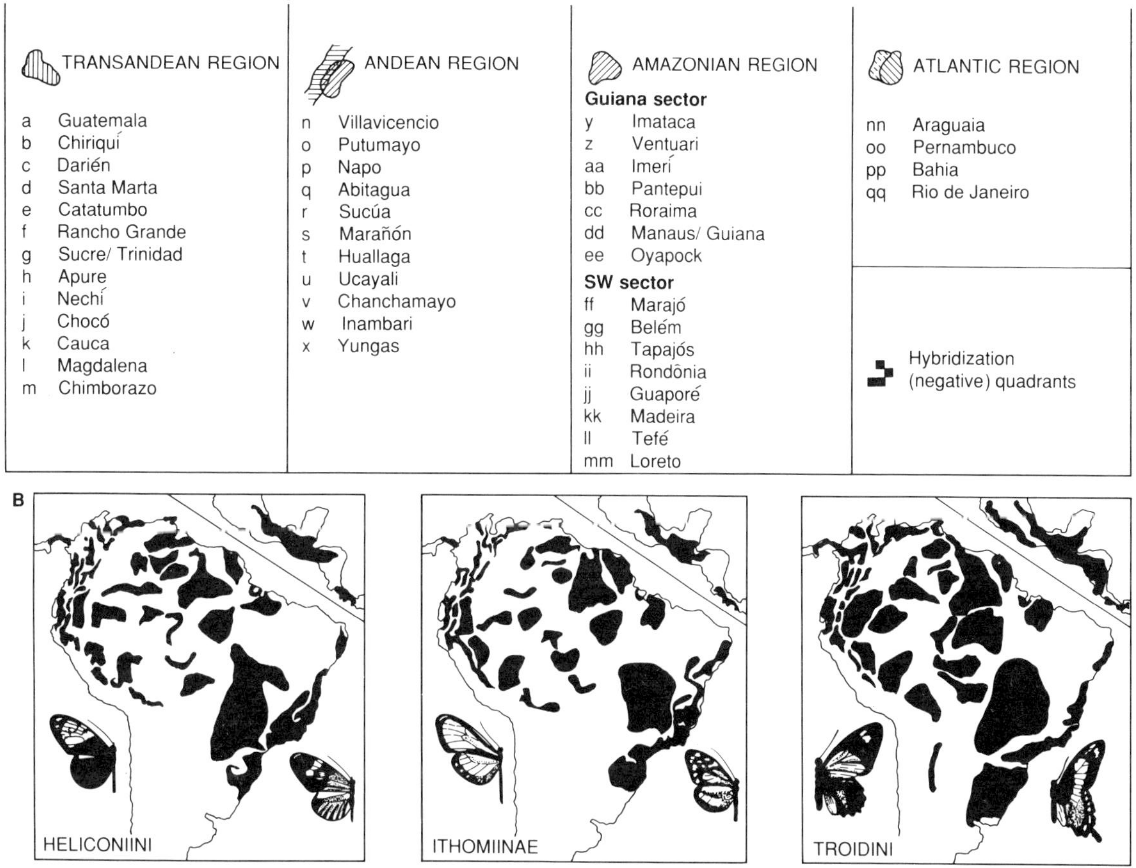

Fig. 2. (A). Four species-endemic regions in the neotropical forests (with number of endemic species considered), and 43 subspecies-endemic centres, one-third of maximum corrected isoline for Heliconiini and Ithomiinae. (B) These two groups separately, compared with Troidini subspecies-endemic centres (total 167 species of the neotropical forest butterfly fauna).

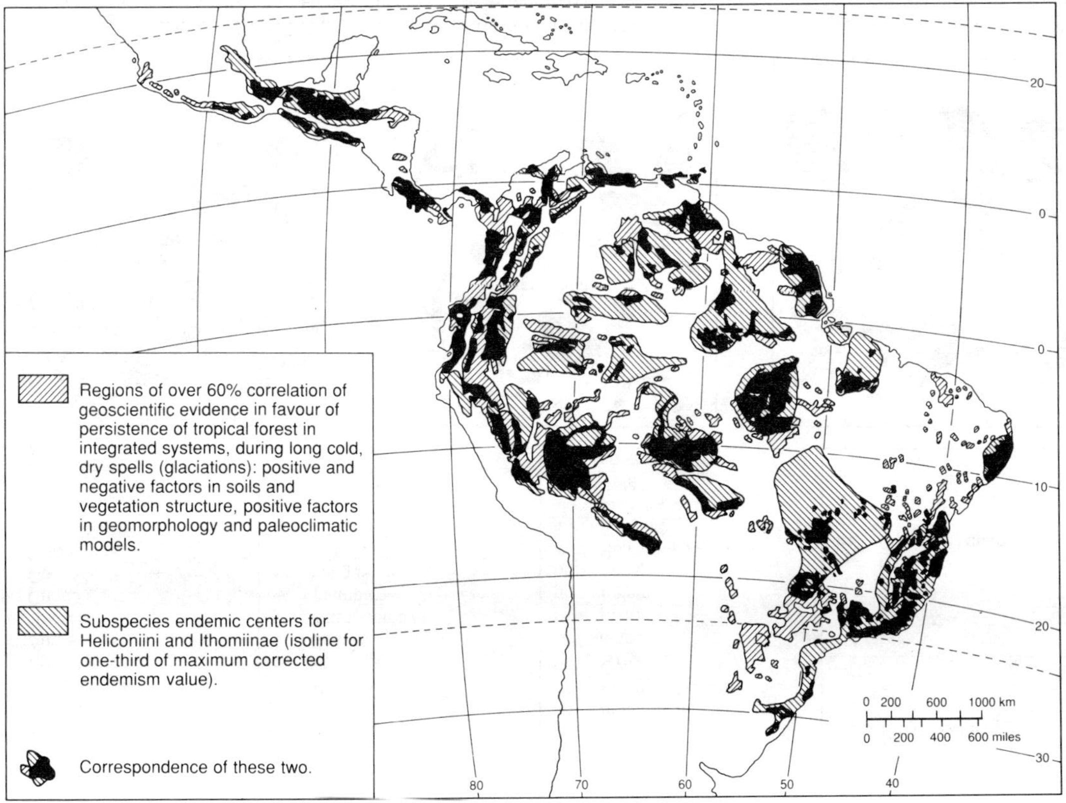

Fig. 3. Geographical overlay of subspecies-endemic centres (Fig. 2A) with a geoscientific model of probability (more than 60% in synthesis of palaeoclimatic, geomorphological, pedological and phytosociological data) for forest permanence during the most recent long cold dry climatic spell (Würm-Wisconsin glaciation, 13 000–20 000 years BP), based on observations of present-day forest in regions of such unfavourable climate in Brazil. Essentially all endemic centres enclose areas of high probability for existence of "paleoecological forest refuges".

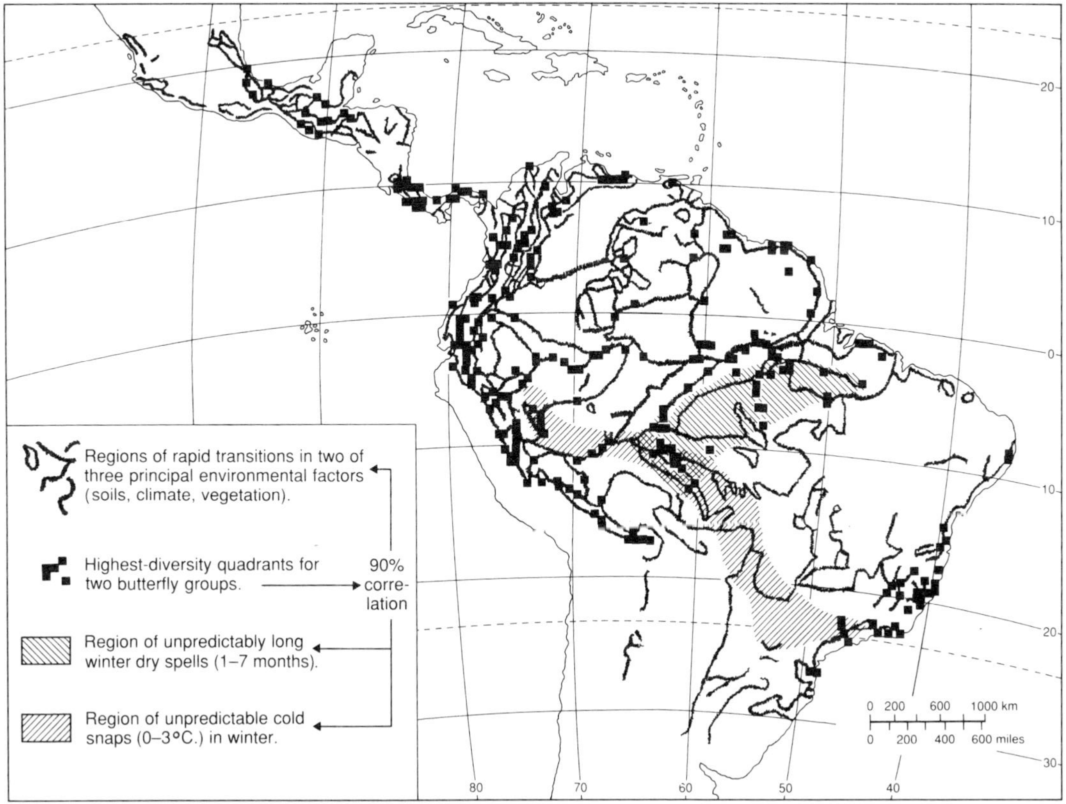

Fig. 4. Geographical correlation between highest diversity quadrants for Heliconiini and Ithomiinae (more than 15 species among those analysed, or 75% or more of the possible species pool present if less than 20) with a model of ecological conformities and rapid transitions in the present, including indication of regions of unpredictable mild disturbance (dry spells or cold snaps in the south-west Amazon) (after Brown, 1982). Over 90% of the quadrants are coincident with the model.

historical or ecological causes, which merely add "the weight of theory" to the proposal and priorities. The divorce of endemism from diversity which became evident with dense field data (Figs 2–4) carries a strong recommendation for conservation units which need to include both: they should be large enough to cover both the nucleus and the peripheries of endemic centres or "refugia", since diversity is concentrated at certain points around the edge (which should be discovered empirically). The "peaks" of the endemic centres can be quite monotonous places, with low-level populations of genetically homogeneous differentiates. More important, many rare primitive species, undoubtedly associated with the endemic centres (by mimetic patterns or host plant affinities), are found only in the microheterogeneous "hot spots" on their peripheries (see below). Thus, a programme of conservation planning and priorities should be strongly supported by empirical field data to identify these "hot spots" and the special, ancient or rare species which help contribute to their diversity.

B. Monitoring the Short-term Effects of Human Intervention: the Biological Dynamics of Forest Fragments Project in Manaus

The butterfly communities present in low-productivity dense tropical forests north of Manaus, Amazonas, Brazil have been followed since 1980 as part of a 25-year project (Biological Dynamics of Forest Fragments, WWF/INPA) designed to test the theories of island biogeography in continental habitat "islands" of different sizes (Lovejoy *et al.*, 1984, 1986). A regional list of fewer than 500 Papilionoidea (flower-addicted Hesperioidea are very scarce in these nearly flowerless forests), and daily lists averaging near 60 species, testify to the poverty of this community, which nonetheless is very rich in metalmarks (Riodininae — almost 200 species, including over 40 in the single genus *Euselasia*) and shows dramatic responses to increase in light levels in the understorey (Figs 5 and 6, "edge effects"), far more than to increased fragment area or total area sampled (Fig. 7, note very small slope of most of the lines). The total butterfly community is strongly bimodal in species distribution and abundance (Fig. 8), including some species that are widespread and common and a majority that are very restricted and rare, with a deficiency in the intermediate classes. Thus, in these habitat fragments, ecological and population–biological effects greatly outweigh the simple area relationships emphasized in island–biogeographical theory.

In the careful monitoring of the effects of human intervention and eventual isolation of various tropical forest "islands" (Fig. 9), elements of

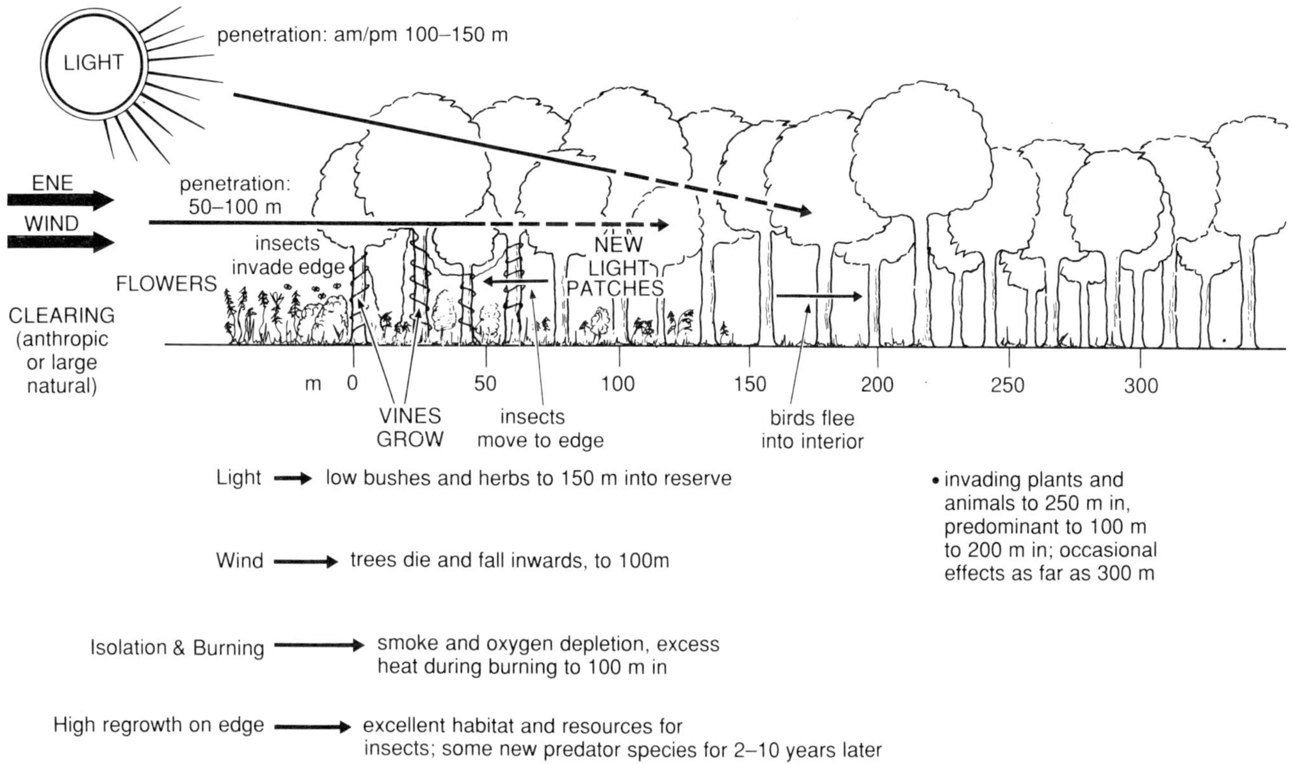

Fig. 5. Different types of edge effects observed in forest isolates in the central Amazon (Biological Dynamics of Forest Fragments project, WWF/INPA).

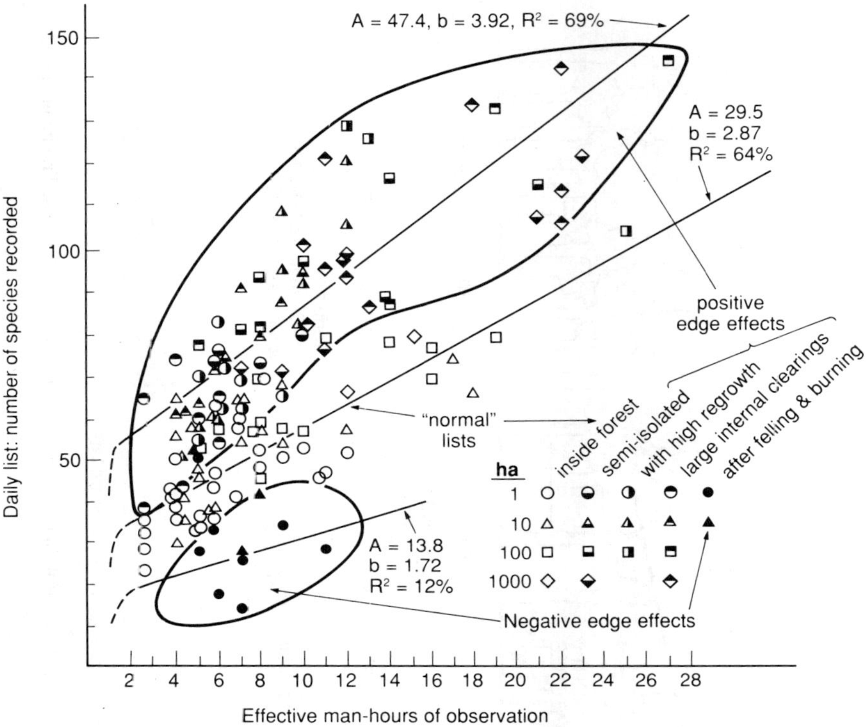

Fig. 6. The results of different edge effects on the daily list of butterfly species observed, as plotted against effective man-hours of observation, including all daily lists since 1980. Note that the artificially linear regression lines apply only to lists of 3 h or more; they must pass through the origin, and therefore are at least partially logarithmic in shape.

the butterfly community which have been of special importance as indicators include:

1. Large species or those with special rare resources, that disappear from small areas after isolation (Morphinae, Brassolinae, Charaxinae).
2. Small shade-loving species with either very uniform (Satyrinae) or very patchy (Ithomiinae) distributions in the forest.
3. Sun-loving nymphalines (including heliconiines) that appear, multiply, show liberated genetic variation, and then may crash, related to fresh growth of their plant resources on newly created edges or in very large internal clearings — very good indicators of edge effects.

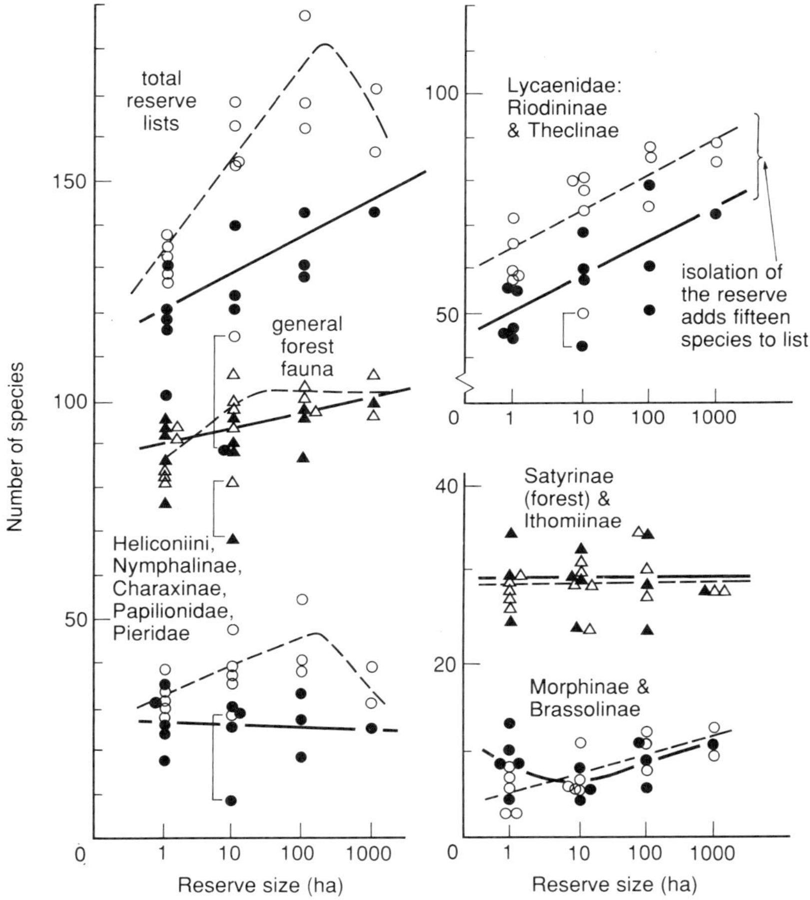

Fig. 7. Species–area curves for the 25 reserves sampled in the BDFF project: cumulative reserve totals, divided into isolated (open symbols and dashed lines) and forest-surrounded (blacked-in symbols and full lines) reserves, such that each total list corresponds to 31–38 man-hours of observations. The total reserve list, and several ecologically separate groups of butterflies, are shown. Lines were drawn by eye; note that the largest reserves (1000 ha), with many internal edges (large clearings), showed a diminished response to area, with relation to smaller, isolated reserves. Smaller shade-loving species (triangles at right) are insensitive to area in both isolated and non-isolated reserves, as are larger mobile sun lovers and very large species in forest-surrounded reserves (lower circles at left and right, solid lines). Widespread species (centre triangles at left) are relatively insensitive to area, but the whole community and the small sun-loving Lycaenids (upper circles at left and right) in both types of reserves, and large species in isolated reserves (lower circles at right) show a noticeable response to larger area. The linked pair of symbols at the bottom of most 10-ha sets are for a single very homogeneous and poor reserve (2206) before and after isolation.

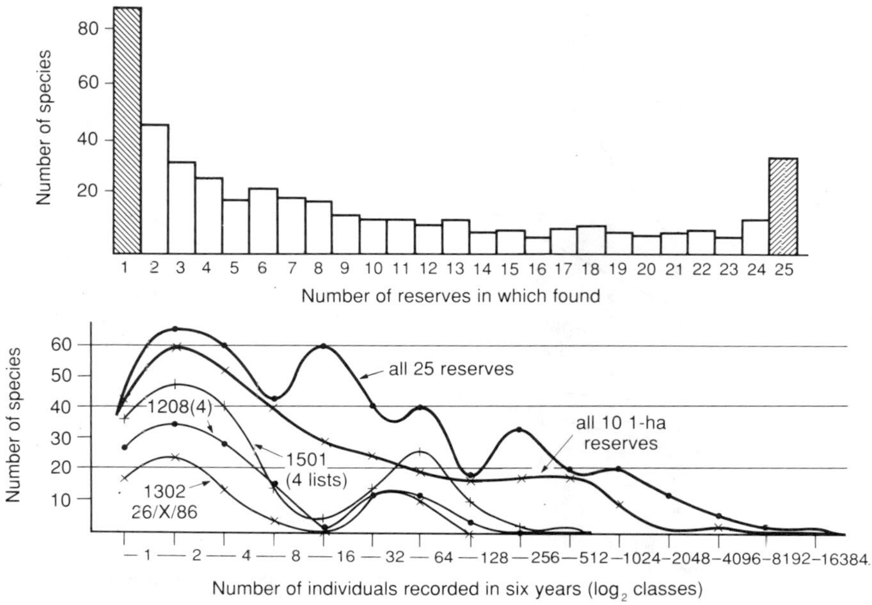

Fig. 8. Some characteristics of the community sampled in the BDFF project. Upper: distribution of the species in the reserves, from 1 to 25 (all); on the left are very localized and rare species, on the right ubiquitous and common ones. Lower: total individuals recorded for each species, on a $\log_2$ scale: all reserves (upper curve), all 10 reserves of 1 ha, two individual reserves (one of 1000 ha, part of the 10 000-ha reserve 1501, and one of 10 ha), and a single day's list in a 100-ha reserve. Of the nearly 65 500 individuals counted, the most abundant are all inhabitants of the dense forest understorey, and all except the three Riodininae prefer the shade: in order, *Hypothyris euclea* (9501 individuals, 15% of all seen) and Riodininae: *Nymphidium aurum*, *Charis cleonus*, *Adelotypa penthia*; Theclinae: *Calystryma keta*; and Satyrinae, *Magneuptychia gera*, *Pierella astyoche* and *P. lamia* (all these 7 between 2000 and 2200, or 3% of all seen). The additional nine species with over 1000 individuals seen (1.5% of all) are all of the forest understorey (one *Morpho*, three Satyrinae, two Riodininae, three Theclinae), all but the first small. The 84 singletons (divided between classes 0–1 and 1–2 in the log-normal plotting method) represent all butterfly groups and are typically widespread, strong-flying, sun-loving species often common in disturbed habitats (45 are Lycaenidae). The bi- or polymodal curves are only partly due to the census method, which uses fixed categories for common species (10–50 recorded, set at 16 in the counting), abundant (50–200, set at 64) and very abundant (over 200, set at 256), which merely emphasizes the underlying pattern of several distinct classes of species, many that are always very scarce (seen only one to three times in six years), others that are usually scarce but occasionally common, and a very few that are usually common to abundant.

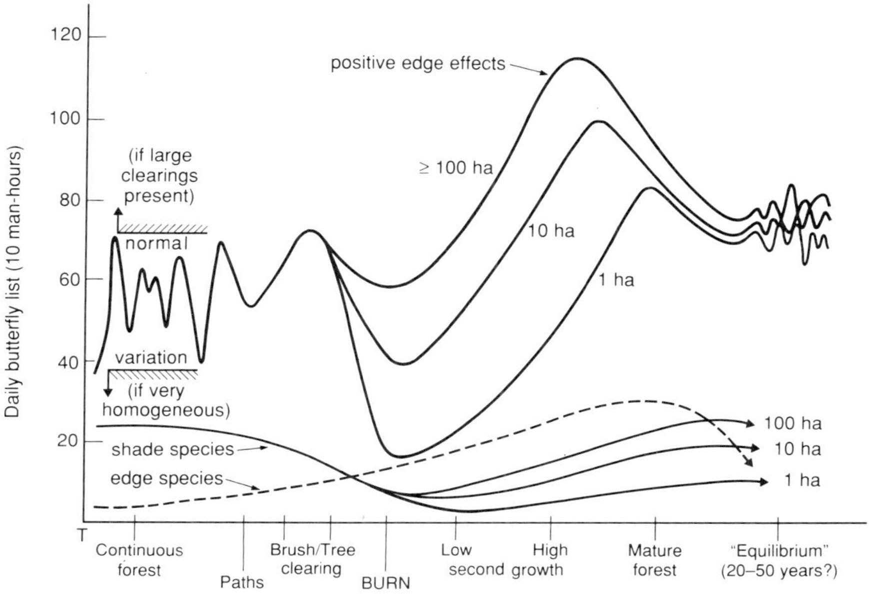

Fig. 9. Expected course of the daily list (corresponding to 10 man-hours of observation, see Fig. 6) before, during and after the isolation process of a reserve in the BDFF project (integrated and averaged from all lists in Fig. 6). The list can be maintained at or returned to any point after first isolation, by reburning or fixation of pasture vegetation around the reserve; the progression follows the list through the complete process of regeneration around the reserve, at which point it will show a new "equilibrium" with a different species composition from that originally present.

4. The highly diversified Lycaenidae (riodinines and theclines), often associated with ants, ranging from very localized deep-shade, primary-forest obligates — some seen only once in 3000 h of field work — through light-gap and edge specialists, to open-field or truly migratory species (*Stalachtis phlegia* and many theclines), thus indicating all possible stages of system degradation.

Many "standard common neotropical forest butterflies", collected everywhere along roads and trails, are exceedingly rare or have not even been recorded yet in the 25 variously isolated reserves of the project, suggesting that an exceptional opportunity has been provided to study and understand the primitive butterfly community of the central Amazon, and the progressive effects of system disturbance on it; most of these species exist less than 50 km to the south (disturbed areas near Manaus) or north (more open forests on richer soils), and it will beinteresting to see when (or if) they arrive in the recently disturbed dense-forest system of the project.

Lessons derived from this study include a renewed appreciation of the delicate nature of the many local populations in these low-productivity systems; the tendency of rare species to concentrate in heterogeneous habitats; the important effects of light, plant growth and flowers on butterfly diversity and abundance (Brown, 1972; Otero and Brown, 1986); the contrasts between different local and regional patterns of distribution and abundance, suggesting an incipient dichotomy in population strategies; the importance of predators, especially spiders and ants, in determining both community structure and population density in isolated systems; the sensitivity of butterflies to special foodplant resources and topography; and the relative instability of local assemblages, forming highly invadable subsaturated communities and favouring appreciable ecological opportunism in these dispersal-adapted winged insects.

C. Long-term Effects of Human Intervention: the Brazilian Atlantic Forests

The Brazilian Shield, comprising the Atlantic Forests (perhumid or semi-deciduous), Araucaria forests (in the south), cerrado (open woodland) and caatinga (north-eastern thorn scrub) as major vegetation systems, and including dozens of additional vegetation types and variations in complex microheterogeneous landscape mosaics, is one of the four principal species-endemic regions in neotropical butterflies (Fig. 2a) and indeed in most animals. Stretching along all the coast from 6° to 32° south and extending well into the interior are the high-endemism tropical Atlantic forests, whose recent mapping by the Projeto RADAMBRASIL (1981a,b, 1982, 1983a,b, 1987; Brasil, 1988) and paleo-environmental synthesis (Fig. 3) make all current maps obsolete, showing a number of effective forest archipelagoes linking the Atlantic coastal systems with those of the Amazon Basin and subandean regions, many even maintained through past periods of unfavourable climate. Although these links are evident in some Atlantic faunal elements including a number of typical Andean and Amazonian butterfly species and subspecies, overall endemism is 45% in the butterfly species of the shield region (42% for Papilionoidea, 48% for Hesperioidea) with 66 endemic genera, mostly skippers (Table IV). The total number of Atlantic butterflies (2124) is over two-thirds of the total

TABLE IV. Diversity and endemism in butterflies of Brazil and of the Atlantic Forest region (Brazilian Shield, including Cerrado, Caatinga, Araucaria), by group.

	Total number of species		Endemic taxa in Atlantic region		
Butterfly group	Brazil	Atlantic	Species	(% of total)	Genera
Nymphalidae					
Morphinae	21	9	7	(78%)	0
Brassolinae	49	36	17	(47%)	4
Satyrinae	194	132	68	(39%)	12
Danainae	6	5	0	(–)	0
Ithomiinae	128	54	22	(41%)	2
Acraeinae	22	21	19	(90%)	0
Heliconiini	44	20	2	(10%)	0
Nymphalinae (other)	251	175	36	(21%)	0
Charaxinae	52	34	9	(27%)	0
Subtotal: Nymphalidae	767	486	180	(37%)	18
Libytheidae	1	1	0	(–)	0
Lycaenidae					
Plebejinae	5	5	2	(40%)	1
Theclinae	424	306	151	(49%)	??
Riodininae	720	350	155	(44%)	4
Pieridae	59	49	19	(39%)	2
Papilionidae	68	43	18	(42%)	0
Subtotal: all of Papilionoidae	2044	1240	525	(42%)	25
Hesperiidae					
Pyrrhopyginae	70	49	28	(57%)	4
Pyrginae (a + b)	450	351	105	(30%)	4
Hesperiinae	566	484	293	(61%)	33
Subtotal Hesperioidea	1086	884	426	(48%)	41
Total: all butterflies	3130	2124	951	(45%)	66

for all of Brazil and about an eighth of the world butterfly fauna, with some lycaenids, hesperids and satyrines still to be described.

This fauna was originally distributed among a great variety of microhabitats, virtual vegetation islands strongly differentiated in topography, soil, microclimate and plant composition, iterated across regional landscapes but strongly interdigitated such that each patch of a certain community might be effectively isolated from other such patches by several very different, potentially "hostile" communities. Thus it is not surprising that:

1. Many Atlantic organisms are known from very few records from only one or two localities.
2. Many of these are constantly being discovered or rediscovered in previously inaccessible sites whose vegetation is similar to that of the type locality; and
3. Although as much as 90% of the Atlantic forests have been strongly disturbed or converted to anthropic systems over the past 490 years of European occupation, among these highly restricted species, little-known, probably endangered and sometimes not seen for over 50 years, *not a single one can be declared extinct*. Typical habitat for all of them, be they butterflies, dragonflies, beetles, frogs, monkeys, birds, fishes, rodents, orchids, palms or other plants, still exists in abundance in the remaining 50 000 km^2 of parks, reserves, and inaccessible areas, or — better yet — in complex successional landscapes where the original mosaic has changed in quality (vegetation components) but not in basic structure, and still retains the resources of these species as part of the complex system, at times even cultivated or multiplied.

Thus, an insect fauna already accustomed to appreciable isolation and disturbance in nature, on both long and short time scales, has met with *savoir faire* the challenge of a human occupation that reduced the systems far below the most restricted levels of the late Pleistocene, through an adaptability selected by surviving many eons of environmental challenges which likewise involved reduction of available habitat both locally and regionally. Essentially all the original species seem to be persisting, and some populations have survived serious, potentially fatal challenges from both abiotic and biotic factors while being monitored (1967–1989), indicating their great resistance and tenacity. They are thus still present to help indicate local environments of high diversity, endemism or rare species concentrations, as discussed in the following section.

D. Identification of Areas With Exceptional Biological Importance: Palaeo-environments

On not infrequent occasions, I have stepped off a bus along a new paved road through a previously unseen or unsampled area of Atlantic forest, and immediately encountered — perhaps right on roadside flowers — one or more individuals of a primitive, rare butterfly, perhaps not seen in many decades or feared extinct. Penetration of remnants of original vegetation near the road, often steeply up or down from it, usually has led to glimpses of other equally primitive or rare species, not only butterflies but also other insects, birds, frogs, monkeys and a host of bizarre and little-known plants. I have called these places "palaeo-environments" since many of the rare species found together in them are recognized by the systematists of their respective taxonomic groups as morphologically primitive or at least very isolated and generalized, or even transitional to adjacent groups; none are present in modern anthropic mosaics, even if patches of supposedly suitable habitat are retained or restored. True islands in time, usually peripheral to endemic centres/refugia and rarely on the best soils or topographies, these environments are among the most significant scientifically in the neotropical mosaics, always full of surprises in any group, and represent first-priority areas for conservation; they are not usually duplicated, they have inordinately high indices of rarity in their biotas, and they are usually rather small in size and surrounded by ever-encroaching pressures for conversion to agricultural uses. Among such places encountered in recent trips in many parts of the neotropics are those described in Table V (from south to north up the coast, then the interior and the Amazon, Andes and Central America, but emphasizing Atlantic forest sites). Most presently are either transformed into reserves or are in the process of preservation (see locations in Fig. 1).

All the areas in this small sample (Table V), and dozens of others throughout the neotropics, were first recognized by the unusual composition or abundance of indicator butterfly species in the first few minutes of exploration, and subsequently shown to contain equally unusual plants and other animals in like proportions. While a few "first-impression rich palaeo-environments" have proven to be more ordinary upon further work, there is little doubt that an early sighting of rare or primitive butterflies is usual in such sites and a faithful indicator of them. It has been very exciting to be able to deliver over the past 20 years (Brown, 1970) reports on biological parameters (endemism, diversity and rarity values) to conservation planners throughout the neotropics, with very high probability that these will be confirmed by more intensive work with other groups of organisms — representing a very reliable, simple, rapid, convenient and useful indication.

E. The Spectre of Extinction: Facts and Myths

Of the 3130 butterfly species described and recorded from Brazil (Tables I and IV), about two-thirds (mostly Lycaenidae and Hesperiidae) are so little known that no evaluation is possible of their present status, though usually they have been found in several scattered areas some of which still have intact vegetation. In the remaining thousand species are about a hundred which may be considered as rare, highly restricted, sparse or directly affected by habitat elimination. Twenty-three of these have been placed, through efforts of Luiz Otero, Olaf Mielke, this author and the Sociedade Brasileira de Zoologia, on the 1989 list of Brazilian animals threatened with extinction (Table VIA). Populations of most of these have been located and followed with mark–recapture techniques, and basic resource and habitat requirements are known; some populations of most of them may be found in official parks and reserves or other preserved areas (Table V), and several have been used to defend the establishment of conservation units. A further 33-subspecies and species are on a second list (Table VIB) which includes some entire genera, awaiting more complete study before being included on the official list. The remaining species are needing further information before being moved into official consideration for protection or management.

Essentially all of these species are good indicators of palaeo-environments or other special habitats, and usually co-occur with other animals and plants on official lists for conservation monitoring and action, in these habitats. Unlike these others, the butterflies are usually rapidly located and easily observed in their environments, thus contributing to the overall habitat preservation efforts presently in intensive expansion in Brazil. As in other countries, a few have been placed on the first list already because they occur in areas under immediate and direct threat, from commercial logging, agricultural expansion, dam-building or other large public works. Since Brazil now requires environmental impact statements including mention of threatened species for any such large-scale intervention in natural systems, it is probable that these butterflies will receive appreciable attention and concern in the near future. Hopefully, they will be properly evaluated and used in conservation planning and encouragement of sustainable and rational long-term use of natural resources.

F. Management of Anthropic Mosaics for Sustainable Yield: Extractive Reserves

Ten years ago, in the first book on Conservation Biology, Larry Gilbert (1980, p. 11) commented that very few opportunities still remained to plan natural reserves, but a very large task was appearing in the management of the pieces left, or of mosaic landscapes. He provided useful bases for the understanding of tropical forest structure and function, and for maintaining these inside and outside of official reserves.

As the pressure on tropical systems grows, the biosphere reserve type

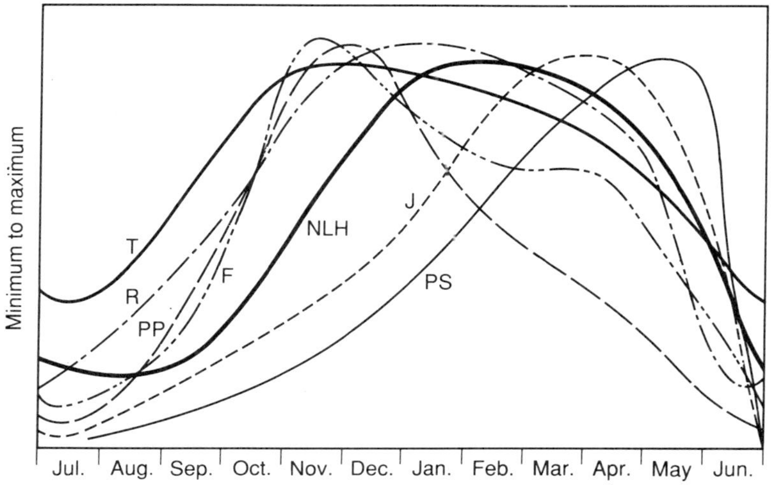

Fig. 10. Yearly cycles of butterfly abundance in a transitional subtropical palaeo-environment (very dry in the past), the Serra do Japi, Jundiaí, São Paulo (from Brown, 1991). Butterflies are very scarce and with few adults present in winter (July). With the rising temperature (curve T) and the onset of the rainy season (R) in September and October, Papilionidae and Pieridae appear in large numbers (PP), some from diapausing pupae, others from immigration, and lay eggs on the abundant fresh new growth of the foodplants (F). These families diminish in mid-summer as the other families (NLH, Nymphalidae–Lycaenidae–Hesperiidae) rise to a peak on the burgeoning foodplant resources, mostly from immigration from all sides. By late summer, juveniles (J) are very commonly seen on the progressively devastated foodplants, but by autumn almost all of them are parasitized (PS); most populations crash by late May with the first cold snaps. Each year, a number of very rarely seen species appear, presumably arriving from inaccessible areas to the south, east and west; these rare species vary from year to year. The resources of Japi may help keep some of these species going (keystones), since individuals probably originating in Japi are sometimes found 50–100 km away, in less suitable habitat but with less rigorous winters.

of conservation unit must become ever more important. Local peoples who use forest resources and base their lives and economy on these, provide the ideal structure for forest preservation and continual renewal of these resources, resisting commercial or industrial exploitation or habitat modification by interests outside the local community. Insects can also be very useful indicators of the best areas for permanent preservation in these mixed-usage schemes, quickly identifying the parts of greater or lesser endemism, diversity or community value — indeed, local populations often use insects intensively to find their resources and manage the natural systems.

TABLE V. Selected neotropical palaeo-environments (Fig. 1).

No.	Country: State/Dept.	Name	Endemism (Fig. 2): Centre, corrected (uncorrected)	Threatened butterflies (Table VI)	Conservation (Fig. 1)/Notes
1	Brazil: Santa Catarina	Serrinha, Joinville	Rio de Janeiro 7	Six species (list 2)	Quite inaccessible
2	Brazil: São Paulo	Serra do Japi, Jundiaí	Rio de Jan. 3(9)	One (list 1), 1(2)	Heritage area; variable fauna (Fig. 10)
3	Brazil: São Paulo	Cajuru region	Araguaia, low	Possibly 5 (list 2)	University study reserve
4	Brazil: São Paulo/MG/RJ	C. Jordão/Itatiaia	Rio de Jan. 4(10)	Two (list 1), 7(2)	Many parks and reserves; 500–3000 m
5	Brazil: Rio de Janeiro	Fazenda União, C. Abreu	Rio de Jan. 6(8)	Poss. 7 (list 1), 5(2)	Preserved by owners
6	Brazil: Rio de Janeiro	Barra de São Joao	Rio de Jan. 4(6)	Four (list 1), 2(2)	Preservation needed
7	Brazil: Espírito Santo	S. Teresa/S. Leopoldina	Bahia 4(10)–RJ (11)	Five (list 1), 9(2)	Several reserves
8	Brazil: Espírito Santo	Linhares region	Bahia 8(10)	One (list 1), 5(2)	Two large reserves
9	Brazil: Bahia	Camacã region	Bahia 10	Six (list 1), 3(2)	Inaccessible; no reserves
10	Brazil: Pernambuco	Tiuma woods	Pernambuco 6	One (list 1), 5(2)	Preserved remnant
11	Brazil: Pernambuco	Agua Azul, Timbaúba	Pernambuco 6	One (list 1), 5(2)	Preserved by owners
12	Brazil: Paraíba	Buraquinho, J. Pessoa	Pernambuco 5	Four (list 2)	Municipal reserve
13	Brazil: Distrito Federal	Parque do Gama	Araguaia 2(4)	Five (list 2)	Municipal reserve

14	Brazil: Mato Grosso	Chapada dos Guimarães	Guaporé 1(3), Ar (4)	Two (list 1), 6(2)	Three reserves; historical locality
15	Brazil: Rondônia	Jaru/Ouro Preto	Rondônia 11(25)	Two (list 1), 7(2)	Small reserves; high diversity
16	Brazil: Acre (Fig. 11)	Cachoeira, Xapuri	Inambari 9(13)	Seven (list 2)	Extractive reserve
17	Brazil: Acre (Fig. 11)	Rio Tejo basin	Inambari 2(22)	Possibly 3 (list 2)	Extractive reserve, very diverse
18	Brazil: Acre	Serra do Divisor	Ucayali 6(12)	Two (list 1)	National Park, high diversity
19	Brazil: Amapá	Lourenço mines	Oyapock 9	One (list 1), 3(2)	Preservation proposed
20	Brazil: Pará	West of Altamira	Tapajós 10(18)	Two (list 1), 2(2)	Indigenous reserves
21	Brazil: Amazonas	Pico da Neblina	Imeri 11(13)	Three (list 1), 1(2)	National Park, tepuis
22	Peru: Junín	Mina Pichita area	Chanchamayo 3(19)	Not defined	No preservation, little access
23	Peru: Cajamarca	Jaen region	Marañón 9	Not defined	Local preservation
24	Ecuador: Azuay, Mor.-Sant.	Inecel to Gualaquiza	Sucúa 12(14)	Not defined	Local preservation
25	Ecuador: Pastaza	Rio Topo region	Abitagua 12(22)	Not defined	Local preservation
26	Colombia: Meta	Cubarral region	Villavicencio 14(20)	Not defined	Local preservation
27	Colombia: Antioquia	Rio Claro-Vallesol	Nechí 9(13)	Not defined	Local preservation
28	Colombia: Valle de Cauca	Rio Tatabro/Aguasclaras	Chocó 11	Not defined	Ecological Reserve
29	Venezuela: Táchira	Rio Negro/Rio Frio	Apure 8(10)	Not defined	National Park
30	Venezuela: Monagas	Cueva El Guácharo	Sucre 7	Not defined	National Park
31	Panamá: Panamá	Cerros Azul, Jefe	Darién 10	Not defined	National Park
32	Panamá/Costa Rica	Talamanca slopes	Chiriquí 18(22)	Not defined	Bi-national Park
33	México: Oaxaca	Sierra de Juárez	Guatemala 16	Not defined	Local preservation

TABLE VIA. Brazilian butterflies threatened with extinction: first list (published in 1989).

Group, genus, species/subspecies	Distribution (states)	IUCN status	Habitat	Notes, population size
Papilionidae				
Eurytides iphitas	Esp. Santo–Rio de Jan.	Endangered?	Steep forest	Not seen in 50 years
Euritides lysithous harrisianus	Rio de Janeiro	Endangered	Restinga forest	Pop. 50–200, one locality
Papilio himeros (*h.h.* and *h.baia*)	Bahia to Rio de Janeiro	Endangered	Seasonal forest	Very few colonies, pops 10–30
Papilio zagreus (*z.neyi*, *bedoci*)	Acre, Rondônia, Amapá, AM	Vulnerable	Perhumid forest	Very few specimens known
Parides ascanius	Rio de Janeiro	Vulnerable	Sandy restinga	Still many strong colonies, 20–100
Parides lysander mattogrossensis	Mato Grosso (central)	Vulnerable	Perhumid forest	Highly restricted, isolated
Pieridae				
Perrhybris flava	Espírito Santo	Vulnerable	Seasonal forest	Highly variable; pops 20–150
Moschoneura methymna	Bahia to Santa Catarina	Vulnerable	Perhumid steep forest	Many colonies
Riodininae				
Joiceya praeclara	Mato Grosso (central)	Vulnerable	Transition forest	Not seen in recent years
Brassolinae				
Orobrassolis ornamentalis	São Paulo/Minas Gerais	Endangered	High-altitude field	Not seen since 1968
Dasyophthalma vertebralis	Espírito Santo (north)	Endangered?	Palmy lowlands	Not seen in 50 years
Ithomiinae				
Eutresis hypereia imeriensis	Amazonas (north-west)	Vulnerable	Tepui forest	Only one colony known
Melinaea mnasias (various sspp.)	Locally throughout	Vulnerable/ Endangered	Marginal forest	Very rarely seen
Mechanitis bipuncta	Amazonas, Roraima	Vulnerable	Tepui forest	Very few colonies known
Scada karschina delicata	Pernambuco, Paraíba	Endangered	Perhumid forest	Only one colony known
Napeogenes cyrianassa xanthone	Bahia (mixed in E. Santo)	Vulnerable	Lowland forest	Only two colonies known
Hypothyris mayi	Pará, west of Altamira	Vulnerable	Liana forest	Very few colonies, local
Hyalyris fiammetta	Bahia, MG, ES, RJ, SP	Endangered?	Transition forest	Not seen in 50 years
Hyalyris leptalina leptalina	Rio de Janeiro	Endangered?	Perhumid forest	Not seen in 50 years
Hypoleria fallens	Minas Gerais, ES, RJ	Vulnerable	Perhumid forest	Several colonies, pops 5–20
Hypoleria mulviana	Pará, Altamira-Itaituba	Vulnerable	Liana forest	Few colonies known
Heliconiniini				
Heliconius nattereri	Bahia, Espírito Santo MG?RJ?	Endangered	Transition forest	Three colonies, pops 10–150
Melitaeini				
Eresia erysice	Bahia (southern)	Vulnerable	Ecotones of forest	Several colonies known

A variant of the biosphere reserve concept is the extractive reserve, where viable and sustainable extractive economies based on fruits, latex, medicines, oils, fibres, gums, spices and (as a lesser item) renewable timber, as well as other natural products, are conserved by the small-scale local harvesters of these economically valuable resources (Hecht and Schwartzman, 1988; Peters *et al.*, 1989; Padoch and de Jong, 1989; Schwartzman, 1989; Fearnside, 1989). A number of such reserves have recently been proposed and established by rubber tappers in the Brazilian Amazon, aided by both national and international agencies (Fig. 11). One of the largest of these, in the basin of the Rio Tejo (alto Rio Juruá), was established with the help of analysis of endemism and diversity in butterflies, which revealed it to show low corrected endemism values (mixture of Ucayali and Inambari centres) and very high diversity including many Andean species otherwise unknown in Brazil (Brown and Cardoso, 1989). Other proposed reserves, as well as the equally important indigenous reserves (Fig. 11), are being evalutaed and monitored in this further effort to preserve as much as possible of the Amazonian biota and human culture, without impeding occupation by humans but not permitting non-sustainable misuse of resources.

Adequate knowledge of the patterns of occurrence, differentiation and resource use of insects can thus lead to their successful and helpful inclusion in both planning and monitoring (Management plan) stages for diverse types of conservation units in the neotropics.

V. A PRACTICAL GUIDE TO CONSERVATION ACTION USING INSECTS (FIG. 12)

A. How to Choose a Group for Use in Conservation

An insect taxon should meet the criteria indicated above under "indicator groups" (Table III) to be useful in conservation planning and monitoring. In particular, it should include a large number of ecologically diversified species whose systematics and ecology are well known, and that are easy to find and evaluate in the field in a short time. Nocturnal, fossorial or other cryptic insects, even if attracted to lights and baits, are of lesser utility, since possibilities for non-destructive population study may be limited. Those whose life histories are incomplete or unknown will carry reduced information content. Species or groups that are migratory, poorly differentiated, highly erratic in time, eurytopic or easily adaptable to anthropic habitats, or phylogenetically inscrutable are of minimal value.

TABLE VIB. Brazilian butterflies possibly threatened with extinction: second list (for more study, 1989–1990).

Group, Genus, species/subspecies	Distribution (states)	Habitat	Notes, population sizes, status
Papilionidae			
Papilio garleppi (*insidiosus, interruptus*)	Amazon Basin, AP-AC	Riparian forest	Very few specimens known
Papilio xanthopleura/diaphora	Amazon Basin, PA-AC	Riparian forest	Very few specimens known
Parides burchellanus	Goiás, Minas Gerais, SP	Riparian forest	Last seen in late 1960s
Parides panthonus (various sspp.)	Amapá to São Paulo, RO	Riparian forest	Few colonies known (includes *aglaope*, *castilhoi*, *numa*, several new sspp.)
Parides hahneli	Pará, Amazonas	Sand forest	Several large colonies still known
Parides pizarro (+ *kuhlmanni*)	Amazonas, Acre	Riparian forest	Very few colonies known
Parides vercingetorix	Amapá	Riparian forest	Uncertain occurrence in Brazil
Pieridae			
Charonias theano	São Paulo, Minas Gerais	Montane scrub	Seen only once in last 50 years
Hesperocharis catasticta	Esp. Santo, Rio de Jan.	Montane hilltops	Very little known; few seen recently
Riodininae			
Nirodia belphegor	Minas Gerais (central)	Montane rockfields	Endemic genus; few seen
Petrocerus catienia	Rio de Janeiro	Montane forest	Endemic genus; very local and rare
Eucorna sanarita	Rio de Jan., MG, SP	Montane streams	Endemic genus; few localities
Helicopis cupido lindeni	Paraíba	Coastal swamps	Very few colonies; southern limit
Genera: *Symmachia, Xenandra, Mesenopsis, Esthemopsis, Colacyticus, Alesa* (all over country, most species very little known)			
The Clinae			
Arawacus aethesa	Bahia, Espírito Santo	Lowland forests	Very rarely seen, restricted
Arcus ducalis	Esp. Santo-Santa Catarina	Mountain tops	Sporadic, rare, easily eliminated

Hesperiidae			
Elbella polyzona polyzona	Rio de Janeiro, S. Catarina	Hilltops	Very rarely seen
Turmada maravilha (2 sspp.)	Esp. Santo, Rio de Janeiro	Lowland forests	Very few specimens known
Drephalys miersi, *mourei*, others	Esp. Santo to S. Catarina	Hilltops	Most species very rarely seen
Morphinae			
North-eastern subspecies of *M. epistrophus*, *M. menelaus*, *M. hercules* and *M. achilles* — habitat greatly reduced			
Brassolinae			
Dasyophthalma delanira	Rio de Janeiro	Montane forests	Only one site known
Dasyophthalma geraensis	Rio de Janeiro, MG, SP	Montane forests	Few colonies known, not common
Charaxinae			
Prepona deiphile	Esp. Santo to S. Catarina	Montane forests	Very rarely seen
Agrias (all four species)	Throughout country	Humid forests	Many colonies are easily eliminated
Ithomiinae			
Tithorea harmonia caissara	Esp. Santo to São Paulo	Montane forest	Very few colonies known
Hypothyris connexa (two sspp.)	Roraima, Amazonas	Tepui forest	Very restricted and rare
Hypothyris leprieuri (many sspp.)	Amazonian region, SE?	Transition forests	Extremely local and little known
Acraeinae			
Actinote zikani	São Paulo, Minas Gerais	Montane forests	No colonies presently known
Nymphalinae			
Callicore hydarnis	Rio de Janeiro, MG, SP?	Montane forests	Very rarely seen
Doxocopa laurona	Santa Catarina	Hilltops	Very rarely seen
Doxocopa zalmunna	São Paulo	Seasonal forest	Not seen in recent decades

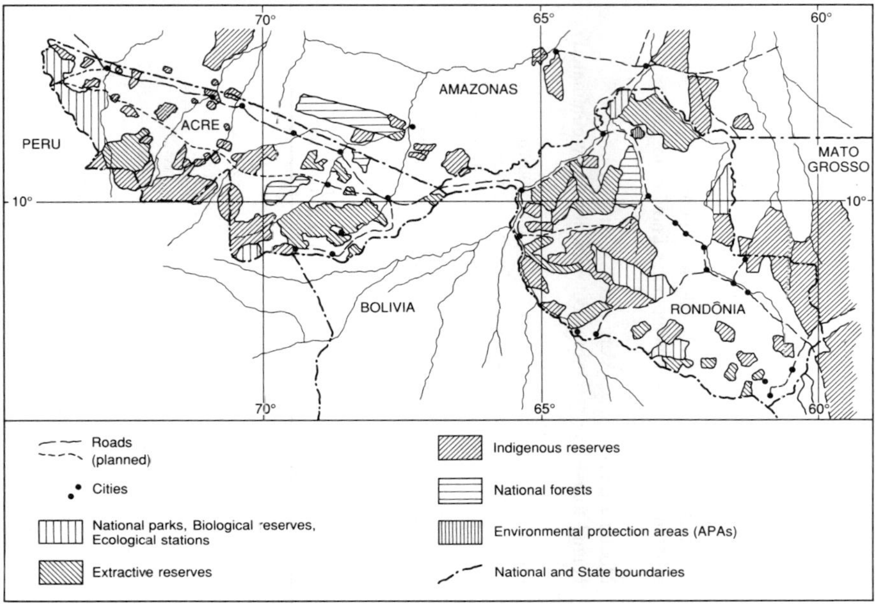

Fig. 11. Extractive reserves, indigenous reserves and other important preserved areas, in the western Brazilian states of Rondônia and Acre. Areas of high species diversity and endemism occur in both states (Guaporé, Rondônia, Madeira, Inambari and Ucayali, plus high-diversity mixed faunas; see Figs 2 and 4).

Ten neotropical groups that are already near to optimum usefulness or could be brought up to this with a few years of intensive biosystematics and geographically widespread field work are: Odonata, especially zygopterans; tabanid and asilid flies; various groups of ants, bees and wasps (Hymenoptera), myrid bugs; and cerambycid, chrysomelid, and curculionid beetles; in addition to several groups of butterflies (especially some hesperids and riodinines) and diurnal or easily observed moths. Obviously, some subgroups of all of these are much more useful than others, and work should be concentrated on these (as long as they have at least several dozen ecologically diversified and faithful species) to bring them into optimal usefulness. As they are added to the roster of insects routinely used in conservation protocols, the likelihood will be increased for well-based priorities, plans and monitoring of preserves in tropical America.

B. How to Predict Areas of Potential Importance

An insect population responds as much to micro-environment as to the general physical characters of a site or region such as are found in coarse-scale maps, and insects are also very sensitive to their immediate biotic resources and the abundance of their enemies. Thus, prediction of insect endemism and abundance must proceed through at least four stages, and always be verified afterwards with on-the-spot checks. The stages that I have used to find palaeo-environments with unusual butterfly communities or rare species are (Table II):

1. Examination of topographic and climatic maps (fine scale if possible), seeking regions of complex landforms and appreciable but variable rainfall, without excessively long dry season or other strongly stressing elements.
2. Use of fine-scale soil maps (broad-scale surveys are less useful) to look for areas of complex soil mosaics including eutrophic soils — these almost always support rich plant and insect communities.
3. Analysis of vegetation maps to define core and mosaic areas with varied plant formations and ecotones, permitting a high density of niches and resources for insects.
4. Preliminary survey, in known collections or published biological information, for localities of rare or threatened species in any plant or animal group.

Because climate, topography, soil and vegetation mosaics and even biotic factors are essentially fractal in nature, continuing to reveal ever more complex patterns of subdivision down to very reduced dimensions which are still recognized as different by small insects, these predictive methods can, at best, suggest larger regions within which the real concentrations of endemism, diversity and rarity must be discovered by traditional footwork.

C. How to Undertake Surveys of Butterflies (or Other Insects)

In the neotropics, many important insect areas are still relatively inaccessible and must be penetrated by walking or riding out from the nearest point served by motor transport. Especially when new data are sought, and new areas suggested by soil or vegetation surveys, one cannot remain in sites along roads, railways or rivers, which have been extensively visited and modified by others. Thus, survey begins with detailed access maps of a region, aerial photos when possible, conversation with local residents and a "nose" for trails into good original habitat. It is usual to go out 3–10 km in the early morning to reach a site where an adequate sample can be made — the same distance instinctively regarded as usual by forest Indian hunters heading out from a village (perhaps a 20–60 km circumference can be guaranteed to provide game to sustain the 10–30 families in the human population). The trail itself usually contains progressively more interesting sites as it diminishes in width and usage; some time can be dedicated to these, and if a circle trip is possible, as many as six major sites (up to 1 km^2) and numerous additional minor sites (a few hectares or less) can be sampled in a day; any one could be the primary key to the value of an area. With a Jeep, horse or helicopter, the sites can be increased in their separation but not in their total number during the day. After the most informative sites are discovered, they can be more extensively explored, on and off the trails, on subsequent days or in different seasons. A general region of hundreds of km^2 (river valley, mountain range, intermontane plateau, lowland plain) may take from a week to a month to explore adequately, and several years really to understand well enough to compare with other sites and know which sectors are more valuable to conserve. This time scale is often impractical in conservation efforts today, but at least a day or two of broad survey and another of intensive observation of adequate indicator groups should precede analysis and recommendation of any area. There are no short cuts, as recent conservation history has confirmed; even large regions preserved on predictive criteria alone have often lacked the target species and biotas upon field examination.

Once in a sampling or survey site, the visitor should intensively seek insects (or, better yet, attract them with appropriate baits) both on and off the trails, although for many groups the ecotones provided by older access routes and clearings provide most species and best opportunities for observation and area coverage. Polymorphic populations should be more thoroughly observed to define their variation and origin; more time should be spent near the normal resources of the organisms being surveyed, also seeking juvenile stages and observing behaviour. The field bag should include adequate food and water (dehydration is surprisingly fast in tropical work, even in rain forest), protection against rain and disease vectors, a compass, a marking pen, a flashlight, measuring aids, plastic bags and dishes, first aid items, and a pair of wide-angle, close-focusing binoculars. Any larger or more numerous animals (especially humans, wasps and hunting ants) should be respected but not feared; large cats and venomous snakes are very rarely encountered, but like the more usual potential threats require a special vigilance, equanimity, and if all else fails, agility!

D. How to Transform Survey Data into Useful Parameters

The biological figures most useful for resource and conservation planners and managers are indices of endemism, diversity, abundance and rarity, and presence of recognized threatened or economically important species. All should be quantitative and comparative to be most useful or convincing. Endemism should be presented as comparable totals of species or subspecies, in the latter case corrected for mixture of regional differentiates (in hybridization or contact zones). Diversity is best compared by evenness (H') in small samples, total species in long-term data. Lists of rare or threatened species should be accompanied by some comparative population data through time and space. All analyses should include as much data as possible on variation, resources, community structure and other organisms, especially large plants and vertebrates; the entomological conservationist should recognize and report on his ferns, vines, palms, orchids, frogs, birds and primates. Starting with an evaluation of the physical environment, the analysis should give ample emphasis to sampling methods, the new biological data and broad ecological considerations, and close with essential human concerns: scenic or archaeological values, support of local populations, regional land-use plans and patterns (including ownership), human use of local water and biotic resources, and integration of the proposed unit into all of these.

E. How to Relate to the Public Sector and the Local Community

The right half of Fig. 12 shows the complex interaction between scientific data and the actual conservation process. In the past, this has often proceeded through the political channels at the far right, but it is now recognized that any conservation unit is safer and less expensive to maintain if substantial sectors of the local population respect and support it as part of their own aspirations, enjoyment, resources and future value (the alternative is to make the unit totally inaccessible, which is always temporary and risky, if not impossible). Thus, ideal conservation starts from these local sectors themselves, who wish to set aside an effective area for resource preservation, education, recreation, tourism and aesthetic appreciation, and contact both scientists and government (upper centre of Fig. 12) to help bring this about. Such an ideal case has recently led to important preserves in at least Mexico, Guatemala, Costa Rica, Panama, Colombia, Venezuela, Ecuador, Peru and Brazil, often with direct involvement of insect surveys and monitoring; the procedure is certainly spreading as rapidly as base democracy throughout the New World. When a scientist wishes to initiate the process (upper left), however, this

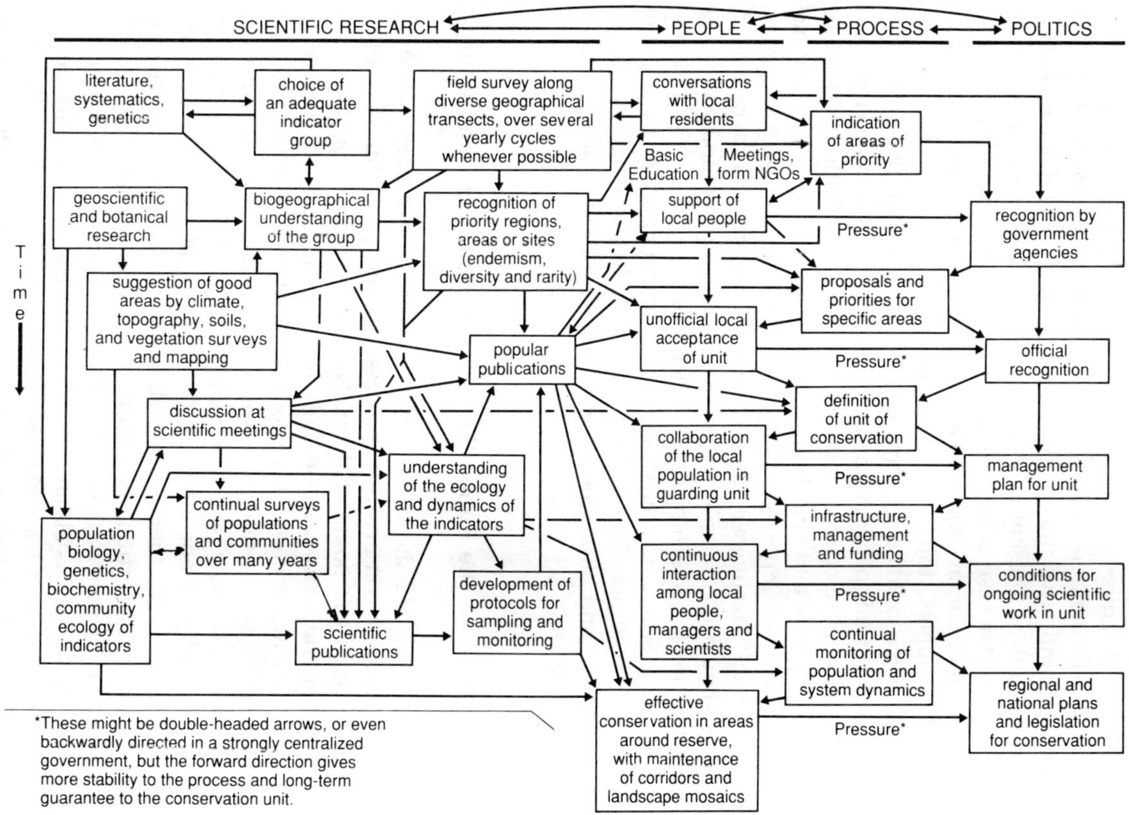

*These might be double-headed arrows, or even backwardly directed in a strongly centralized government, but the forward direction gives more stability to the process and long-term guarantee to the conservation unit.

Fig. 12. The conservation process: interaction between science and society.

should always involve conversations with local people, living around or within the surveyed area, seeking to discover and disseminate broad awareness of the value of the area to them and their future. Negative opinions should especially be sought, respected, and delicately brought into discussion with those of other residents. Educators, religious leaders and other influential people in the local population should be contacted, befriended, listened to and praised for their interest (at times still latent, but usually expressed) in the community and its future; their own problems and dilemmas should be understood and helped in any way possible. This is the true and continual basis for any successful conservation programme of species, habitats, watersheds or resources, and also the trump card for the actions at levels more removed from the local scene.

The higher public sectors may be much more interested in immediate money and power than in scientific data or resources for the future. Thus, a direct approach to government agencies should only be attempted by politically trained scientists — even when called in for consultation, the scientist should expect all the exact data and logical recommendations to be reshaped or even inverted through channels necessarily constrained by expediency, priority and compromise. At this level — through which an effective conservation unit should probably pass at some time — the most useful implements are generalizations (scientific theory, even if quite fuzzy) and votes; in the latter case, local support for a proposal, especially by numerous and vociferous non-governmental organizations (NGOs), is very important. A careful, very condensed and tightly organized mix of these can be most convincing, while a careless, wordy or ambiguous proposal will quickly be set aside. Patience, calm confidence and insistence, well-controlled flexibility, very broad knowledge and humility are fundamental tools in the interaction with official channels; in a scientist, they may be honed through teaching undergraduate core courses, participating in local heterogeneous non-scientific communities or having many children. Fame, wealth, intelligence, number of publications and academic rank of the proponent hinder as often as help in bringing about the conditions for correct decision making on conservation cases by the public sector.

Local, private or NGO-sponsored conservation units can bypass the complexities of interaction needed for large public parks and reserves, and if well chosen and managed may be just as effective as the latter, even though usually smaller. Many large national or international (biosphere or world heritage) reserves grow out of small, intensely educational local reserves such as are now springing up in all parts of the neotropics, often (as in Brazil) with local or national tax incentives to leave natural vegetation untouched or to replant heterogeneous arboreta or gardens. The real hope for future conservation probably lies in these very local and basic initiatives for multiple land-use planning, stimulated and aided by scientific data and good friendship, which in any case will greatly facilitate action at higher and more distant levels.

F. Insects and the Conservation Transition

Just as the demographic transition can bring considerable tension and anguish to a country or people in the explosive growth in the period between falling death rates and the corresponding drop in birth rates, so the conservation transition can lead to appreciable loss of natural resources in the gap between accessibility of habitats with economic practicability of their conversion and public recognition of the importance of saving these resources, which provide along with the associated systems so many fundamental services to life (Fig. 13). In most industrialized countries, the time lag between the rise of these curves has given an intermediate curve of non-sustainable removal of natural resources leading to high asymptotes of conversion of natural systems to anthropic ones (Fig. 13). Typically, the transition has been completed with only 2–5% of natural vegetation remaining in reserves, and another 5–20% relatively intact in private or public domain, in pieces of various sizes. The rapidity of the response to excessive devastation with accompanying deterioration of local climate, air quality, soils, water sources, biological diversity and ecosystem services depends strongly on educational systems and local NGOs who defend nature in a wide variety of ways. Exceptionally, the Brazilian state of Acre (Fig. 11) has declared a moratorium on all conversion, with over 95% of primitive habitat still intact, while they study economic zoning and the benefits of extractivism; this example and the very low, later carefully planned asymptote (Fig. 13) should be widely praised and publicized.

Some insects are well known as useful to humankind; farmers have always admired their ability to increase fruit crops through cross-pollination (Hymenoptera, Diptera, skippers and sphingids are most prominent in this), and everyone appreciates good honey. Butterflies invariably appear in primitive and modern mythologies as symbols of freedom, movement, colour, beauty, transformation and renewal (one of the most ancient Christian symbols for the Resurrection), and can be easily used to support campaigns for nature preservation at a popular level; in the neotropics, they are so ubiquitous and beautiful that all people know and like them, and are eager to know how they can be kept around. Thus, they have an important potential in accelerating the conservation transition and arriving

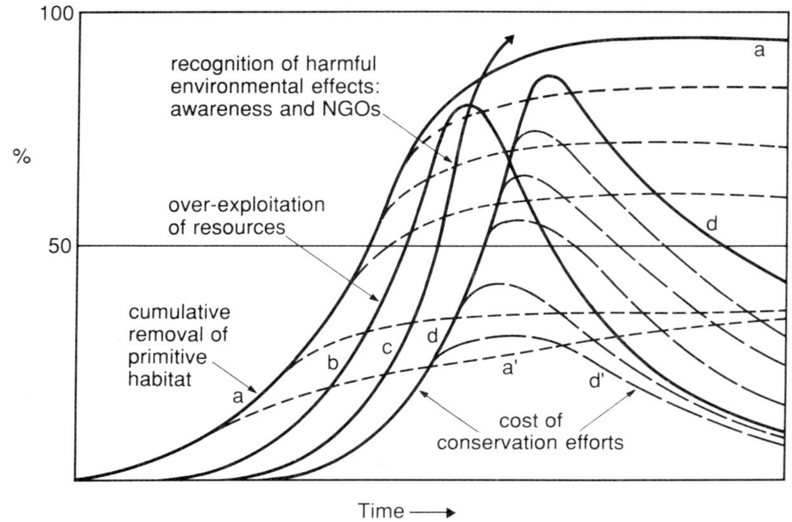

Fig. 13. The conservation transition: the total conversion asymptote (curves a to a′) and the cost of conservation efforts (curves d to d′) increase with the time lag between accelerated removal of natural systems (a) and the appearance of environmental defence groups (c), controlling the unsustainable overexploitation of natural resources (b). The uppermost curves (a, d) represent the case of the Brazilian Atlantic forests, about 90% converted before 1970 when environmental awareness became strong; the conservation cost is high for the remaining systems. The lowest dotted lines (a′, d′) represent the projection for the Brazilian Amazon, presently about 10% converted, with the final amount to be kept under 33%, most areas being protected and cared for by local communities, resulting in a minimal cost for effective conservation.

at a lower value for the conversion asymptote, after which the people will no longer tolerate further removal of natural vegetation.

VI. EPILOGUE: IMPORTANT CONSIDERATIONS FOR THE AGENDA, 1990–2010

A. The Primacy of Education and Grass-roots Movements: Community and Conservation

The preceding sections have argued the fundamental importance of the local community in effective conservation efforts. While not all members of the community will share the entomologist's love for insects, some will even be eager to prevent survey collecting in "their" reserves, a noble attitude which should always be immediately and rigorously respected and cultivated (understanding of the monitoring process can be sought at later stages, if necessary). In general, this community will be more likely to know which local resources are preserved and where than the entomologist can guess even after years of survey. The hoped-for coming heyday of base democracy in the neotropical countries (Brazil is already there) gives unlimited opportunities for scientific collaboration with local groups to help achieve true "conservation landscapes" where all resources are effectively saved within the anthropic mosaic, including islands and corridors of natural vegetation to reduce the effects of habitat attrition by occasional disasters. Many such mosaics already exist in some parts of the Brazilian Atlantic forests, and others are being proposed and established in various regions with information already in hand. Further areas, many still almost unoccupied (over 90% of the 4 900 000-km^2 Brazilian Amazon region is still untouched by man, other than well-equilibrated indigenous groups; Brown and Prance, 1987; Cunha, 1989) are being surveyed as soon as they become accessible, and various options are being tried in order to discover optimum regimes of development with landscape planning and management, including maintenance of large areas as pristine reserves, indigenous lands, extractive reserves, sustainable-production forests or integrated mosaics of agriculture and primary vegetation. Older proposals based on preliminary information are being updated with more widely based amphibian, bird, primate and insect surveys and vegetation studies. Fiscal incentives are no longer available for unsustainable use of the forest areas (like massive timber extraction, reforestation, pasture, peasant farming on poorer soils or non-native agroindustry), and illegal wildcat attempts at cutting and conversion are often successfully blocked in stand-offs with local indigenous, rubber-tapping, fruit-gathering or other resident communities. Important recent studies revealing the direct economic advantages of sustainable economies based on extractivism over any conversion-based activity, even in the shorter term (Hecht and Schwartzman, 1988; Browder, 1989; Anderson, 1990; Hecht and Nations, 1989; Peters *et al.*, 1989) are completing the information necessary to permit effective resource preservation for further use — with participation of local communities in the planning, management and benefits of resource mobilization. This transition has depended greatly upon conservation education at all levels in all parts, carried through by local teachers (ecology and natural resource courses are now required continuously from first grade through the University in Brazil). A true consensus is thereby in the making, hopefully resistant to the limited immediatist and profit-motivated forces which still seek to dominate natural resource use in some areas.

B. Biosphere Reserves as Ideal Insect Habitat

This consensus, a product of the demographic and conservation transitions, could eventually transform much of the landscape into mosaic patterns not greatly different from those presently existing in some parts of the world with old, stable cultures in equilibrium with their resources. Technology is now available to permit this equilibrium to exist at a higher human population density (up to 0.5 ha^{-2} = 50 km^{-2}) than that of present Amazonian indigenous groups (only about 0.1 person km^{-2}), with part isolated into cities at even higher densities – which should not be located in the more important endemism or diversity areas, or in limited palaeoenvironments. The biosphere reserve concept (Maldague, 1984) and the second World Conservation Strategy (McNeely *et al.*, 1989; Prescott-Allen, 1989, in preparation) provide for peaceful coexistence between humankind and the rest of the biosphere through careful planning and management of landscapes and economies combining all. Because insect diversity is closely tied into habitat heterogeneity and mild disturbance (Fig. 4), it may be presumed that such reserves and landscapes will be optimal for the maintenance of all species and the genetic heritage, especially of insects, unless excessive insecticides, hydrocarbons or particulate pollution are permitted to overly affect non-industrial or agricultural mosaics. The provisions for education, research, training and ecosystem restoration, which form part of the biosphere reserve programme, should guarantee its multiplication and continuity, thereby providing for adequate insect conservation in the future.

C. Economic, Cultural and Scientific Imperialism

The "remnants of authoritarianism" which are still amply evident in the industrialized world and in some cases dominate the developing nations, are the chief antagonists to the "community control of resources" ideal presented above. They are typically observed as the attempts of one community, large or small, to control and derive immediate profit from the resources of other communities, without the necessary bilateral agreements which guarantee equitable division of the benefits and long-term sustainability of the process. This was formerly called "imperialism", and the term has been expanded in recent times to cover a broad range of actions, from brute force to technology, rock music and psychological persuaders, through which a dominant group controls subdominant ones with unequally divided benefits. It has been emphatically declared that by the year 2000 such "imperialistic" actions, on any scale, must be universally outlawed if the human race is to survive. The conservation scene also reflects (and generates) these situations and supports this declaration, as discussed above. Hopefully, all conservationists will be able to work together in the coming decades to avoid the economic, political, social and environmental cataclysms that must result from any increase in over-dominance-directed actions at local, regional or international levels.

D. The Importance of Optimism

In an increasingly overcrowded, polluted, undereducated and destroyed urban environment, it is easy to lose optimism and resort to egotism or fatalism in relation to conservation of resources (including insects and their habitats). Some even tend towards pessimism, exaggerating the rate or extent of destruction or focusing attention on supposedly "unavoidable" impending catastrophes, subtracting their efforts from the imperative and immediate actions necessary to avoid them. The biosphere is still healthy and even exuberant in the neotropics, and its energy and homeostatic capacity far exceed any "contributions" made or possible by humankind. While it is conceivable that exponentially increasing insults could lead to a severe punishment of the arrogant insulting, presumptuously over-dominant species, by adequate expression of the homeostatic mechanisms of nature, it seems more sensible that we learn to be friendly to ourselves and to nature before such becomes necessary. In the Israelite Genesis tradition, shared by Middle Eastern and Occidental cultures, humans were but a part of Creation (a unifying term; the language did not contain a term for a separate nature), placed within it as a representative ("image") to help care for it and guarantee its productivity (in spite of other current interpretations caused by poor translation, fuelled by the technological mentality; see Granberg-Michaelson, 1984, pp. 54–104). Recent tendencies in the neotropics suggest that this integrative relationship, also central to many Eastern traditions, can be creatively recovered by modern peoples, perhaps with the help of older cultures which still persist in harmony with their natural resources. There are ample bounds for optimism — that very fundamental human trait which ironically has led to technological development without full consideration of its consequences — to coexist with realism; both are especially necessary for this time. Insect scientists, who deal with the vast majority of natural diversity, should be willing to participate in exploring these bounds and contributing their vision to others in the coming decades.

Acknowledgements

The Royal Entomological Society of London, through the Symposium Convenors N. M. Collins and J. A. Thomas, provided support to present this paper in London. The late Dr David Azambuja originally stimulated the use of insect data to propose a conservation unit in the Chapada dos Guimarães in 1969 (finally declared a National Park in 1989). Drs Paulo Vanzolini, Aziz Ab'Sáber, Igo Lepsch and G. T. Prance helped greatly in the palaeo-ecological synthesis; the last gave access to the Flora Neotropica map used in Figs 1–4, and the Projeto RADAMBRASIL in Rio de Janeiro permitted access to unpublished data for these syntheses. The Brazilian IBDF (especially Maria Teresa Jorge Pádua) and SEMA (Dr Paulo Nogueira Neto), now combined to the IBAMA, provided travel funds and stimulus for many surveys and transects in the Amazon region. Drs T. E. Lovejoy and R. O. Bierregaard invited me to participate in the BDFF Project north of Manaus and guaranteed travel funds and infrastructure for this work, in which Roger W. Hutchings helped in later stages. Drs Olaf Mielke, Curtis J. Callaghan, Robert K. Robbins and Lee D. Miller helped develop the Brazilian and Atlantic butterfly lists, stimulated by Dr Angelo B. Machado and the Sociedade Brasileira de Zoologia. Drs Ernesto Schmidt-Mumm and Gerardo Lamas provided comparable data on Colombian and Peruvian butterflies. Gonzalo Halffter of México, and Jorge Orejuela and Emílio Constantino of Colombia, first provided information, stimulus and effective field tests of methods for working with local communities and forming biosphere reserves. Recently, new concepts and data have been furnished by Dr R. B. Mittermeier of the WWF-US, Drs Paul R. Ehrlich, Bruce A. Wilcox and Dennis D. Murphy of the Centre for Conservation Biology (Stanford University), Dr Kenton R. Miller of the World Resources Institute, Elizabeth Nagagata of the Reserva Biológica Poço das Antas, and many other Brazilians young and old. Research fellowships and grants have been received from the Brazilian CNPq and FAPESP. I am deeply grateful to all of these persons and agencies for their support.

REFERENCES

Anderson, A. B. (ed.) (1990). "Alternatives to Deforestation: Steps Towards Sustainable Use of the Amazon Rainforest". Columbia University Press, New York.

Brasil, SEPLAN/IBGE and Min. Agric./IBDF (1988). Mapa de Vegetação do Brasil, 1: 5 000 000.

Browder, J. O. (ed.) (1989). "Fragile Lands of Latin America: Strategies for Sustainable Development". Westview, Boulder, Colorado.

Brown Jr, K. S. (1970). Proposta: uma reserva biológica na Chapada de Guimarães, Mato Grosso. *Brasil Flor.* **4**, 17–29.

Brown Jr, K. S. (1972). Maximizing daily butterfly counts. *J. Lepid. Soc.* **26**, 183–196.

Brown Jr, K. S. (1978). Heterogeneidade: fator fundamental na teoria e prática de conservação de ambientes tropicais. "Encontro Nacional sobre a Preservação da Fauna e Recursos Faunisticos, Brasília 1977", pp. 175–183. IBDF. Brasília.

Brown Jr, K. S. (1979). "Ecologia Geográfica e Evolução nas Florestas Neotropicais", XXXI + 265 + 120 pp. Universidade Estadual de Campinas, São Paulo.

Brown Jr, K. S. (1982). Historical and ecological factors in the biogeography of aposematic Neotropical Lepidoptera. *Am. Zool.* **22**, 453–471.

Brown Jr, K. S. (1984). Species diversity and abundance in Jaru, Rondônia (Brazil). *News Lepid. Soc.* 45–47.

Brown Jr, K. S. (1987). Biogeography and evolution of Neotropical butterflies. *In* "Biogeography and Quaternary History in Tropical America" (T. C. Whitmore and G. T. Prance, eds), pp. 66–104. Clarendon Press, Oxford.

Brown Jr, K. S. (1991). Borboletas da Serra do Japi: diversidade, habitats, recursos alimentares e variação temporal. *In* "A Serra do Japi" (L. P. M. Fonzar, ed.). UNICAMP Press, Campinas, São Paulo.

Brown Jr, K. S. and Cardoso, A. J. (1989). Aspectos ecológicos da proposta reserva extrativista do Tejo, Acre. Report to Brazilian Government, 15 pp.

Brown Jr, K. S. and Prance, G. T. (1987). Soils and vegetation. *In* "Biogeography and Quaternary History in Tropical America" (T. C. Whitmore and G. T. Prance, eds), pp. 19–45. Clarendon Press, Oxford.

Brown Jr, K. S., Trigo, J. R., Francini, R. B., Morais, A. B. B. and Motta, P. C. (1991). Aposematic insects on toxic host plants: coevolution, colonization, and chemical emancipation. *In* "Evolutionary Ecology of Plant/Animal Interactions: Tropical and Temperate Perspectives" (P. W. Price, T. M. Lewinsohn, G. W. Fernándes and W. W. Benson, eds). Wiley, New York.

Coope, G. R. (1978). Constancy of insect species versus inconstancy of Quaternary environments. *In* "Diversity of Insect Faunas" (L. A. Mound and N. Waloff, eds), Royal Entomological Society Symposium No. 9, pp. 176–187. Blackwells, Oxford.

Cunha, R. P. da (1989). Deforestation estimates through remote survey: the state of the art in the Legal Amazonia. Anais Simpósio sobre Amazônia: Fatos, Problemas e Soluções, São Paulo.

Erwin, T. L. (1988). The tropical forest canopy: the heart of biotic diversity. *In* "BioDiversity" (E. O. Wilson, ed.), pp. 123–129. National Academy Press, Washington, D.C.

Fearnside, P. M. (1989). Extractive reserves in Brazilian Amazonia: an opportunity to maintain tropical rain forest under sustainable use. *BioScience* **39**, 387–393.

Futuyma, D. J. and Slatkin, M. (eds) (1983). "Coevolution". Sinauer, Sunderland. Massachusetts, X + 555 pp.

Gilbert, L. E. (1980). Food web organization and the conservation of Neotropical diversity. *In* "Conservation Biology: An Evolutionary–Ecological Perspective" (M. E. Soulé and B. A. Wilcox, eds), pp. 11–33. Sinauer, Sunderland, Massachusetts.

Granberg-Michaelson, W. (1984). "A Worldly Spirituality". Harper and Row, New York, XIX + 210 pp.

Haffer, J. (1969). Speciation in Amazon forest birds. *Science* **165**, 131–137.

Hecht, S. B. and Nations, J. D. (eds) (1989). "The Social Dynamics of Deforestation: Processes and Alternatives". Cornell University Press, Ithaca, New York.

Hecht, S. B. and Schwartzman, S. (1988). The good, the bad and the ugly: Amazonian extraction, colonist agriculture, and livestock in comparative perspective. UCLA Grad. Sch. Architecture & Urban Planning, Working Paper.

Lamas, G. (1985). Los Papilionoidea (Lepidoptera) de la Zona Reservada de Tambopata, Madre de Diós, Peru. I. Papilionidae, Pieridae y Nymphalidae (en parte). *Rev. Per. Entomol.* **27**, 59–73.

Landres, P. B., Verner, J. and Thomas, J. W. (1988). Ecological uses of vertebrate indicator species: a critique. *Conserv. Biol.* **2**, 316–328.

Lovejoy, T. E., Rankin, J. M., Bierregaard, R. O., Brown Jr., K. S., Emmons, L. H. and Van der Voort, M. E. (1984). Ecosystem decay of Amazon forest remnants. *In* "Extinctions" (M. H. Nitecki, ed.), pp. 295–325. University of Chicago Press, Chicago.

Lovejoy, T. E., Bierregaard, R. O., Rylands, A. B., Malcolm, J. R., Quintela, C. E., Harper, L. H., Brown Jr, K. S., Powell, A. H. Powell, G. V. N., Schubart, H. O. R. and Hays, M. B. (1986). Edge and other effects of isolation on Amazon forest fragments. *In* "Conservation Biology: The Science of Scarcity and Diversity" (ed. M. E. Soulé), pp. 257–285. Sinauer, Sunderland, Massachusetts.

Lugo, A. (ed.) (1988). Diversity of tropical species: questions that elude answers. *Biol. Int.*, Special Issue **19**, II + 37 pp.

Maldague, M. (1984). The Biosphere Reserve concept: its implementation and its potential as a tool for integrated development. *In* "Ecology in Practice" (F. di Castri, F. W. G. Baker and M. Hadley, eds), pp. 376–401. Tycooly, Dublin.

McNeely, J. A., Miller, K. R., Reid, W. V., Mittermeier, R. A. and Werner, T. B. (1989). "Conserving the World's Biological Diversity". IUCN, Gland, Switzerland, VIII + 265 pp.

Murphy, D. D. and Wilcox, B. A. (1986). Butterfly diversity in natural habitat fragments: a test of the validity of vertebrate-based management. *In* "Wildlife 2000: Modeling Habitat Relationships of Terrestrial Vertebrates" (J. Verner, M. L. Morrison and C. J. Ralph, eds), pp. 287–292. Univ. Wisconsin Press, Madison.

Myers, N. (1988). Tropical forests and their species: going, going ....? *In* "BioDiversity" (E. O. Wilson, ed.), pp. 28–35. National Academy Press, Washington, DC.

Oldfield, M. (1989). "The Value of Conserving Genetic Resources", revised edn. Sinauer, Sunderland, Massachusetts, XVII + 379 pp.

Otero, L. S. and Brown Jr, K. S. (1986). Biology and ecology of *Parides ascanius* (Cramer, 1775) (Lep., Papilionidae), a primitive butterfly threatened with extinction. *Atala* **10–12**, 2–16.

Padoch, C. and de Jong, W. (1989). Production and profit in agroforestry: an example from the Peruvian Amazon. *In* "Fragile Lands of Latin America: Strategies for Sustainable Development" (J. O. Browder, ed.), pp. 102–113. Westview, Boulder, Colorado.

Peters, C. M., Gentry, A. H. and Mendelsohn, R. O. (1989). Valuation of an Amazonian rainforest. *Nature* **339**, 655–656.

Pielou, E. (1975). "Ecological Diversity". Wiley-Interscience, New York, viii + 165 pp. See also Chapter 19, pp. 291–311 in "Mathematical Ecology" (1977).

Pimm, S. (1986). Community stability and structure. *In* "Conservation Biology: The Science of Scarcity and Diversity" (M. E. Soulé, ed.), pp. 309–329. Sinauer, Sunderland, Massachusetts.

Prescott-Allen, R. (1989). "Lessons of the World Conservation Strategy". Abstracts of the Annual Meeting of the Society for Conservation Biology, Toronto, Ontario, p. 1.

RADAMBRASIL (Projeto; MME-DNPM) (1981–1987). "Levantamento dos Recursos Naturais." Vol. 23 (1981a), Folhas SB 24/25, Jaguaribe/Natal, 740 pp.; Vol. 24 (1981b), Folha SD24, Salvador, 620 pp; Vol. 29 (1982), Folha SD23, Brasília, 655 pp.; Vol. 30 (1983a), Folhas SC 24/25, Aracaju/Recife, 852 p.; Vol. 32 (1983b), Folhas SF23/24, Rio de Janeiro/Vitória, 775 pp.; vol. 34 (1987), Folha SE 24, Rio Doce, 544 pp. Rio de Janeiro.

Schwartzman, S. (1989). Extractive reserves: the rubber tappers' strategy for sustainable use of the Amazon rainforest. *In* "Fragile Lands of Latin America: Strategies for Sustainable Development" (J. O. Browder, ed.), pp. 150–165. Westview, Boulder, Colorado.

Sheppard, P. M., Turner, J. R. G., Brown Jr., K. S., Benson, W. W. and Singer, M. C. (1985). Genetics and the evolution of muellerian mimicry in *Heliconius* butterflies. *Phil. Trans. R. Soc. Lond. B* **308**, 433–610.

Spencer, K. C. (ed.) (1988). "Chemical Mediation of Coevolution". Academic Press, San Diego, XV + 605 pp.

Vanzolini, P. E. (1970). Zoologia sistemática, geografia e a origem das espécies. *Inst. Geog. São Paulo, Sér. Teses Monogr.* **3**, 1–56.

Vanzolini, P. E. (1973). Paleoclimate, relief and species multiplication in tropical forests. *In* "Tropical Forest Ecosystems in Africa and South America: a Comparative Review" (B. J. Meggers, E. S. Ayensu and W. D. Duckworth, eds), pp. 255–258. Smithsonian Institution Press, Washington, DC.

Wilcox, B. A., Murphy, D. D., Ehrlich, P. R. and Austin, G. T. (1986). Insular biogeography of the montane butterfly faunas in the Great Basin: comparison with birds and mammals. *Oecologia (Berlin)* **69**, 188–194.

Wilson, E. O. (1988). The current state of biological diversity. *In* "BioDiversity" (E. O. Wilson, ed.), pp. 3–18. National Academy Press, Washington, DC.

15

Insects and Tropical Forest Conservation

S. L. SUTTON AND N. M. COLLINS

I. Introduction 405
II. The size of the Insect Resource 406
III. Do Insects Matter? 408
IV. Pattern and Process 409
V. Deforestation 409
VI. Identification 411
A. Indicator Groups 412
B. Standard Techniques 413
VII. Extending the Protected Areas Network 414
A. Selection of Conservation Areas 415
B. Latent Extinction and Transient Diversity 418
C. Conservation of Disturbed Forest Mosaics 419
D. External Threats 420
VIII. Insects as Tools for Conservation 421
IX. Conclusions 421
References 422

I. INTRODUCTION

Fears are running high that our planet is currently at the foot of an exponentially increasing slope of biological impoverishment. The evidence is fragmented, more circumstantial than factual and sometimes contradictory, but nevertheless persuasive. Impoverishment is not yet happening in terms of substantial species extinctions, but in the massive reduction of species populations to rare or vulnerable status. We believe that the time to save a species is when it is common. By taking action as outlined in this paper it should still be possible to save the vast majority of organisms that concern us here — the insects of tropical forest. We are also concerned with the part that insects — or rather insect science — can play in the conservation of the forests themselves. In other words, how insects can operate as tools of conservation.

In the first part of this paper we examine the size of the insect resource, consider the question "Do insects matter?", look at the extent to which both ecological patterns and processes have been studied in tropical forest ecosystems, and set the threat to the insect resource in the context of rampant deforestation. We then consider how to alleviate problems of species identification, and the choice of criteria for choosing indicator groups and standard techniques of study. This is followed by consideration of the protected areas network and conservation area selection. We then introduce the ideas of latent extinction and transient diversity, and discuss the conservation of insects in disturbed forest. We end with proposals for the use of insects as tools for conservation.

II. THE SIZE OF THE INSECT RESOURCE

Recent published estimates of the number of species of insects and other terrestrial arthropods range between 10 and 30 million (Stork, 1988; Erwin, 1982, 1983), with an estimated 1 million so far described (Wolf, 1987). May (1986, 1988) has reviewed the question and put forward the challenging proposition that, to a good approximation, all organisms are

TABLE I. A history of estimates of the described and total number of insects world-wide.

Date	Number of described species	Estimated total number of insects worldwide	Author
1690s		10 000–20 000[a]	John Ray quoted in Westwood (1833)
1760s	3000		Linnaeus, *Systema Naturae*, 12th edn
1821		100 000	McLeay (1819–1821)
1815–1826		400 000–600 000[b]	Kirby and Spence (1815–1826)
1924	450 000		Imms (1925)
1950s		3–5 million[c]	C. B. Williams (attributed)
1952	700 000		USDA estimate
1980	900 000		BM(NH) annual report
1982		30 million[d]	Erwin (1982)

[a] Estimated as 10 times the British fauna of the day.
[b] Estimated at six times the number of known plant species.
[c] Extrapolated from species/area curves.
[d] Total is for arthropods, extrapolated from beetles as a proportion of the fauna in tropical forest tree canopies.

insects, because they so overwhelm other taxa in species number! At least this puts the importance of insects in a clear perspective.

Past trends in the estimation of both the number of described species and the total number of species extant are given in Table I. There seems every prospect that a figure of 10 million is conservative for terrestrial arthropods, and Erwin's claim seems less extravagant now than when first made.

This cornucopia of species is not, of course, evenly spread around the globe. There is a strong latitudinal gradient, with almost all insect groups increasingly speciose towards the equator. Figure 1 shows this clearly for termites in south-east Asia and Fig. 2 for species of swallowtail butterflies (Papilionidae). The work on the latter (Collins and Morris, 1985) appears

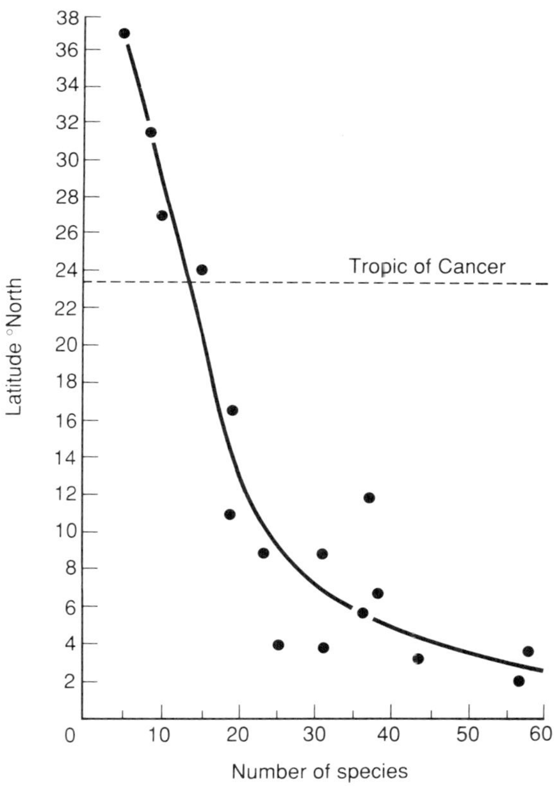

Fig. 1. The number of termite species according to latitude, north of the equator. Data from Collins (1983, 1989).

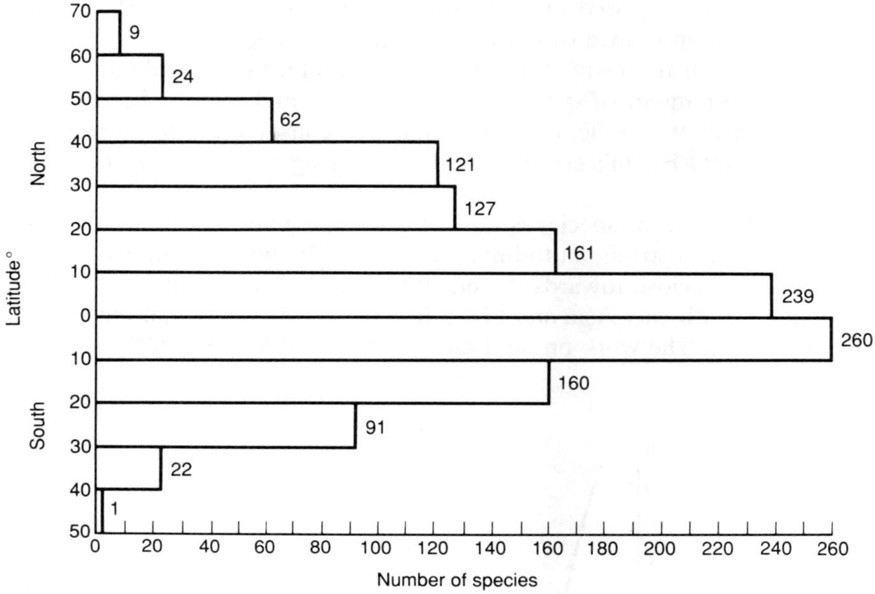

Fig. 2. Latitudinal gradients in species richness of swallowtail butterflies (Lepidoptera: Papilionidae). Data from Collins and Morris (1985).

to be the only complete published study of the world distribution of an insect family.

Insects are not only concentrated in the tropics, but more importantly in the wet tropics. Seventy-five to ninety per cent of all insects are thought to occur in tropical moist forests, i.e. those with a closed canopy and broadleaved evergreen, semi-evergreen or deciduous foliage. We thus include rain forests, mangroves and monsoon forests, but exclude open canopy formations such as savanna woodlands.

III. DO INSECTS MATTER?

Conservation of this mighty diversity poses a formidable challenge, and there can be few who have not wondered if insects matter that much, or at least if some are anything like as important as others. Janzen (1987), in a classic account of the operation of insects within a tropical forest (at Santa Rosa in Costa Rica), makes it clear that many insect relationships are highly specific, and highly complicated. So much so, in fact, that it would be a foolish person, in our present state of ignorance, who would attempt to predict which could be "let go". Wilson (1987) has summarized the key nature of insect activities in ecosystems, and Greenwood (1987) has outlined the involvement of insects in tropical forest food webs. Our limited knowledge suggests that insects are nowhere more important than in tropical forests, given their great diversity, the prevalence of tight one to one associations between each other and with other organisms, and the way in which they pervade all the functions of forest ecosystems. Until we have better information, the only safe course is to adopt a policy to conserve all tropical forest insects.

IV. PATTERN AND PROCESS

It has been said that conservation biologists are graduating from a study of pattern to a study of process (Hanski, 1989). This is true to some extent for tropical forest insects, but is held back because the study of pattern is still in the early stages and is a necessary preparation for the study of process. With a few honourable exceptions (e.g. Elton, 1975; Brown, 1982; Lovejoy *et al.*, 1986; Janzen, 1987) the conservation of insects in tropical forests has received very little attention or application of resources. It has been held back by an ignorance of pattern. This means that effective management prescriptions for disturbed systems have been slow to appear — in fact for most tropical forests they do not exist. Until processes involving insects have been worked out such prescriptions cannot be generated. For large pristine systems, of course, no such formultion should be necessary, for all they should need is protection against encroachment. Fortunately, large blocks of moist forest still exist in the Amazon and Zaire basins, and in parts of south-east Asia.

V. DEFORESTATION

However, there is no room for complacency. FAO/UNEP's 1981 figures for the deforestation rates are given in Table II. They suggest that the rates of deforestation in the first half of the 1980s were broadly similar in the three main continental blocks, but that the actual area being affected was greatest in Latin America, where the forests are largest. The figures given in Table II are widely believed to underestimate the current situation. Indeed, previews of data being gathered by FAO to a 1990 baseline indicate

TABLE II. Estimated rates of deforestation and logging of closed, broadleaved forests, 1981–1985 (FAO/UNEP, 1981 as revised in FAO, 1988).[a]

	Total area of forest (km^2)	Deforestation per year		Logging per year[b]	
		Area (km^2)	% of total	Area (km^2)	% of total
Latin America	7 037 300	40 045	0.57	19 600	0.28
Africa	2 163 500	13 253	0.61	6 350	0.29
Asia	3 116 840	17 617	0.57	16 940	0.54
Oceania	378 040	301	0.08	470[c]	0.12
Totals	12 695 680	71 216	0.56	43 360	0.34

[a] FAO is currently reassessing tropical (and temperate) forest cover and deforestation rates to a 1990 baseline. Data are due to be released in full in 1992, but early indications are that deforestation rates are now at least 50% higher

[b] Note that an error arises if deforested and logged areas are summed, because these processes often occur on the same sites. The proportion of deforestation that occurs in logged-over lands is higher in Africa (77%) and Asia (70%) than in Latin America (43%) (Lanly, 1982).

[c] Papua New Guinea only.

that deforestation rates have increased by about 50% over the decade. In Indomalaya, already severely affected, the pace of logging has speeded up on the island of Borneo, while rates in countries such as the Philippines and Thailand have necessarily slowed as the resources have become depleted. In the Philippines, for example, the production of timber has declined by 90% in less than 20 years. Only a fraction of the forest remains pristine, perhaps 10 000 km^2 out of the 65 000 km^2 total. In Africa the situation is relatively stable in the central African nations, but deforestation is severe in the west (FAO/UNEP, 1981; FAO, 1988).

Pressures on forests will continue to increase. The world's population is set to double to 10 billion in the next half century, by which time a rise in global sea level is predicted (Dobson *et al.*, 1989), substantive change in regional climate in tropical forest areas is almost certain and at least locally the forest will be affected by acidification (Rodhe *et al.*, 1988).

The next decade is critical for putting in place a comprehensive scheme for the conservation of tropical forest insects if the worst effects of the coming global modifications are to be avoided. That is, if they can be. Hartshorn (quoted in Dobson *et al.*, 1989) points out the hazards to the continued existence of forests if rainfall patterns are substantially altered.

In addition, the possible rise of 1–2 m in sea level within 60 years would be too much for low-lying forests to survive except possibly for mangroves.

It would be a mistake, however, to assume that these grim scenarios will necessarily be realized, and planning for insect conservation must be made on the basis that the physical environment will continue to support forest growth, at least in most areas.

The assumption has so far been made that deforestation means loss of insect diversity. This needs to be qualified. Permanent clearance does lead to massive loss. There are rather few detailed studies, but it has been quantified for scarab dung beetles (Howden and Nealis, 1975), soil invertebrates (Lasebikan, 1975), moths (Holloway, 1987), termites (Collins, 1980, 1989) and a variety of Central American insects attracted to light traps (Wolda, 1987). However, a small degree of agricultural disturbance, by opening up the canopy, can actually increase insect diversity (Janzen, 1987). Where a forest has been selectively felled with removal of only a small proportion of big trees, and where small islands of primary forest have been left, the same may apply (data for insects are lacking). Such a forest may return, over one or two decades, to something which looks similar to primary forest. The insect fauna, however, may be very different from that of primary forest because of the lack of regeneration of some tree species. Although diversity may have been increased, in this case the essential character of the primary forest insects will have been lost, (see also Chapter 14).

VI. IDENTIFICATION

The fundamental problem which has affected the study of both pattern and process and which has led to delay in the conservation of insects in tropical forests is the difficulty of identifying species.

It is often said that there are far too few taxonomists to identify tropical insects, to revise genera and overall to build an orderly taxonomy, even if the insects themselves could be collected (see Greenslade and New, Chapter 3). We agree with Dourojeanni (1989) that the problem is not so much a lack of taxonomists as a lack of resources to direct them to tropical forest work. Dourojeanni lays out an argument in detail for the redeployment of insect taxonomists in developing countries at a fraction of the cost arrived at in other estimates, which are generally based on US scales of pay. The feeling that the taxonomy problem cannot be tackled because it is too expensive is both mistaken and short-sighted, in our view. In any case, as May (1988) has said "Future generations will find it blankly incomprehensible that we are devoting so little money and effort to the study of these questions".

A. Indicator Groups

Our fundamental problem can be much alleviated by putting to one side any ambition to describe *all* the species of insects in tropical forest and instead concentrating on taxa which can be used as indicator groups. These will be taxa which have already attracted a lot of attention, and which are also taxonomically straightforward, easy to collect and reasonably (but not too) diverse. They should also show a fair degree of endemism. Brown (Chapter 14) has examined these criteria in more detail.

First and foremost among the target taxa are the butterflies, or at least the larger ones. The work of Collins and Morris (1985) on papilionids has already been mentioned. Comprehensive work has also been done on the Danainae (Ackery and Vane-Wright, 1984). Both studies provide a sound base for future study, using these two groups as indicators of species richness generally. Just how representative they are of other insects is debatable, but this can be checked as studies proceed.

Dragonflies are another promising group, except that the sources of reference are less extensive (but see Davies and Tobin, 1984, 1985). They fulfil the criteria laid out above and could be particularly useful because endemism in dragonflies is often tied to watersheds, which are commonly used in forestry operations to delimit areas of coupe. As forests become increasingly fragmented this pattern of endemism will grow in importance.

Sphingid moths (hawk moths) have been collected extensively in the past, are diverse, have a marked degree of endemism and are common enough to be important in food webs (e.g. see Janzen, 1987). They are also relatively easy to collect using light traps and flower baits. Holloway (1984) advocates caution in using them as indicators, because certain generalist species are highly dispersive and may appear anywhere, even in deep forest. While bearing this in mind, we feel they merit priority attention. Certain other groups of moths have been highlighted as indicators (Holloway, 1984).

Other target groups should include bush crickets and dung beetles. The latter are particularly easy to collect at suitable baits and extensive collections already exist. The song patterns of bush crickets have been used to study their adaptations to different elements of the habitat mosaic outside the tropics (Samways, 1989). This neat approach could be easily adapted to tropical forest and would provide a powerful tool for studying the influence of the habitat mosaic found in disturbed forest on insect distribution.

Along with a concentration of effort on well-known taxa, we need to develop other groups as indicators as well. Trapdoor spiders have been studied in Australia and a persuasive general case made to justify their use (Main, 1987). Termites, leafhoppers and carabid beetles would between them cover a wide range of ecological activity and are abundant throughout the tropical forest. This is also true of pyralid moths, which would bring in a representation of fruit and seed eaters.

B. Standard Techniques

Coupled with a focusing of resources on key taxa, we need to find and develop a number of simple techniques for sampling these taxa which can be applied in a standard manner throughout the tropics. We also need to adopt standard procedures for analysis of the data obtained. Too often in the past differences in field technique and method of analysis have precluded best use of the results. One has only to look at the varied use of diversity indices.

Techniques which are already tried and tested should be more widely deployed. Pitfall trapping, vacuum sampling, canopy fogging and the use of malaise and light traps are such techniques. In each case one standard design of equipment and technique of use might be adopted. Efforts are needed to capitalize on expertise in sampling gained in temperate habitats and to adapt techniques to tropical forest conditions. One potentially valuable method of assessing butterfly abundance is the "transect walk" (Pollard, 1977; Pollard *et al.*, 1986). Applying this to a tropical forest situation is fraught with hazard, but a low-technology approach like this has tremendous advantages. The software developed by the Rothamsted Insect Survey (Taylor *et al.*, 1978) to analyse catches from a national network of light traps in the UK could be easily adapted to deal with data from an array of solar-powered light traps in tropical forest. These could be produced at no great cost and could operate at low efficiency so as to trap manageable numbers of insects. Suitable prototypes exist. By thus exploiting work done on other ecological systems it should be possible in the next decade to provide the basic knowledge of pattern essential to pinpoint areas most in need of conservation.

VII. EXTENDING THE PROTECTED AREAS NETWORK

The first aim of tracing out the distribution patterns of selected taxa must be to extend the existing protected areas network in tropical forest to protect more insects. The network has been set up largely on the basis of protecting vertebrates, with very little consideration of insect distributions, mainly because these are so inadequately known.

It is vital to include as many insect species as possible within the protected areas system. Virtually all of the projected growth of the

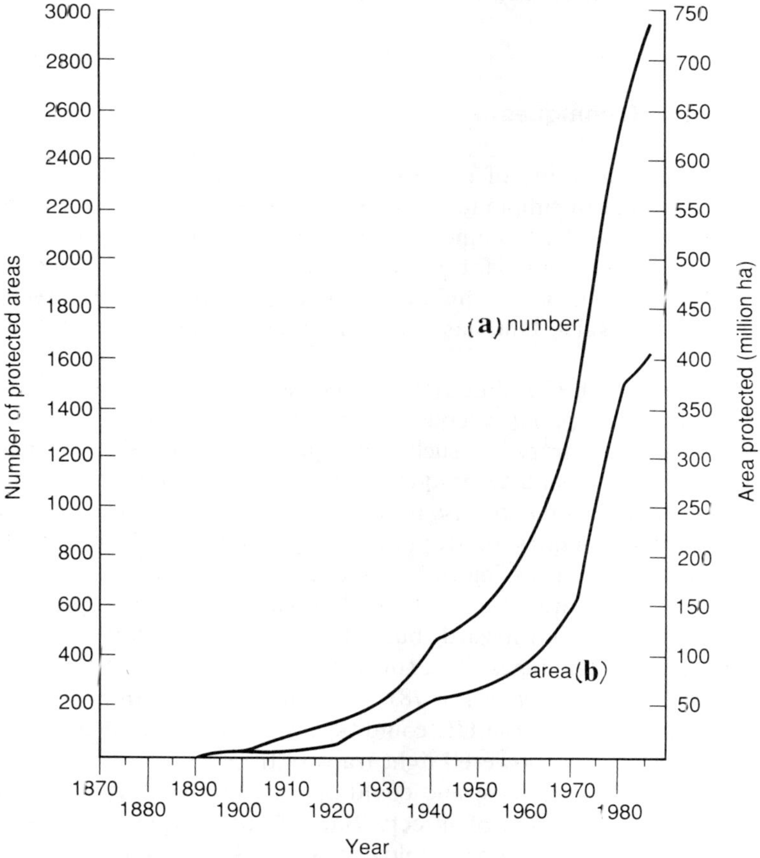

Fig. 3. Growth in world coverage of protected areas (a) Number of protected sites. (b) Total area of protected sites. Data from the protected areas database, World Conservation Monitoring Centre.

world's population to 10 billion over the next half century will take place in the developing world, most of it in countries with tropical forests. Such growth makes the continuing deforestation of tropical lands an inevitability (Sayer and Anadu, 1989).

The maintenance of insect diversity, in this situation, depends on setting aside the richest pristine habitats for conservation, and the appropriate management of the lands in between. The Bali Action Plan prepared by IUCN in 1982 has as its first objective "To establish by 1992 a world-wide network of national parks and protected areas, exemplifying all terrestrial ecological regions" (McNeely and Miller, 1984). Considerable, indeed gratifying progress has been made towards this first objective, as the rapid growth in size and number of protected areas through the 1980s testifies (Fig. 3). Something like 4% of the tropical forest biome is now protected (Sayer and Anadu, 1989). The original target of 10% may be unrealistic, but a doubling of the present coverage is certainly possible by the end of the century, and it is in the delineation of this that insects must figure more prominently.

Sayer *et al.* (1990), in reviewing the extent to which tropical forest vertebrates are covered by the protected areas network, can note with satisfaction the high or even very high level of protection already secured. A similar analysis of insects (if we had the detailed knowledge to produce it), would almost certainly make alarming reading. Although, yet again, detailed evidence is lacking, it seems highly likely that insects by no means slavishly follow the patterns of vertebrate distribution. A policy of protecting arthropods under the "umbrella" of vertebrate sites is therefore unsound. Plants are better indicators of insect-rich habitats. Nevertheless insects and their allies are so much more diverse and specialized that endemism is far more widespread and localized. The protected areas network needs to reflect these properties as much as possible.

A. Selection of Conservation Areas

A variety of systems is in use for the selection of conservation areas. Most of these have not involved any application of insect science. Four are discussed below.

1. Analysis using Biogeographical Provinces and Vegetation Types

The papilionid analysis of Collins and Morris (1985) included a breakdown of latitudinal gradients of species richness in the Americas, Africa and eastern Asia and Australia, as shown in Fig. 4 The data have been further

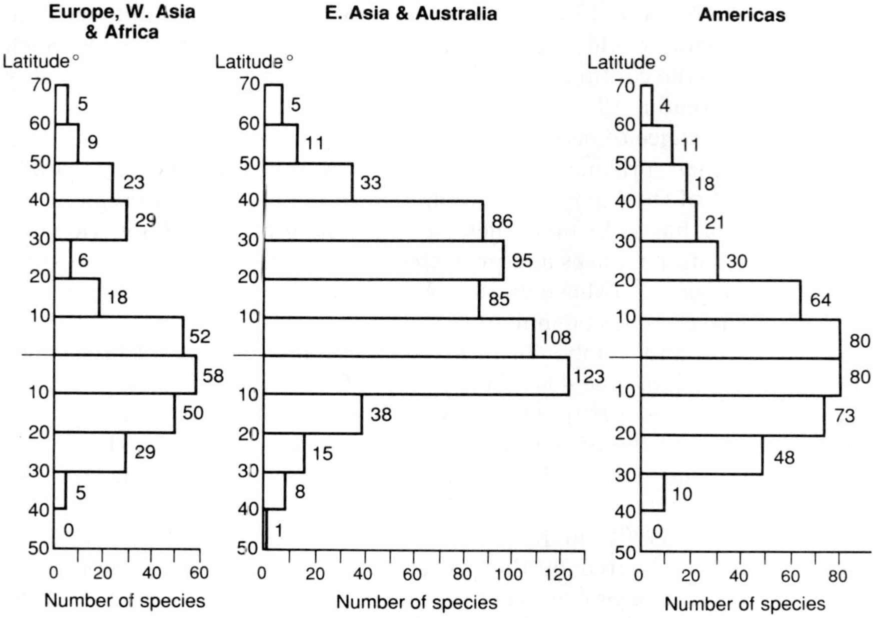

Fig. 4. Latitudinal gradients in species richness of swallowtail butterflies (Lepidoptera: Papilionidae) in Eurasia and Africa, east Asia and Australia and the Americas. Data from Collins and Morris (1985).

broken down to biogeographical realms (Table III). These show the prime importance, in terms of species richness, of the Indomalayan and Neotropical regions, both with large numbers of forest butterflies. The poor showing of the Afrotropical region as compared to the Palearctic may seem surprising, but it is widely recognized that the tropical forest of Africa is neither as extensive, nor as well developed, nor as rich as that of the other two main blocks (Richards, 1973).

The same table shows a preliminary analysis of dragonflies carried out for this chapter by D. A. L. Davies and N. M. Collins. Here the total for the Afrotropical region is higher than that for the Palearctic, though still lower than for other realms with tropical forest. The moral of this comparison is not to rely too heavily on one group. Both analyses provide a first approximation for assessing where most conservation effort should be directed in order to preserve the greatest number of species. Clearly this differs with the taxon being considered. A good range of taxa is needed.

TABLE III. Swallowtail butterfly (Papilionidae) and dragonfly (Odonata) species assigned to biogeographical realms.

	Number of species		%		Rank	
	Papilionidae	Odonata	Papilionidae	Odonata	Papilionidae	Odonata
Indomalayan	225	1435	40	26	1	2
Neotropical	166	1473	29	27	2	1
Palearctic	60	662	11	12	3	4
Afrotropical	53	881	9	16	4	3
Oceania	44	626	8	11	5	5
Nearctic	19	454	3	8	6	6
Total species	567	5531	100	100		

Once the overall richness of a biogeographic realm has been assessed, a search must be made for centres of endemism and species richness within it. Brown (1982) has shown that these two do not necessarily go together.

Biogeographical analysis seems the only rational way of deciding where conservation effort is most merited, and is the foundation of the three other approaches for the selection of conservation sites. Its disadvantage is the amount of time it takes to assemble the necessary information and the fact that as biogeographical realms cross national boundaries it is more difficult to apply the results.

2. *Identification of "Hot Spots"*

This is a generalized approach applying a qualitative assessment of threat as well as species richness (Myers, 1988). Involving the concept of threat is highly necessary, since some species-rich areas are much more threatened than others and require much more rapid conservation action. This approach has, at present, very little insect science input, which limits its value.

3. *Analysis of "Critical Faunas"*

The approach here is to recognize the significance of national boundaries in conservation work, and to analyse the extent of endemism in different countries for a particular taxon. This has been done for papilionids as an extension of the biogeographical analysis mentioned already (Collins and Morris, 1985), as shown in Table IV. It has the advantage of alerting those countries which have a particularly rich heritage. There is a possible disadvantage in that those which are low in the endemism "league table" may think they have less to conserve, even though they may have a species-rich fauna. Such a case is Myanmar, Burma (Table IV), with 68 species but no endemics. The critical faunas approach looks a very valuable one, and there are quite a number of taxa for which it could be used, without recourse to much field work. It provides an objective base for making hard choices in diversity.

4. Within-nation Identification of Areas of Particular Conservation Value

This again takes biogeographical data and applies them to nation-states, so that a country can see where its natural riches lie. Thus the Ornithoptera (birdwing butterflies) of Papua New Guinea, which are commercially valuable (Morris *et al.*, Chapter 13) have been the subject of intensive study and a number of reserves have been proposed (Parsons, 1983).

B. Latent Extinction and Transient Diversity

Brown (Chapter 14) has pointed out that although their habitat has been heavily fragmented for some time, none of the butterflies of the Atlantic coast rain forest of Brazil has yet become extinct. As we know so little about the population dynamics of insects in tropical forest, we could perhaps be underestimating their powers of persistence. Let us hope that this is so.

An alternative, and rather more plausible view from the theoretical standpoint, is that the persistence of the Atlantic forest butterflies is temporary — that they represent examples of latent extinction which has

TABLE IV. Swallowtail butterfly (Papilionidae) critical faunas. A selection of countries ranked according to the number of endemic species. Data from Collins and Morris (1985).

			Non-endemics		
	Rank	Endemics	New to list	In higher-ranked countries	Total species
Indonesia	1	53	68	0	121
Philippines	2	21	4	24	49
China	3	15	61	28	104
Brazil	4	11	63	0	74
Madagascar	5	10	3	0	13
Myanmar	44	0	2	66	68

resulted in a transient diversity. They are the "living dead" of Janzen (1986a). We believe that latent extinction and transient diversity will become key concepts in understanding future changes in species diversity of tropical forest insects. As forests become more fragmented, insect populations outside the protected areas network will become latently extinct if fragments are allowed to become too small and too isolated.

A good example of local latent extinction is shown by the Brazil nut tree, *Bertholletia excelsa*. This relies on euglossine bees for pollination and seed setting (Prance, 1983) while the bees rely on other resources in the forest to complete their life cycle. If any of these resources are removed (as can happen in forest disturbance), the bees disappear and the local population of *Bertholletia* is unable to reproduce itself. It may continue to live for another 100 years or more, but as a local population it is latently but effectively extinct from the time of the disappearance of the bees. Insect life cycles are very much shorter than those of trees, so the time span over which latent extinction might operate for them would be shorter — unless the insect is dependent in some way on a long-lived resource like a tree and will persist until the tree itself becomes locally extinct. Given the complexity of interactions between insects and other tropical forest biota, time spans are impossible to predict, but could be very long.

A good example of a forest island which probably contains a good many taxa subject to latent extinction is Bukit Timah on Singapore Island. This still has an extensive insect fauna (Murphy, 1973), but it is so isolated from other primary forest, and so small, that is is unlikely that many of the tree species can set seed, and that local extinction of these species, and the insects that depend upon them, is steadily taking place.

The danger with latent extinction and transient diversity is that they may engender a false sense of optimism when the fragmentation process fails immediately to cause a marked fall in species diversity.

C. Conservation of Disturbed Forest Mosaics

The primary aim of tropical forest insect conservation must be to establish an effective protected areas network in pristine habitat. However, there needs to be a secondary aim of saving as many as possible of the insects that are able to survive in disturbed forest areas. Given the degree of local endemism implicit in many studies (particularly in those of Erwin, 1983) many insects will not be contained within the protected areas network, even on the most optimistic predictions. It may be possible to save a good number of them despite this. Most tropical forest is going to be converted in the next few decades into a mosaic of primary forest fragments, secondary forest, pasture and tree plantations. In some parts the mosaic will be coarse grained and consist largely of pasture or bush crops (e.g. in Costa Rica or parts of Brazil). There is little hope for forest insects in these. In other places (e.g. Sabah) the mosaic may turn out to be fine grained, with primary forest fragments, smallholdings and tree plantations embedded in a matrix of secondary growth. Some of this will be on a 60-year cutting cycle, which is long enough to develop quite mature forest. The rest may be on a 30-year cycle, or even less in accessible places (C. Marsh, personal communication). In Sabah there is now not enough lowland dipterocarp forest left to create a very effective protected areas network, so the only hope for saving forest insects is to develop a conservation policy for this disturbed mosaic.

The pristine forest islands are the key component in this mosaic. If each is surrounded by an extensive stand of long cutting-cycle second growth the chances for effective conservation are probably quite good, although insect species tied to the big dipterocarp trees will decline greatly, as these trees have very limited future prospects in Sabah. The provision of linking corridors of second growth between islands should be encouraged, to aid the dispersal of insects which will not cross open ground.

D. External Threats

The disturbance of forest offers not only a threat to species richness through decrease of habitat. There is a whole range of so-called external threats (Janzen, 1986b). One of the most alarming of these is the invasion of pristine habitat by exotic wild or domesticated organisms which have escaped from human supervision. African honey bees offer a classic example. Aggressive invaders like these can disrupt normal pollination processes and shift plants into latent extinction through inhibition of seed setting. Greenslade and New (Chapter 3) note the possible threat to some Australian habitats through the encouragement of domestic bees to forage in native vegetation. Introductions are discussed in detail by Howarth and Ramsay (Chapter 4). Another kind of threat is the build-up of large populations of herbivorous insects of second growth. Some of these may then spread into pristine forest and greatly increase the mortality of (for example) tree seedlings, again possibly pushing the tree species into latent extinction.

External threats will vary uniquely from one reserve to another. Janzen emphasizes that the best defence is an intimate knowledge of the ecological processes within each particular reserve and how these may be threatened. The implication is for the need for well-trained field ecologists on reserve management staff.

VIII. INSECTS AS TOOLS FOR CONSERVATION

With one or two conspicuous exceptions (see for example Brown, Chapter 14) insect science has never been used in the building up of the protected areas network. Particular birds and mammals have, by and large, been the "umbrella species" under which everything else has sheltered, the assumption being (if any thought was cast at all) that insects (and plants) would be adequately catered for by this arrangement. In fact, the umbrella might have been better stitched together with insects. Insects have close and often extremely precise relationships with plants, particularly through herbivory, seed eating and pollination. They are therefore very good indicators of the diversity of plant life. They are also bound by a multitude of relationships to virtually all the rest of the biota. In addition they show a higher degree of endemism than any other group. These facts make insects ripe for consideration as indicators of areas of high conservation value for other organisms. Exactly the same procedures can be used as suggested for extending the protected areas network for the conservation of insects themselves (see Section VI).

Furthermore, insects are so central to ecosystem function, and so finely tuned to change that they make excellent monitors of the impact of disturbance. Choosing selected taxa as suggested earlier, insects could be used to follow the consequences of fragmentation and to monitor a programme of environmentally sound exploitation. In this latter case, management of the disturbed mosaic would involve maintaining a diversity of genetic resources, as well as supporting a local human population and providing economic revenue. Again, the techniques and taxa mentioned in the earlier part of this paper could be used. By using insects as management tools a great deal might be salvaged from disturbed forest systems by ensuring that the mosaic patterns provide support for (perhaps) a major part of the primary forest fauna and flora.

IX. CONCLUSIONS

Insects are the most diverse group of living organisms on earth. The number of species may exceed 10 million. They are particularly abundant in tropical forest, where they are central to the function of these ecological systems. Perhaps 95% of all insect species are confined to tropical forest.

Tropical forests are under threat. They are disappearing so fast that very few extensive tracts of lowland forest may be left by the year 2050. If action is taken now to extend the existing protected areas network to cater for the conservation needs of insects, much potential loss of diversity can be avoided. Additionally, if the mosaic of habitats resulting from fragmentation of forest is managed with conservation as a subsidiary aim (exploitation to meet human needs being the primary aim), much diversity of insects and other organisms can be saved.

Latent extinction and transient diversity emerge as key concepts which need to be quantified and assessed.

Successful conservation of insects turns on adoption of simple techniques of sampling and analysis, together with the selection of a small number of indicator groups, of which butterflies and dragonflies are prime candidates.

We argue that insects themselves can be put to use as tools for conservation of other tropical forest organisms (plants as well as animals).

Acknowledgements

The authors are grateful to Prof. D. Allen L. Davies for advice on the distribution of Odonata.

REFERENCES

Ackery, P. R. and Vane-Wright, R. I. (1984). "Milkweed Butterflies: their Cladistics and Biology". British Museum (Natural History), London.

Brown, K. S. (1982). Historical and ecological factors in the biogeography of aposematic neotropical butterflies. *Am. Zool.* **22**(2), 253–275.

Collins, N. M. (1980). The effect of logging on termite (Isoptera) diversity and decomposition processes in lowland dipterocarp forests. *Trop. Ecol. Devel.* 113–121.

Collins, N. M. (1983). Termite populations and their role in litter removal in Malaysian rain forests. *In* "Tropical Rain Forest; Ecology and Management" (S. L. Sutton, T. C. Whitmore, A. C. Chadwick eds), pp. 311–325. Blackwells, Oxford.

Collins, N. M. (1989). Termites. *In.* "Tropical Rain Forest Ecosystems. Ecosystems of the World Vol. 14B" (H. Lieth and M. J. A Werger, eds), pp. 455–472. Elsevier Science Publishers, Amsterdam.

Collins, N. M. and Morris, M. G. (1985). "Threatened Swallowtail Butterflies of the World". International Union for the Conservation of Nature and Natural Resources (IUCN), Gland, Switzerland, 401 pp.

Davies, D. A. L. and Tobin, P. (1984, 1985). "The Dragonflies of the World". Societas Internationalis Odonatologica Rapid Communications, Supplements Nos, 3 and 5, 125 and 151 pp.

Dobson, A., Jolly, A. and Rubenstein, D. (1989). The greenhouse effect and biological diversity. *Trends Ecol. Evol.* **4**(3), 64–68.

Dourojeanni, M. J. (1989). "Entomology and Biodiversity Conservation in Latin America". Plenary address: 1988 Entomological Society of America, National Conference, Louisville, Kentucky.

Elton, C. (1975). Conservation and the low population density of invertebrates inside neotropical rain forest. *Biol. Conserv.* **7**, 3–15.

Erwin, T. L. (1982). Tropical forests; their richness in Coleoptera and other arthropod species. *Coleopt. Bull.* **36**(1), 74–75.

Erwin, T. L. (1983). Beetles and other insects of tropical forest canopies at Manaus, Brazil, sampled by insecticidal fogging. *In* "Tropical Rain Forest: Ecology and Management" (S. L. Sutton, T. C. Whitmore and A. C. Chadwick, eds), pp. 59–75. Blackwell Scientific Publications, Oxford.

FAO (1988). "An Interim Report on the State of Forest Resources in the Developing Countries". FAO, Rome, 18 pp. + tables.

FAO/UNEP (1981). "Tropical Forest Resources Assessment Project", 3 vols. FAO, Rome.

Greenwood, S. R. (1987). The role of insects in tropical forest food webs. *Ambio* **16**(5), 267–271.

Hanski, I. (1989). Metapopulation dynamics: does it help to have more of the same? *Trends Ecol. Evol.* **4**(4), 113–114.

Holloway, J. D. (1984). Moths as indicator organisms for categorizing rain-forest and monitoring changes and regeneration processes. *In* "Tropical Rain-Forest: The Leeds Symposium" (A. C. Chadwick and S. L. Sutton, eds), pp. 235–242. Special Publication of the Leeds Philosophical and Literary Society, Leeds, UK.

Holloway, J. D. (1987). Macrolepidoptera diversity in the Indo-Australian tropics: geographic, biotopic and taxonomic variations. *Biol. J. Linn. Soc.* **30**, 325–341.

Howden, H. F. and Nealis, V. G. (1975). Effects of clearing in a tropical rain forest on the composition of the coprophagous scarab beetle fauna (Coleoptera). *Biotropica* **7**(2), 77–83.

Imms, A. D. (1925) "*General Textbook of Entomology*", Methuen, London.

Janzen, D. H. (1986a). The future of tropical biology. *Ann. Rev. Ecol. System.* **17**, 305–23.

Janzen, D. H. (1986b). The eternal external threat. *In* "Conservation Biology: the Science of Scarcity and Diversity" (M. E. Soulé ed.); pp. 286–303. Sinauer.

Janzen, D. H. (1987). Insect diversity of a Costa Rican dry forest; why keep it, and how? *Biol. J. Linn. Soc.* **30**(4), 343–356.

Kirby, W. and Spence, W. (1815–1826). "An Introduction to Entomology", 4 vols., Longman, Hurst, Rees, Orme and Brown, London.

Lanly, J.-P. (1982). "Tropical Forest Resources". FAO Frestry Paper 30. FAO, Rome, 106 pp.

Lasebikan, B. A. (1975). The effect of clearing on the soil arthropods of a Nigerian rain forest. *Biotropica* **7**(2), 84–89.

Lovejoy, T. E., Bierregaard, R. O., Rylands, A. B., Malcolm, J. R., Quintela, C. E., Harper, L. H., Brown, K. S., Powell, A. H., Powell, G. V. N., Schubart, H. O. R. and Hays, M. B. (1986). Edge and other effects of isolation on Amazon forest fragments. *In* "Conservation Biology: the Science of Scarcity and Diversity" (M. E. Soulé, ed.), pp. 257–286. Sinauer.

Main, B. Y. (1987). Persistence of invertebrates in small areas: case studies of trapdoor spiders in Western Australia. *In* "Nature Conservation: The Role of Remnants of native Vegetation" (D. A. Saunders, G. W. Arnold, A. A. Burbidge and A. J. M. Hopkins, ed), pp. 29–39. Surrey Beatty, Pty, with CSIRO & CALM.

May, R. M. (1986). How many species are there? *Nature* **324**, 514–515.

May, R. M. (1988). How many species are there on earth? *Science* **241**, 1441–1449.

McLeay, W. S. (1819–1821). "Horae Entmologicale", second part, p. 469.

McNeely, J. A. and Miller, K. R. (eds) (1984). "National Parks, Conservation and Development: the Role of Protected Areas in Sustaining Society". Smithsonian Institution Press, Washington, DC 838 pp.

Murphy, D. H. (1973). Animals in the forest ecosystem. *In* "Animal Life and Nature in Singapore" (S. H. Chuang, ed.), pp. 53–73. Singapore University Press, Singapore.

Myers, N. (1988). Threatened biotas: 'Hotspots' in tropical forests. *Environmentalist* **8**(3), 1–20.

Parsons, M. (1983). A conservation study of the birdwing butterflies, *Ornithoptera* and *Troides* (Lepidoptera: Papilionidae) in Papua New Guinea. Final report to Department of Primary Industry, Papua New Guinea, 111 pp.

Pollard, E. (1977). A method for assessing changes in the abundance of butterflies. *Biol. Conserv.* **12**, 115–134.

Pollard, E., Hall, M. L. and Bibby, T. J. (1986). "Monitoring the Abundance of Butterflies", 1976–1985. Nature Conservancy Council, Peterborough, UK, 280 pp.

Prance, G. J. (1983). Pesquisas botanicas e a conservaçao da floresta Amazonica. *In* "Anais do XXXIV Congresso Nacional de Botanico" Vol I: 63–71. Puerto Alegre, Gio grande do Sul, Brazil.

Richards, P. W. (1973). Africa, the 'Odd Man Out'. *In* "Tropical Forest Ecosystems in Africa and South America: a Comparative Review" (B. J. Meggers, E. S. Ayensu and W. D. Duckworth, eds), pp. 21–26. Smithsonian Institution Press, Washington, DC.

Rodhe, H., Cowling, E., Galbally, I., Galloway, J. and Herrera, R. (1988). Acidification and regional air pollution in the tropics. *In* "Acidification in Tropical Countries" (H. Rodhe and R. Herrera, eds). SCOPE. John Wiley & Sons, Chichester.

Samways, M. J. (1989). Insect conservation and landscape ecology: a case history of bush crickets (Tettigoniidae) in southern France. *Environ. Conserv.*, **16**, 217–226.

Sayer, J. and Anadu, P. A. (1989). The role of forestry in conserving biological diversity. Paper for presentation at the Commonwealth Forestry Congress September, 1989.

Sayer, J. A., McNeely, J. A. and Stuart, S. N. (1990). The conservation of tropical forest vertebrates. *In* "Vertebrates in The Tropics" (G. Peters and R. Hutterer, eds) pp. 407–419. Museum Alexander Koenig, Bonn.

Stork, N. E. (1988). Insect diversity: facts, fiction and speculation. *Biol. J. Linn. Soc.* **35**, 321–337.

Taylor, L. R., French, R. A. and Woiwod, I. P. (1978). The Rothamsted Insect Survey and the urbanization of land in Great Britain. *In* "Perspectives in Urban Entomology" (G. W. Frankie and C. S. Koehler, ed), pp. 31–65. Academic Press, New York.

Westwood, J. O. (1833). On the probable number of species of insects in the creation; together with descriptions of several minute Hymenoptera. *Mag. Nat. Hist.* **6**, 116–123, 414–421, 880.

Wilson, E. O. (1987). The little things that run the world (the importance and conservation of invertebrates). *Conserv. Biol.* **1**(4), 344–346.

Wolda, H. (1987). Altitude, habitat and tropical insect diversity. *Biol. J. Linn. Soc.* **30**(4), 313–323.

Wolf, E. C. (1987). "On the brink of extinction: conserving the diversity of life". Worldwatch Paper 78, 53 pp. Worldwatch Institute, Washington, DC.

16

Poster Presentations

MANAGING PLANTATION WOODLAND ON ANCIENT WOODLAND SITES FOR BUTTERFLIES AND OTHER INSECTS

J. NICK GREATOREX-DAVIES AND MARNEY L. HALL

Introduction

In the past, many butterflies associated with British woodland flourished under traditional coppice management. The rich ground flora characteristic of coppice woods provided ideal conditions for a number of butterfly species for 2–5 years after coppicing. With the virtual demise of coppicing, and the corresponding increased shade in woodland, butterfly numbers have declined. The violet-feeding fritillaries have been particularly severely affected. However, on some nature reserves where coppice management has been reinstated (e.g. Gait Burrows in Lancashire and Blean Woods in Kent), there has been a dramatic increase in the numbers of some species, notably fritillaries.

Since 1945, about 30% of ancient woodland has been planted with conifers. At first many of these new plantations created conditions similar to newly cleared and young coppice, with the result that many butterflies flourished. Indeed, according to the Nature Conservancy Council's (NCC) Invertebrate Site Register, many of the finest butterfly woods are conifer plantations on ancient woodland sites. Some of these woods retain most, if not all, of their original butterfly fauna, though in rapidly declining numbers as the conifers grow.

This poses the question: can these commercial plantations be managed in such a way as to maintain their butterfly and insect fauna? In 1985, the NCC contracted the Institute of Terrestrial Ecology to provide management guidelines for the conservation of invertebrates, especially butterflies, in plantation woodland on ancient woodland sites (Hall and Greatorex-Davies, 1989).

Methods and Results

The relationship between shade and butterfly numbers was studied in rides of eight plantation woods in southern and eastern England. Butterfly monitoring scheme transect data were available for these sites. Some other insect groups were also sampled, notably Heteroptera and chrysomelid beetles.

Both butterflies and Heteroptera showed a highly significant negative correlation with shade. Chrysomelid beetles were also significantly negatively correlated with shade. However, numbers of chrysomelid species per sample were small compared with both butterflies and Heteroptera. Relatively few species showed a preference for shaded conditions.

Few of the rarer butterflies often associated with woodland, such as pearl- and small pearl-bordered fritillaries (*Boloria selene* and *B. euphrosyne*) and the Duke of Burgundy (*Homeosis lucina*), were recorded in rides with more than 25% shade.

A simple method was devised to predict the shadiness of rides of different widths and adjacent tree crop heights, on relatively flat sites. It could be predicted that a ride would need to be 30 m wide when the average adjacent crop height was 20 m in order to be only 25% shaded. Insect species diversity and richness declined rapidly in rides more shaded than this.

Apart from such abiotic factors as light, temperature and humidity, vegetation structure is an important determinant of the oviposition sites selected by certain butterflies. Observations were made on a number of butterfly species, but particularly on the Duke of Burgundy. This butterfly nearly always selected *Primula* plants in vegetation intermediate between very open sites and areas with dense surrounding herbaceous vegetation.

In addition to studying and recording butterflies and other invertebrates, botanical transects were set up in many of the rides. Again there was a highly significant negative correlation between the number of plant species and shade. In addition, it was apparent that almost all plant species found in the wood could be found in or immediately adjacent to the rides, indeed in some conifer plantations herbaceous vegetation was virtually eliminated under the crop trees, and true woodland ground flora was restricted to the edges of rides.

Conclusions

In managing a plantation woodland for insects, a number of factors need to be taken into account. For example we need to:

1. Maintain open areas.
2. Maintain temporal and structural diversity.
3. Take into account individual species' requirements.
4. Select the best areas for management.
5. Plan management for the long term.
6. Adjust management in the light of experience.

Because of the nature of plantation woodland, most conservation management must inevitably be concentrated on the edges of the plantations. There may also need to be a compromise between management for particular species and general management likely to benefit a wide range of species.

In general, a practical compromise for foresters seeking to manage their woodlands for insects may be a series of box junctions interlinked with wide rides. Both the ride edge scrub and the grassland of the rides needs to be managed on cutting rotations, in order to maintain the floristic and structural diversity necessary to maintain a rich and diverse insect fauna. Box junctions need to be at least 0.5 ha in size to be effective, and some rides at least 30 m wide.

REFERENCE

Hall, M. L. and Greatorex-Davies, J. N. (1989). "Management Guidelines for the Conservation of Invertebrates, especially Butterflies, in Plantation Woodland", 110 pp. Unpublished Nature Conservancy Council/Natural Environment Research Council contract report (HF3/08/12).

THE ROLE OF FIELD BOUNDARIES IN HOVERFLY (DIPTERA: SYRPHIDAE) FORAGING ON ARABLE LAND

SUE COWGILL

Observations were made of syrphid behaviour in two floristically different field boundary types. In particular, the use of arable weeds by adult hoverflies was studied, to facilitate the development of selective weed control strategies for field boundaries which maximize natural enemy action.

The larvae of certain species of hoverfly are important cereal aphid predators (Chambers and Adams, 1986). Adult flies feed upon pollen and nectar; females require proteins and amino acids available from pollen for the maturation of their reproductive systems (Schneider, 1948). This has implications for cereal aphid predation since the provision of appropriate floral resources has the potential to influence both the timing of oviposition and the number of eggs laid per fly.

Ten common species of hoverfly were recorded at both sites and the relative contribution of individual syrphid species to the site totals varied with the field boundary type.

Flowers were not visited according to their abundance; particular species of fly showed a preference for certain weeds. The "forage ratio" was used to quantify this preference, which varied with the species of hoverfly studied and according to habitat composition.

Work currently being undertaken is an attempt to determine whether rich floras (of preferred flowers) have an effect on the number of syrphids within an area and whether field boundary floral composition can influence syrphid oviposition. That is, are more eggs laid in cereal fields adjacent to weedy field boundaries?

REFERENCES

Chambers, R. J. and Adams, T. H. L. (1986). Quantification of the impact of hoverflies on cereal aphids in winter-wheat: an analysis of field populations. *J. Appl. Ecol.* **23**, 895–904.

Schneider, F. (1948). Beitrag zur Kenntnis der Generation-sverhaltnisse und Diapause rauberischer Schwebfliegen. *Mitt. Schweiz. Entomol. Ges.* **21**, 249–285.

THE BERN CONVENTION AND THE SPANISH LEPIDOPTERA

J. L. VIEJO, J. MARTÍN, E. GARCÍA-BARROS AND M. L. MUNGUIRA

The aim of this work is to provide a preliminary survey to assess the status of the 13 Spanish Lepidoptera species (butterflies and moths) included in the Bern Convention.

The records of *Maculinea teleius* and, perhaps, *Coenonympha oedippus* are probably due to misidentification and should be eliminated from the Spanish census.

Parnassius mnemosyne, *Lopinga achine*, *Maculinea nausithous* and *Plebicula golgus* are the most threatened species. Each occurs in fewer than ten 10 × 10 km squares, and *P. golgus* is endemic to Sierra Nevada. *Parnassius apollo* and *Maculinea arion* occur in a large number of sites although they are locally endangered. *Euphydryas aurinia* is widely distributed and is out of danger.

Graellsia isabelae is believed to be out of danger. Although its range seems to be increasing, there are threatened populations, for example in the Sierra de Guadarrama.

The status of the three other moths is more poorly known. *Proserpinus proserpina* may be present in a relatively large area of the Iberian Peninsula, but the two other species, *Hyles hippophaes* and *Eriogaster catax*, seem to be rare and restricted to a few sites in the north of Spain.

CAPTIVE BREEDING — A VITAL LINE OF DEFENCE

P. PEARCE-KELLY

With a meagre 3% of terrestrial Earth accorded any degree of protected status, it is evident that a reliance on habitat protection is often inadequate to ensure the maintenance of biotic diversity. As clearly stated in the IUCN Policy Statement on Captive Breeding (15 October 1986), the establishment of self-sustaining captive populations and other supportive intervention will be needed to avoid the loss of many high-risk species. Furthermore, it is vital that captive-breeding programmes be established before species are reduced to critically low numbers. They also need to be co-ordinated internationally according to sound biological principles with a view to the maintenance or re-establishment of viable populations in the wild.

Invertebrates are ideally suited to captive breeding programmes due to their small size (enabling large numbers to be kept in small areas), adaptability to laboratory conditions, and dependence upon low-budget maintenance systems.

A diverse range of invertebrate species has been successfully maintained in captivity for many years and with little effort. However, it is only in recent years that attention has been focused on applying these cultural methods to the ever-increasing number of invertebrate species that are under threat in the wild. An excellent example is the international breeding programme for the Moorean *Partula* snails.

Due to habitat destruction and introduced pests, the St Helena giant earwig (*Labidura herculeana*) and Lord Howe Island stick insect

(*Dryococelus australis*) are now extinct. It would have been easy to have set up captive breeding populations of both these species. We must not allow a similar fate to befall other fascinating invertebrate species when captive breeding is such a simple alternative.

The Invertebrate Department of London Zoo is fully committed to the above conservation policies.

INSECTS AND KRAKATAU: A CONSERVED HABITAT OF LONG-TERM RESEARCH INTEREST

I. W. B. THORNTON

Krakatau's 1883 eruption destroyed two-thirds of the island leaving its remnant (Rakata) and two adjacent islands, all effectively sterilized, around its submerged caldera. Now clothed in secondary forest, the islands are part of Indonesia's Ujung Kulon National Park, in Sunda Strait.

Insect recolonization has been monitored by surveys in 1908, 1919–1923, 1929–1933 and 1982–1986. Although some groups (Orthoptera, Blattodea, nymphalid and hesperiid Lepidoptera) may be approaching equilibrium species numbers, others (Thysanoptera, Neuroptera, aculeate and braconid Hymenoptera, three dipterous families) have at least doubled in species numbers since the closure of the forest canopy (*ca* 1915–1935), the period of maximum turnover for all but Blattodea. It is clear that plant succession has differentially affected the approach to equilibrium by insect groups.

In 1930 Anak Krakatau emerged from Krakatau's submarine caldera. Its 1952 self-devastating eruption and subsequent activity has resulted in only 10% (17 ha) of its area being vegetated, with early stages of vegetation broadly analogous to those on Rakata before the 1920s, when zoological surveys were minimal.

The first fruiting of Anak Krakatau's two fig species in 1985 was correlated with the activity and establishment of frugivorous vertebrates; fig chalcids, captured in aerial fall-out in 1984, were present in fig syconia in 1986. An aeolian scavenger–predator guild, dominated by a partly nocturnal, cryptic, flightless cricket practically confined to barren areas, exploits the substantial arthropod fall-out and parallels communities on Hawaii's high bare lava flows. This conduit for extra-island energy, short-circuiting the usual plant–herbivore–carnivore succession, may have evolved to exploit this first post-volcanic stage in areas of persistent volcanism.

THE AMATEUR ROLE IN INSECT CONSERVATION

DAVID LONSDALE

Amateur entomologists have long been concerned about the decline of insect species and the destruction of their habitats. The Amateur Entomologists' Society (AES) is committed to seek the reversal of these trends in two ways: by calling on the skills and interests of its members, and by acting as a body which promotes insect conservation. Individual amateurs can play an essential role in surveys of species distributions and of individual sites where protection may be needed. Amateurs also work with more generalist organizations which need a continuing entomological input. The AES as a whole works for conservation in three main ways: dissemination of ideas and information, committee work and representation in respect of individual sites. Dissemination takes place through the newsletter "Insect Conservation News", through special publications on conservation-related subjects, through the promotion of (JCCBI) Joint Committee for the Conservation of British Insects codes of conduct and through exhibition displays. The AES Conservation Committee is actively represented on the JCCBI and plays a leading role in encouraging better co-ordination of effort amongst the JCCBI's constituent organizations. Representation on individual site protection has influenced decisions in a number of important cases. In all aspects of its work, the AES stresses the need for the conservation movement to give insects the attention which they deserve relative to other forms of wildlife.

THE NATURE CONSERVANCY COUNCIL APPROACH TO INVERTEBRATE CONSERVATION

R. KEY AND THE NCC TERRESTRIAL INVERTEBRATE ZOOLOGY BRANCH

The Nature Conservancy Council (NCC) is the government body which promotes nature conservation in Great Britain. Most of the invertebrate conservation activities are co-ordinated by a small group of entomologists, based in the Great Britain headquarters at Peterborough. Currently, we are seeking to overcome the historical neglect of invertebrate conservation in Britain, with the objective of integrating practical measures to conserve invertebrates within the day-to-day work of NCC, and other conservation organizations.

The poster summarizes the current status of NCC's national data base, the Invertebrate Site Register, which holds information on important sites and rare species. Additionally, the relevance of ongoing survey and research projects to practical conservation issues on the ground is illustrated. Another aspect of our work is the widespread promulgation of general principles for tackling invertebrate conservation amongst reserve wardens and other land managers. These guidelines can help conservationists who are not entomologists to conserve invertebrates, and this is one way of improving implementation. This approach needs to be coupled with specific measures to cater for rare and threatened species, when their particular requirements are known. Invertebrates are challenging to conserve, and their specialized and distinctive needs are not yet generally recognized amongst conservationists. There remains the need to improve monitoring of species and their critical habitat if conservation management is to take invertebrates fully into account.

CONSERVATION OF BRITISH TRICHOPTERA

I. D. WALLACE

There are 195 caddis species in Britain. The rarest 16 species (RDB1, RDB2, RDBK, on criteria established by the Nature Conservancy Council) deserve active national conservation. In order of importance, the threats are peatland destruction, pollution and drainage improvements to large rivers. Several of the 16 species are common elsewhere in Europe and other species which are regarded as requiring conservation in some European countries are common in Britain. Perhaps nationally, species which are common in a country but generally threatened elsewhere deserve as much conservation consideration in that country as species which are rare but common abroad.

The poster display categorizes the abundance of the various caddis species in Britain into: 110 species common in at least a quarter of Britain; 40 species which are rare or local in at least a quarter of Britain; 12 species which are very local, or uncommon generally; 14 species which are rare generally; 14 species which are so rare or decreasing as to be threatened with extinction in Britain; 5 species which are believed to be extinct.

Index

Numbers in italics refer to figures and tables

A

Acacia, associated phytophagous insects, 40
Acanthobrahmaea europaea, 125
Acari, *267*
Achatinella, Hawaii, 93
Acidification, 246
Acraeinae, possible extinction threat, *389*
Acrodipsas illidgei, 61
Actinote zikani, possible extinction threat, *389*
Aculeata
  NCC habitat conservation, *290*
  Red Data lists, wetland insects, *243*
Adelocosa anops, 98
Adelotypa penthia, *376*
Adonis blue butterfly *see Lysandra bellarqus*
Aeshna isosceles, 248
Aeshna juncea, *273*
Africa, deforestation, 408
Agabus uliginosus, *290*
Agabus undulatus, 248
Aganippe castellum, 47
Agaontidae, 327
Agelastica alni, 185
Aglais urticae
  colliery spoil reclamation, 269, 270
  fully sprayed land *vs* conserved headlands, *302*
Aglia tau, 127
Agonum bogemanni, 116
Agriades, 127
Agrias, possible extinction threat, *389*
Agricia artaxerxes, *289*
Agrilus pannonicus, 170
Agrodiaetus, 127, 134, 136
Albulina orbitulus, 127
Alesa, possible extinction threat, *388*
Alfalfa aphid, 43
Alien species, introduction, 98
Alkali bee, 326
Alystryma keta, *376*
Amateurs/naturalists, and existing biological information, 360
America, tropical, conservation units and palaeo-environments, *353*
Ampedus nigrinus, 170
Amphipods, 98
Andricus kollari, 336
Anillinid coleoptera, 140
Anisolabis, 79
Anisotominae, 178, 179
Anisozygoptera, 5
Anobiidae, 180
Antheraea pernyi, 128
Antheraea yamamay, 128
Anthocharis cardamines, fully sprayed land *vs* conserved headlands, *171*, 303
Ants
  Australia, 41–42
    numbers of species, *48*
  numbers and diversity, 48–49
Apamea anceps, 113
Apamea epomidion, 184
Apamea scolopacina, 184
Apatura iris, *171*, 172
Aphantopus hyperantus, 171
  fully sprayed land *vs* conserved headlands, 303
  movement patterns, 310
  relative index of abundance, *305*
Aphids, and insecticides, 297
Apodemia mormo, 26
Apodemia mormo langei, 19, 21, 26, *27*
Apollo *see Parnassius apollo*
Aptinus, 125
Aquatic and wetland habitats
  British Red Data Book, 240–248
  colourful species, 257
  environmental education, 258

food for other animals, 258
ignorance, 256
insects list, Red Data book, *243*
international species protection, 238–240
new and unstable habitats, 259
non-entomologists, acceptance of need for conservation, 257–260
pond creation, 271–272
Ramsar Convention, 136
site quality criteria, 251–252
species rarity and quality scores, 249–250
species rarity and site evaluation, 249–255
surface-active insects, 258
tactics, 259
threats and causes of loss, *243*
threats not mentioned in RDB, 245–247
water beetle site list evaluation, 252–255
world conservation strategy, 255–256
Aradidae, 180
Aradus, 116
Araneae, *291*, 285
Arawacus aethesa, possible extinction threat, *386*
'Area of search', 4
Argynnis adippe, 171, 172–174, 275
management operations on reserve, *289*
Argynnis aglaja, *171*, 186, 272
management operations on reserve, *286*
Argynnis elisa, 140
Argynnis pandora, 140
Argynnis paphia, 171, 185, 186
fully sprayed land *vs* conserved headlands, *302*
Aricia agestis, *171*, 224
Arthropods *see* Insects; named groups
Asia, deforestation, 410
Asida, 127
Assemblage, defined, 250
Atheta, 127
Athous mutilatus, 113
Atlases and mapping, North America, 16
Atrichobrunettia, 125
Atrytonopsis hianna, 11
Attelabidae, 185
Auchenorrhyncha, 342
Aulonium trisulcum, 170, 189
Australia
Cainozoic history, 38–39
cave-adapted arthropods, 72
Chrysomelidae, 40
climatic zones, *35*
Cupedidae, 42
current threats to insects, 49–55
distribution of rain forests, *37*
environments and their history, 34–39
federal and state initiatives, 58–62
greenhouse effects, 54–55
habitat templet, *64*
history and status of insect conservation, 55–63
insect fauna, 39–49
legislation, 57–58
phytophagous insect groups, *41*
potential natural vegetation, *36*
rural parks, 59
screw worm fly, 54
soil degradation severity index, *52*
status of some insects of conservation interest, *44–45*
status of taxonomic study, *40*
Tasmanian taxa, classification, *64*
vegetation clearance and degradation of habitat, 49–53
vegetation types, *50*
Victoria, changes in tree cover, *51*
Axymyiidae, Diptera, 180

B

Bali Action Plan, 413
Bay checkerspot butterfly, 23, 24–25, *25*
Bees *see* Hymenoptera
Beetles *see* Coleoptera
Belize, 331
Bemisia tabaci, 325
Bern Convention, 238
and Spanish Lepidoptera, *poster presentation*, 426
Bessa remota, Hawaii, 84
Bhutanitis, 334
Biodiversity, one-day survey, 359
Biological controls
classical, threat to ecological balance, 100
genetic resources, 323–325
Biological Monitoring Working Party score, 255
Biosphere reserves concept, 397

Birds, food insects and chick survival, conservation headlands, 299–301
Birdwing butterflies, Papua New Guinea, 418
Black hairstreak *see Strymonidia pruni*
Black veined moth *see Siona lineata*
Blastophaga psenes, 327
Boloria aquilonaris, 117
Boloria euphrosyne, 171, 172, 174, 186, 275, *289*
Boloria selene, *171*, 172, 173, 188
Bombyx mori, 335
Boreus, 127
Botanic gardens, role in conservation biology, and risks, 98
Boxworth Project, pesticides, 322
Brachypodium pinnatum, changes in population size, and foodplants, *223*, 224
Brachystomella parvula, *268*
Brassolinae
  extinction threat, *386*, *389*
  as indicators, 372
  species-area curves, *375*
Brazil *see* Neotropics; Tropical forest
Breckland Environmentally Sensitive Area, 314
Brillouin diversity index, *267*
British Nature Conservation Review, 342
Brunettini, 125
Bupestridae, 55–56, 125, 170, 181
Burning *see* Fires
Bush crickets, as indicator groups, 412
Butterflies *see* Lepidoptera
Byctiscus betulae, 185
Byctiscus populi, 185

C

Caenis horaria, 271
Caithness Biological Records Centre, 247
Caledonian pine forests, 161
Callicore hydarnis, possible extinction threat, *389*
Callippe silverspot butterfly, 21
Callophrys rubi, *171*
Calosoma inquisitor, 183
Camponotus, Australia, 41
Canada
  conservation programmes, 11–12
  see also North America
Capsodes flavomarginatus, 184
Captive breeding, *poster presentation*, 429
Carabidae, 125, 126, 138, 180, 183, 322
  larval, *300*
Carabus alysidotus, 130
Carabus clathratus, 130
Carabus intricatus, 112
Carabus italicus, 130
Carabus olympiae, 130
Carterocephalus palaemon, *171*, 172, 184, *290*
Catocala fraxini, 111
Cave-adapted arthropods, Australia, 72
Caves, threats to fauna, 134–135
Celastrina argiolus, *171*
  fully sprayed land *vs* conserved headlands, *302*
Cerambycidae, 180, *285*, *286*
  Hawaii, 85
Cerambyx cerdo, 112
Ceratocystis ulmi, 170
Ceratopogonidae, Diptera, 271
Cereal aphids, insecticides, 297
Chalcidoidea, 327
Chaoborus, 271
Charaxes jasius, 125, 126, 140
Charaxinae
  as indicators, 372
  possible extinction threat, *389*
  species-area curves, *375*
Charis cleonus, *376*
Charonias theano, possible extinction threat, *388*
Charpentiera, host to *Mapsidius*, 85
Cheilosia carbonaria, 183
Cheilosia semifasciata, 176, 184
Chequered skipper *see Carterocephalus palaemon*
Chionea, 127
Chironomidae, Diptera, 271
Chloeon dipterum, 271
Chrysomela tremula, 174, 176, 185
Chrysomelidae, 125, 183, 185
  Australia, 40
  larval, *300*
Chrysomya bezziana, 54
Chrysops sepulcralis, 244

Cicadas, Australia, 42
Cicadellidae, 85
Cicindelidae, 15, 138
CITES *see* Convention of International Trade in Endangered Species of Wild Fauna and Flora
Cixiidae, 85
Coenagrion hastulatum, 6
Coenagrion mercuriale, 7, 248
Coenagrion puella, *273*
Coenagrion tenellum, 4
Coenonympha oedippus, 137, 428
Coenonympha pamphilus
  colliery spoil reclamation, 269, 270
  fully sprayed land *vs* conserved headlands, 303
Coenonympha tullia, 285
Colacyticus, possible extinction threat, *388*
Coleoptera
  as indicators, *364*
  NCC habitats, *290*
  Red Data lists, *111*
    wetland insects, *243*
  saproxylic species, 168–170
  US Animal Notice of Review, *17*
  water beetle site list evaluation, 252–255
  see also Aquatic and wetland habitats
Colias palaeno, 117, 127
Collection and trade
  deadstock trade, 332–335
  decorative goods, 336
  'entomological livestock', 330–332
  Mediterranean insects, 130
Collembola
  ecological succession, *267–268*
  as indicators, *364*
  pesticides, 322
  sminthurids, 41
  Tasmania, 53
    generic diversity and distribution, *46*
  US Animal Notice of Review, *17*
Colombia *see* Neotropics
Colorado beetle, 128
Colydiidae, 170
Comma *see Polygonia c-album*
Communication among entomologists, 6–7
Community, defined, 250
Conservation, cultural biases, 91–92
Conservation areas
  growth in world sites, 412
  key areas
    identification, 341–342
    maintenance, 342–343
Conservation headlands, 293–314
  background to development, 294–296
  butterflies, 301–312
  chick food insects and chick survival, 299–301
  design and agronomy, 296–299
  as tool in conservation, 312–314
  wildlife corridors, 313
Conservation process
  interaction between science and society, *394*
  Nature Conservancy Council, management operations for arthropod conservation, 283–291
  reserve management, 95–97
  transition, 395–396
Convention on International Trade in Endangered Species of Wild Fauna and Flora (CITES), 239, 333–334
Coppice
  loss, 162–164
  process, 159–160
  rotational management of coppiced woodland, 193–194
Cordulegaster bottonii, *273*
Corsican swallowtail, 333
Crickets, bush, as indicator groups, 410
Criorhina, 182
Cryptocephalus bipunctatus, 185
Cryptocephalus coryli, 185
Cryptocephalus querceti, 183
Cryptocephalus sexpunctatus, 185
Cryptostigmata, *267*
Ctenophora oranata, 179
Ctesias serra, 180
Cucujidae, 179
Cucujus cinnaberinus, 112, 113
Culcididae, Diptera, 271
Cultural biases, hindering conservation, 91–92
Cupedidae, Australia, 42
Curculionidae, Coleoptera, 127, 170, 180, 185, 189
  larval, *300*
Cyanostolus aeneus, 180
Cydnidae, 184
Cynipid galls, 336

Cynthia cardui, *171*
Cyrotocephalus colryli, 176
Cyrotocephalus nitidus, 176
Cyrtobagous salvineae, water fern weevil, 324

D

Damselfly, 4, 5, 6
Danaus plexippus, 11, 14, 19
Dark bordered beauty *see Epione parallelaria*
Dark green fritillary *see Argynnis aglaja*
Dasyophthalma, possible extinction threat, *389*
Dasyophthalma vertebralis, extinction threat, *384*
Decomposition, 328–330
 see also Saproxylic species
DECORANA, 252–254
 ecological gradients, *254*
Decorative trade, 332
 see also Collection and trade
Decticus verrucivorus, *288*
Deforestation, tropical forests, estimated rates (FAO/UNEP), 409–411
Deinacrida heterocantha, 96
Deinacrida rugosa, *79*, 96
Delias, 355
Delphacidae, 85
Delta green ground beetle, 13
Denmark, conservation law, 113
Dermaptera, Red Data lists, *111*
Dermestidae, 180
Dialectica scalariella, 43
Dicranocephalus medius, 184
Didae fasciata, 183
Diplura, Red Data lists, *111*
Diptera
 Ceratopogonidae, 271
 Chironomidae, 271
 Culcididae, 271
 Dixidae, 271
 as indicators, *365*
 Psychodidae, 125
 Red Data lists, wetland insects, *243*
 US Animal Notice of Review, *17*
Diversity, 338–341
Dixidae, Diptera, 271
Dolomedes fimbriatus, 285
Doriphora, 128
Dorset heaths, habitat loss, 144
Doxocopa, possible extinction threat, *387*
Dragonflies *see* Odonata
Drainage, and peat loss, 246
Drephalys, possible extinction threat, *387*
Driapriidae, 140
Drosophila, 73
Dryococelus australis, 53
Dryophtorus corticalis, 170
Drypta, 125
Duke of Burgundy butterfly *see Hamearis lucina*
Dung beetles, 329
Dyscritomyia, Hawaii, 93
Dytiscid water beetles, *290*
 guilds, 251
Dytiscus latissimus, 239

E

Earwigs, 79
Eastern Europe
 protection of insects and their habitats, 112–114
 Red Data lists, 110–112
Ecological myths, 87–91
Ecological succession
 review, 264–266
 Wharram Quarry Nature Reserve, 266–269
Ecosystems, replacement, factors, 89
Edge effects, tropical forests, *371*
Education, in conservation, 94, 201
El Segundo blue butterfly, 21, 25
Elaeidobius kamerunicus, 326
Elaphrus viridis, 13
Elateridae, 170
Elbella polyzona polyzona, possible extinction threat, *387*
Elmid beetles, 247
Eltham Copper, 56, 60
 'crisis management', 65
Embioptera, Red Data lists, *111*
Emus hirtus, 245
Enallagma cyathigerum, 273
Endangered species, (CITES), 239, 333–334

Endemism and diversity, neotropics, 362
Entomobrya nicoleti, *268*
Entomological livestock
trade, 330–332
see also Collection and trade
Environmental impact report, New Zealand, 94
Environmental impact statement, 57
Ephemeroptera, 271
Red Data lists, *111*
US Animal Notice of Review, *17*
Epione parallelaria, *290*
Epiophlebia laidlawi, 5
Epiophlebia superstes, 5
Epuraea angustula, 183
Erastria cruentaria, 11
Erebia, 136
Erebia aethiops, *171*
Erebia eriphyle, 218
Erebia melampus, 230, *230*
Erebia pharte, 218
Eresia erysice, extinction threat, *386*
Eriodictyon californicum, 266
Eriogaster catax, 429
Ernoporus caucasicus, *286*
Erynnis tages, *171*
Esthemopsis, possible extinction threat, *388*
Eucalyptus
associated phytophagous insects, 40
optimal size of reserve, 47
Euchirines, 125
Eucorna sanarita, possible extinction threat, *386*
Euglandina rosea, Hawaii, 84
Euglossine bees, pollination of Brazil-nut tree, 418–419
Euophryum confine, 189
Enphilotes battoides allyni, 21, 25
Euphydryas
ecological succession, 266
metapopulation dynamics, 273
Euphydryas aurinia, 429
Euphydryas chalcedona, 266
Euphydryas editha, 266
Euphydrays editha bayensis, 23, 24–25, *25*
Euphydrays intermedia wolfensbergeri, 218
Eupithecia, 75
Euproserpinus wiesti, 13
Eurodryas aurinia, *171*, *285*, *289–290*
Europe
grasslands of lowland and upland Europe, 213–233
Lepidoptera, 213–233
see also Eastern Europe; Finland; Mediterranean; Northern Europe
European nations, national parks, 132
Eurydemia dominulus, 184
Eurymelidae, Australian, 40
Eurytides, extinction threat, *386*
Eutresis hypereia imeriensis, extinction threat, *386*
Existing biological information, 360–362
Expeditions, necessary kit, 390
Extensification scheme, 299
Extinction
latent, and transient diversity, 418–419
local, and recolonization, 152

F

Farming and Wildlife Advisory Groups, 5–6
Farming and Wildlife Trust, 5–6
Farmland, conservation headlands, 293–314
Fenlands, peat loss, 246
Fennoscandia, glacial history, 109
Fig insects, 327
Finland
causes of threat and similarity of species lists, 114–116
endangered species, reasons for decline, *115*
glacial history, 109
Red Data lists, 110–112
comparisons, *116*
total area of reserves, 113
Fires
and adapted species, 166–167
effect on species richness, 275
Flow Country, Scotland, investment priorities, 247
Folsomides parvulus, *268*
Food insects, 336–338
Forcipomyia hardyi, Hawaii, 90
Forests *see* Neotropics; Tropical forests; Woodlands
Formica rufa, 176
Formicidae, 183

Fragmentation, isolation and mobility of insect populations, 143–151
Frankliniella, 325
Freshwater habitat *see* Aquatic and wetland habitats
Friesea mirabilis, *268*
Fungicides, insecticidal activity, 297–298
Furcula furcula, 185
Fynbos, species richness, after burning, 275

G

Galerca interrupta, 244
Gamebirds
  brood sizes and population trends, 299
  population trends, mortality and pesticides, 294–295
Genetic resources, biological controls, 323–325
Geological information, neotropics, 360
Geometridae, 184
Glaciations, 109, 367
Glaucopsyche xerces, 20
Global warming, rising sea levels, *245*
Gnorimus nobilis
  dependence on orchards, 161
  moisture for pupation, 179
Gonepteryx rhamni, *171*
  fully sprayed land *vs* conserved headlands, 303
Gonimbrasia belina, 337
Goods from insects, 330–338
Graellsia isabelae, 125, 130
Graphoderus bilineatus, 239, 244
Graphoderus zonatus, 248
Graptodytes flavipes, *290*
Grasslands
  abandoned, 217–218
  British lowland, 220–222
  intensively cultivated, 215–217
  lowland and upland Europe, Lepidoptera, 213–233
  reasons for rapid change in insect populations, 222–230
  subalpine meadows, 218–220
  summary, 231–233
Grey partridge *see* Gamebirds
Guild, defined, 250
Gyrinus substriatus, 258

H

Habitat continuity indicator, 192
Habitat creation, review, 269–270
Habitat loss, 339
  Britain, 143–144
Habitat mosaics, different successional age, 275–278
Habitat piles, 196
Habitat re-creation
  butterfly habitats, 273–275
  review, 272–273
Hadramphus stilbocarpae, *82*
Hallodopus montadoni, *288*
Hamearis lucina, *171*, 172, 275
Hammerschmidtea ferruginea, 168
Hawaii
  alteration of habitats by land use changes, 85–86
  biodiversity crisis, 86–87
  island conservation, 73
  research, 93–94
Heath fritillary *see Mellicta athalia*
Hedgerows
  early, 161–162
  loss, 167
Heliconiini
  extinction threat, *384*
  geographical correlation, *369*
  as indicators, *365*, 372
  neotropical forests, 366
  species-area curves, *373*
Helicopis cupido lindeni, possible extinction threat, *386*
Helicoverpa armigera, 53
Helicoverpa punctigera, 53
Helioconius nattereri, extinction threat, *384*
Heliocoverpa see also Heliothis
Heliothis, 79
Helophorus lapponicus, 239
Helophorus longitarsis, 259
Hemiphlebia mirabilis, 5, 47
Hemiptera
  Australian, 40
  as indicators, *364*
  management operations on reserve, *288*
  Red Data lists, *111*
  US Animal Notice of Review, *17*
Henderson Island, 86
Heodes virgaureae, *230*, 230

Hesperia comma, 221, 222, 224, 227, 274
Hesperia leonardus montana, 20, 27, *28*
Hesperiidae
  fully sprayed land *vs* conserved headlands, *302*
  lowa, 15
  possible extinction threat, *389*
Hesperocharis catasticta, possible extinction threat, *388*
Heteropan dolens, Hawaii, 84
Heteroptera
  Cydnidae, 184
  larval, *300*
  Red Data lists, wetland insects, *243*
High brown fritillary *see Argynnis adippe*
Highland glens, 161
Hipparchia, 140
Hipparchia semele, 224, 270
Hippocrepis comosa, *223*, 225
Histeridae, 180
Homerus swallowtail, 333
Homoptera
  as indicators, *364*
  US Animal Notice of Review, *17*
Hoverflies
  role of field boundaries, *poster presentation*, 425
  see also Syrphidae
Human intervention
  coppicing, 159–160
    loss, 162–164
  effects, neotropics, 372–378
  hedgerows, 161–162, 167
  invasive sampling, 167
  long-term effects, 378–380
  orchards, 161
  public access to sites with ancient trees, 165–166
  short-term effects, 372–376
  touristic development, Mediterranean area, 131
  wood pasture
    loss, 164–165
    pollarding, 160
Hydrochara caraboides, 248
Hydroporus elongatulus, 239
Hydropsyche tobiasi, 79
Hylaeus, Hawaii, 93
Hyles hippophaes, 429
Hymenoptera
  Aculeata, NCC habitats, *290*
  alkali bees, 326
  Australian, 40
  euglossine bees, pollination of Brazil-nut tree, 418–419
  fig insects, 327
  as indicators, *365*
  leafcutter bees, 326
  Red Data lists, *111*
    wetland insects, *243*
  semi-domesticated bees, threat to native species, 418
  US Animal Notice of Review, *17*
Hypebaeus flavipes, 177, 189, *286*
Hypogastrura purpurescens, 53
Hyposmocoma, 73
Hypothyris, extinction threat, *386*, *389*
Hypothyris euclea, *376*

I

Imperialism, economic, cultural and scientific, 399–400
Inachis io
  colliery spoil reclamation, 269, 270
  fully sprayed land *vs* conserved headlands, *302*
Incisalia mossii bayensis, 20, 21, 21
Incurvariidae, 327
Indicator groups
  neotropics, 363–365
  timber habitats, 192
  tropical forest, 412
Indomalaya, deforestation, 410
Insecticides
  cereal aphids, 297
  effects on environment, 321
Insects
  conservation biology, nascent principles, 29
  history of estimates of described and total numbers, worldwide, 404
  mobility, 187–190
  NCC workshops on conservation, 201
  number of species found and threatened, *351*
  populations, isolation, fragmentation and mobility, 143–151

as a resource, educative effect, 339
tools for conservation, 421
International Union for Conservation of Nature and Natural Resources
British contribution, 255
Invertebrate Red Data Book, 55
Statement on Captive Breeding, *poster presentation*, 429–430
and World Wide Fund for Nature, 7
Introductions, and Red Data Book, 248
Invertebrate Site Register, NCC, 249
Investment priorities, Flow Country, 247
Iridomyrmex, Australia, 41
Ischnomera cinerascens, *285*
Ischnura elegans, *273*
Island conservation, 77–92
alteration of habitats by land use changes, 85–86
biodiversity crisis, 86–87
controlling introduction of alien species, 99
cultural biases, 91–92
ecological myths, 87–91
education, 94–95
impacts of invasive alien species, 80–85
legal and legislative actions, 97–98
pest control programmes, 100
research, 93–94
reserve management, 95–97
role of invertebrates in ecosystem functioning, 88
Isolation
insects responding to increased isolation, 146–151
population dynamics, 144–146
Isoptera, *364*, 407
Isotoma notabilis, *268*
Isotoma viridis, 267, *268*
Ithomiinae
extinction threat, *384*, *387*
geographical correlation, *369*
as indicators, *365*, 372
neotropics, 366
species-area curves, *373*

J

Japan, Dragonfly Kingdom of Japan, 6
Joiceya praeclara, extinction threat, *386*

K

Karner blue butterfly, 22
Karocolens pittospori, *77*
Katianninae, 41

L

Laccobius simulatrix, 259
Ladoga camilla
fully sprayed land *vs* conserved headlands, *302*
increase in numbers, 172, 185
Lange's metalmark butterfly, 19, 21, 26, *27*
Large blue butterfly *see Maculinea arion*
Large copper *see Lycaena dispar*
Lasiommata megera
colliery spoil reclamation, 269, 270
fully sprayed land *vs* conserved headlands, 303
Lasius brunneus, 183
Lathridiidae, 178, 179
Latin America
deforestation, 410
see also Neotropics
Laurisilya, 125
Lauro-Prunetea, Laurisilya, 125
Leafcutter bee, 326
Leafminer *see Liriomyza trifolii*
Legal and legislative actions, 97–98
Leiodidae, 178, 179, 180
Lepidoptera
ancient sites, plantation woodland, *poster presentation*, 425
Australian, 40
birdwing butterflies, Papua New Guinea, 418
breeding in British woods, main habitats preferences, *171*
butterfly farming, 335
butterfly house industry, 330–332
in conservation planning, 358
diversity and endemism, *377*
extinction, facts and myths, 380–381
as indicator groups, 412
as indicators of change, grasslands of lowland and upland Europe, 213–233
larval, *300*

management operations on reserve, *288*
neotropical, endemism and diversity, 366–370
numbers and ecological status
cultivated meadows, *216*, *221*
different stages of abandoned grassland, *221*
surveys, 391–393
threatened, *351*
Red Data lists, *111*
wetland insects, *243*
US Animal Notice of Review, *17*
Victoria, ten minute quadrats, *61*
yearly cycles, abundance, subtropics, *381*
Leptidea sinapis, 171, 172, 173, 184, 186
Leptura thoracica, 113
Lestes dryas, 7
Lestes sponsa, *273*
Lethocerus indicus, 337
Levuana iridescens, Hawaii, 84
Libellula quadrimaculata, *273*
Limnebius crinifer, 259
Limoniscus violaceus, 177, 183, 189
Liriomyza trifolii, 325
Lopinga achine, 112, 429
Lorax, role, 101
Lord Howe Island stick insect, 53
Ludius ferrugineus, 179, 181
Luzon peacock butterfly, 333
Lycaena dispar, 112, 137, 248, *289*
Lycaena phlaeas, *171*, *302*
Lycaenidae, 185, 188, 220
fully sprayed land *vs* conserved headlands, *302*
species-area curves, *373*
Lygaeid seed bug, 75
Lyperobius huttoni, 76, *78*
Lysandra bellargus, 220–221, 227, *284*
changes in population size , and foodplants, *223*, 224, 225
distribution of eggs, 224, *225*
Lysandra coridon, *230*, 230
distribution of eggs, 224, *225*
Lytta vesicatoria, 336

M

Macro-invertebrate index, Shannon-Wiener index of diversity, 255
Maculinea arion
changes in population size, and foodplants, *223*, 224
extinctions post grazing changes, 221
niche, 226
response to isolation, 148–151
Spain, 429
Maculinea nausithous, 225, 427
Maculinea teleius, 225, 426
Magneuptychia gera, *374*
Mallota cimbiciformis, 189
Management operations for arthropod conservation, 283–291
Manaus, forest fragments project, 370–376
Mangrove butterfly, 61
Maniola jurtina
coaltip reclamation, 269, 270
fully sprayed land *vs* conserved headlands, 303
movement patterns, 310, *311*
relative index of abundance, *305*
Maniola nurag, 140
Mannerheimia, *126*, 127
Mapsidius, Hawaii, 85
Marbled white butterfly *see Melanargia galathea*
Mark-release-recapture (MRR) experiment, 309
Marsh fritillary *see Eurodryas aurinia*
Mayflies *see* Ephemeroptera
Meadow brown butterfly *see Maniola jurtina*
Mechanitis bipuncta, extinction threat, *386*
Mecoptera
Red Data lists, *111*
US Animal Notice of Review, *17*
Mediterranean insects, 123–140
collection and trade, 130
conservation measures, 129–132
endangered environments, 136–140
origin and composition of insect fauna, 123–128
physical background, 123
Pleistocene refugia, 128–129
threatened environments, 135–136
total insect fauna, 123
vulnerable environments, 133–135
Megachile rotundata, 326
Megachilidae, 176
Megalagrion, Hawaii, 5, 93
Megalanura, 46
Melanargia galathea
colonization of new sites, 268

fully sprayed land *vs* conserved headlands, 303
Melinaea mnasias, extinction threat, *386*
Melitaea, 136
Melitaea cinxia, 224, 226, 269
Melitaeini, extinction threat, *384*
Mellicta athalia, 171, 172, 185, 188, 273, *289*
Melophorus, Australia, 41, 42
Melyridae, 177, *286*
Meranoplus, Australia, 41
Mesites tardii, 180
Mesostigmata, *267*
Mestolobes miniscula, Hawaii, 90
Mexico, conservation programme, 14
Millipede, in Australia, 53
Minoa murinata, 184
Minucia lunaris, 185
Miridae, 184, 87
Mission blue butterfly, 20, 21
Mobility of insects, 187–190
Molytine weevil, 76
Monarch butterfly, 11, 14, 19
Mormon metalmark butterfly, 26
Morphinae
  as indicators, *365*, 374
  species-area curves, *375*
Moschoneura methymna, extinction threat, *386*
Mutillidae, 138
Mycetophilidae, 180
Mygalomorphs, 47
Myrmeleontidae, Australia, 41
Myrmica sabuleti, 148, 221, 222, 230
  changes in population size, and foodplants, *223*, 224

N

Nacerdes melanura, 180
Napeogenes cyrianassa xanthone, extinction threat, *386*
National Trust, conservation policies, 166
Nature Conservancy Council
  approach to invertebrate conservation, *poster presentation*, 431
  management operations, targeting, survey of the 15 regions, *283–291*
  saproxylic invertebrates, 191
  SSSIs, 191
  workshops on insect conservation, 201
Nectar/pollen, saproxylic invertebrates, 183
Neelus minimus, *268*
Nemotaulius punctatolineatus, 247
Neotropics
  biosphere reserves, 399
  Brazilian Amazon region, 398
  Brazilian Atlantic forests, 378–380
  butterflies threatened with extinction, *386*
  case histories, 366–382
    endemism and diversity, 366–370
  defining options, 357–359
  education and grass-roots movements, 397–399
  effects of human intervention, Manaus project, 370–376
  endemism and diversity, 362
  exceptional biological importance, Palaeo-environments, 378–380
  extinction, facts and myths, 382–383
  extractive reserves, indigenous reserves and other important preserved areas, *390*
  geological information, 360
  habitat and biodiversity conservation, 350–352
  imperialism, 397–398
  indicator groups, 363–367
  insects as indicators of their biotas, 349–399
  management of anthropic mosaics for sustainable yield, 381–385
  palaeo-environments, *284–286*
  practical guide to conservation action, 385–395
  preserves, 352
  urban, rural and natural habitats, 352–356
  see also Tropical forest
Nesiomiris, Hawaii, 87
Neuronia clathrata, management operations on reserve, *286*
Neuroptera
  Red Data lists, *111*
  US Animal Notice of Review, *17*
New Guinea, insects, 43
New Zealand
  environmental impact report, 94
  island conservation, 74
  legislation, 98
Nirodia belphegor, possible extinction threat, *388*

Noctuid moths
extinctions, 79
niche breadth, 184, 185
pests, 53
Nomia melanderi, 326
Normandia nitens, 247
North America
atlases and mapping, 16
conservation programmes, 9–14
management and recovery, 20–28
progress in listing candidates, 16–18
protection and habitat acquisition, 18–20
status surveys, 14–16
atlases, 16
governmental efforts, 14–15
private and volunteer, 15
see also Canada; USA
Northern brown argus *see Agricia artaxerxes*
Northern and Eastern Europe
protection of insects and their habitats, 112–114
Red Data lists, 110–112
Northern Europe, biogeographic history, 109
Norway
Red Data lists, 110–112
total area of reserves, 113
Nothomyrmecia, 55–56, 60
Notodontidae, 185
Nuclear power stations, 244
Nymphalidae
fully sprayed land *vs* conserved headlands, *302*
pollen/nectar requirements, 185
Nymphalinae
as indicators, *365*, 372
possible extinction threat, *389*
species-area curves, *373*
Nymphalis polychloros, *171*, 172
Nymphidium aurum, *376*
Nysius wekiuicola, 97

O

Oceania, deforestation, 408
Ochlodes venata, *171*
fully sprayed land *vs* conserved headlands, *302*
Ochthebius lenensis, 244
Odacantha, 125
Odonata
colonization of newly created pond in Loch Lomond, *273*
conservation sites, 3–4, 6–7
Dragonfly Kingdom of Japan, 6
as indicators, *364*, 410
management operations on reserve, *286–288*
number of species assigned to biogeographical realms, 417
pond creation, 270, 271–272
Red Data lists, *111*
wetland insects, *243*
regional thresholds, 257
US Animal Notice of Review, *17*
Odynerus, Hawaii, 93
Oecophoridae, Australian, 40
Oedemeridae, 170, 180
Oil-related development, 244
Omiodes blackburni, Hawaii, 90
Ommatoiulus moreletii, in Australia, 53
Oncochila simplex, 184
Onychiurus ambulans, *268*
Onychiurus armalus, *268*
Onychophorans, 60
Orchards, 161
Oregon silverspot butterfly, 21
Oreodytes alpinus, 247
Ornithoptera, Papua New Guinea, 418
Ornithoptera alexandrae, 333
Ornithoptera 'allottei', 332
Orobrassolis ornamentalis, extinction threat, *386*
Orthetrum nitidinerve, *126*
Orthoptera
management operations on reserve, *288*
Red Data lists, wetland insects, *243*
(*sensu lato*), Red Data lists, *111*
US Animal Notice of Review, *17*
Osmia pilicornis, 176
Osmoderma eremita, 113
Osmylidae, Australia, 42
Oxygastra curtisi, 4

P

Palaeo-environments, *282–284*, 380–382
defined, 378–380

Paleochrysophanus hippothoe eurydame, 220
Papilio alexanor, 130
Papilio aristodemus ponceanus, 19
Papilio chikae, 333
Papilio garleppi, possible extinction threat, *388*
Papilio himeros, extinction threat, *386*
Papilio homerus, 333
Papilio hospiton, 130, 333
Papilio laglaizei, 337
Papilio machaon, 5, 111, 220, 227, *289*
  autecological research, 220
  changes in population size, and foodplants, *223*
  number of species
    by latitude, 407
    threatened, *351*
  response to isolation, 146–148
Papilio xanthopleura diaphora, possible extinction threat, *388*
Papilio zagreus, extinction threat, *386*
Papilionidae
  critical faunas, 418
  as deadstock on open market, *333*
  extinction threat, *386*, *388*
  as indicators, *365*
  number of species
    assigned to biogeographical realms, 417
    latitudinal gradients, 416
  species-area curves, *375*
Papua New Guinea, Ornithoptera, 418
Pararge aegeria, 171, 172
  fully sprayed land *vs* conserved headlands, 303
Parides, possible extinction threat, *388*
Parides ascanius, extinction threat, *386*
Parides burchellanus, possible extinction threat, *388*
Parides lysander mattogrossensis, extinction threat, *386*
Parnassius, 127
Parnassius apollo, 112, 134, 220, 334, 427
Parnassius mnemosyne, 112, 427
Paropsis, Australian, 40
Partridge *see* Gamebirds
Pawnee montane skipper, 20, 27, *28*
Peacock butterfly *see Inachis io*
Pearl-bordered fritillary *see Boloria euphrosyne*
Peat loss, fenlands, 246
Pentatomidae, 184
Pergidae, Australian, 40
Perrhybris flava, extinction threat, *386*
Pesticides
  Boxworth Project, 322
  Collembola, 322
  and environmental effects, 321–322
  Mediterranean area, 131
  selective use, *296*
Pests
  classical biological control, 100
  natural control, 322–323
  pest control programmes, 100
Petroceris catenia, possible extinction threat, *388*
Peucedaenum palustre, changes in population size, and foodplants, *223*
Phalacrognathus muelleri, 58
Pheasant *see* Gamebirds
Pheidole megacephala, 83
Philanthus triangulum, 113
Philippines, deforestation, 10
Photedes fluxa, 184
Phylosamia cynthia, 128
Phytophagous insect groups, Australia, *41*
Pierella astyoche, *376*
Pierella lamia, *376*
Pieridae
  Australia, 43
  behaviour in cereal field margins, conservation and fully sprayed headlands, *307*
  extinction threat, *386*
  flower visit frequency, *308*
  fully sprayed land *vs* conserved headlands, 303
  host plants, 305
  as indicators, *365*
  population structures, 304
  possible extinction threat, *388*
  relative index of abundance, 305
  species-area curves, *375*
Pieris brassicae, fully sprayed land *vs* conserved headlands, 303
Pieris callidice, 127
Pieris napi, fully sprayed land *vs* conserved headlands, 303
Pieris rapae
  Australia, 43

fully sprayed land *vs* conserved headlands, 303
Pieris virginiesis, 11
Pigs, threats to oak forests, 140
Plantation woodland, ancient sites for butterflies and others, *poster presentation*, 425
Platypezidae, 180
Plebejus argus, 224, 226–229, 274, *289*
Plebejus icarioides missionnensis, 20, 21
Plebicula golgus, 429
Plecoptera
Red Data lists, *111*
US Animal Notice of Review, *17*
Pleistocene refugia, Mediterranean insects, 128–129
Pocota personata, 182
Pollarding, wood pasture, 160
Pollination, specialised relationships, 326–328
Polygonia c-album, fully sprayed land *vs* conserved headlands, *302*
Polyommatini, colonization, 127
Polyommatus, 134
Polyommatus icarus, *171*
colliery spoil reclamation, 269, 270
fully sprayed land *vs* conserved headlands, *302*
Population dynamics, isolation, 144–146
Poster presentations, 425–430
Powelliphanta, New Zealand, 91
Prepona deiphile, possible extinction threat, *389*
Proctotrupoidea, 140
Prolasius, 42
Propomacrus bimucronatus, 125
Proserpinus proserpina, 429
Prostomis mandibularis, 179
Pselactus spadix, 180
Pselaphidae, 140
Pseudochazara, 127
Pseudocistela ceramboides, 179, 181
Pseudosinella, *268*
Psilocephala melaleuca, 180
Psychodidae, Diptera, 125
Psyllidae
Australian, 40
gall-forming lice, 185
Psyllopsis fraxini, 185
Ptiliid beetles, 189
Public sector and local community parameters, 393–396
Pug moth caterpillars, 75
Pyralid moths, Hawaii, 90
Pyrazophos, toxicity to insects, 297
Pyrgus malvae, *171*
Pyrochroidae, 170, 181
Pyronia tithonus
behaviour, *307*, 308
flower visit frequency, *309*
fully sprayed land *vs* conserved headlands, 303
movement patterns, 310, *312*
relative index of abundance, *305*
Pyrrhidium sanguineum, *286*
Pyrrhosoma nymphula, *273*

Q

Queen Alexandra's birdwing, 333
Queensland stag beetle, 58
Quercusia quercus, *171*

R

RADAMBRASIL project, 378
Rain forests *see* Neotropics; Tropical forest
Ramsar Convention, wetlands, 136, 240
RDB *see* Red Data Book
Red Data Book
conservation measures specified, 247–248
saproxylic invertebrates, 169
species selection, 240–242
threats and causes of loss reports, 242–244
threats not mentioned, 245–247
wetland insects list, *243*
Northern and Eastern Europe, 110–112
Reservoir population, defined, 273
Resource management
conservation programmes, 95–97
planning, 338–341
Rhantus aberratus, 244
Rhizophagidae, 180
Rhytididae, *Powelliphanta*, 76
Riodininae
extinction threats, *386*, *388*

neotropics, *376*
species-area curves, *375*
Rosalia alpina, 127
Rutidosoma globulus, 185

S

Saldula longicornis, 79
Salpingidae, 180
Salvinea, water fern weevil, 324
San Bruno elfin butterfly, 21
Sandflies, NCC habitat conservation, *290*
Saproxylic invertebrates
beetles
larvae, upper lethal temperature, 181
richest assemblages, 178
habitat
fallen dead wood, 158
location of niches, 182
microclimate, 178–181
niche specialization, 177–184
orchards, 161
habitat loss, 165–166
Windsor Forest and Great Park, 166
interdependence between species, 182–183
invasive sampling methods, 167
long-term survival of isolated assemblages, 197
need for nectar/pollen by imagos, 183
niche specialization, 177–184
nutrient status of dead wood, 181–182
Red Data books, 169
Satellite populations, defined, 273
Saturnia pavonia, 111
Satyridae
fully sprayed land *vs* conserved headlands, 303
movement patterns, on farmland, 309–312
Satyrinae
as indicators, *365*
neotropics, *376*
species-area curves, *375*
Scada karschina delicata, extinction threat, *386*
Scarabaeidae, 43, 125, 138
grass grub, New Zealand, 90
Scaritini, 140
Scelionidae, 140
Schaus swallowtail butterfly, 19, 21
Schinia gloriosa, 11
Schizotus pectinicornis, 169–170
Scolia unifasciata, 112
Scolytidae, 178, *286*
Scolytus scolytus, 183
Screw worm fly, 54
Scydmaenidae, 140
Sehirus beguttatus, 184
Set-aside schemes, 299
Shannon-Wiener index of diversity, macro-invertebrate index, 255
Shipstern Wildlife Reserve, 331
Siagonium quadicorne, 170
Silver-spotted skipper *see Hesperia comma*
Silver-studded blue *see Plebejus argus*
Singa hamata, *291*
Siona lineata, *290*
Sitalcina tiburona, 13
Site evaluation
existing biological information, 360–362
and monitoring, surveys, *361*
Sites of Special Scientific Interest (SSSI), 4
Nature Conservancy Council, 191
Small heath butterfly *see Coenonympha pamphilus*
Small tortoiseshell butterfly *see Aglais urticae*
Sminthurids, 41
Snails, New Zealand, 91
Society, and the conservation process, *394*
Soviet Union, Red Data lists, 111
Spanish Lepidoptera, and Bern Convention, *poster presentation*, 428
Species quality score, 250
Spelaeorchestia koloana, 98
Speyeria callippe callippe, 21
Speyeria zerma hippolyta, 21
Sphecidae, 138
Sphindidae, 178, 179
Sphingidae
as indicator groups , *367*, 412
Wiest's sphinx moth, 15
Spider communities, upland heather moorland, 276–278
Spray drift, conservation headlands, 313
SSSIs *see* Sites of Special Scientific Interest
Stalachtis phlegia, 375
Staphylinidae, 127, 170

Staphylinoidea, 125, 126
Stenelmis canaliculata, 247
Stenocephalidae, 184
Stenopelmatid weta, 75–76
Stephens Island weta, *79*
Stethophyma grossum, 248
Stick insect, Lord Howe Island, 53
Stilbocarpa weevil, *82*
Stratiomyidae, 180
Strymonidia pruni, *171*, 188, *289*
Strymonidia w-album, *171*
Surface-active insects, high quality habitats, 258
Surveys
  data, transforming into useful parameters, 391
  how to predict areas of potential importance, 391
  site evaluation and monitoring, *361*
Swallowtail butterfly *see Papilio machaon* and Papilionidae
Sweden
  causes of threat and similarity of species lists, 114–116
  Red Data lists, 110–112
    comparisons, *116*
  total area of reserves, 113
Sycorax, 125
Symmachia, possible extinction threat, *388*
Sympetrum danae, *273*
Sympetrum sanguineum, *288*
Syrphidae, 170, 176, 180, 182, 183, 184, 189
  role of field boundaries, *poster presentation*, 427

T

Tabanidae, 15
Tachinid parasite, Hawaii, 84
Taphropeltus hamulatus, *288*
Taxonomy, problems, 39, 86, 357–358, 411
Tegeticula, 327
Teinopalpus imperialis, 334
Tenebrionidae, Coleoptera, 127, 138, 179
Tenthredinidae, larval, *300*
Termites
  as indicators, *364*
  number of species, north of equator, by latitude, 407
Tetratoma fungorum, 182
Tetratomidae, 182
Tettigarcta tomentosa, Australia, 42
Tettigoniidae, *288*
Thecla betulae, *171*, 185
Theclinae
  neotropics, *376*
  possible extinction threat, *388*
  species-area curves, *375*
Therioaphis trifolii F *maculata*, 43
Thrips *see Frankliniella*
Thymelicus acteon, changes in population size, and foodplants, *223*, 224
Thymelicus lineola, *171*
  fully sprayed land *vs* conserved headlands, *302*
Thymelicus sylvestris, *171*, 270
  fully sprayed land *vs* conserved headlands, *302*
Thymus praecox, changes in population size, and foodplants, *223*, 225
Tiburon blind harvestman, 13
Tineidae, 43
Tingidae, 184
Tipulidae, 179, 180
Tithorea harmonia caissara, possible extinction threat, *389*
Tobias' caddisfly, 79
Trade *see* Collection and trade
Trapdoor spiders, 47
Trechid beetles, 125
Trechus quadristriatus, 322
Trichoptera
  British, conservation, *poster presentation*, 430
  Red Data lists, *111*
    wetland insects, *243*
  US Animal Notice of Review, *17*
Troidini, 366
Tropical forest, 403–420
  conservation of disturbed forest mosaics, 419–420
  deforestation, 409–411
  edge effects, *373*
  extending the network of protected areas
    critical faunas, 417–418
    selection, 415–417
  external threats, 420
  forest islands, 420
  identification of species, 410

importance of insects, 408–409
indicator groups, 411
latent extinction, and transient diversity, 418–419
sampling, standard techniques, 413
studies, pattern and process, 409
see also Neotropics
Tullbergia affinis, 267, *268*
Tullbergia callipygos, *268*
Turmada maravilha, possible extinction threat, *389*
TWINSPAN, 252–254
Tymbopiptus valeas, *80*

U

Unesco, World Heritage Sites, 57
United Kingdom
arable farmland, conservation of insects, 293–315
creating new and successional habitats, 263–291
Flow Country, investment priorities, 247
fragmentation, isolation and mobility of insect populations, 143–151
grasslands, Lepidoptera, 213–233
poster presentations, 425–432
threats to species and habitats
wetlands *see* Aquatic and wetland habitats
see also Northern Europe; specific habitats
United States
conservation programme, 12–14
Endangered Species Act (1973), insects, *10*, 18–23
Fish and Wildlife Service, Animal Notice of Review, *17*
see also North America
USSR, Red Data lists, 111

V

Vacciniina optilete, 127
Vanessa atalanta, *171*
fully sprayed land *vs* conserved headlands, *302*
Vanessa cardui, fully sprayed land *vs* conserved headlands, *302*
Vespidae, 138
Vespula germanica, 83
in Australia, 53
Vespula pensylvanica, 83
Vespula vulgaris, 83

W

Wall butterfly *see Lasiommata megera*
Wasps
Australia, 53
see also Hymenoptera
Water, extraction of ground water, 247
Water beetles
site list evaluation, 252–255
species lists, area and number of sites, *254*
Water fern weevil, *Cyrtobagous salvineae*, 324
Watsonian vice-counties, 175
Weeds, selective weed control, 296
Weevils, *77*, *80*
New Zealand, 76, *77*
stilbocarpa weevil, *82*
water fern weevil, 324
Wekiu bug, 97
Wellington speargrass weevil, 76, *78*
Wesmaelius mongolicus, *126*
West Virginia white butterfly, 11
Wetas
giant weta, 96
stenopelmatid weta, 75–76
Stephens Island weta, *79*
Wetland habitats *see* Aquatic and wetland habitats
Wharram Quarry Nature Reserve, ecological succession, 266–269
Whitefly *see Bemisia tabaci*
Wicken Fen, *Papilio machaon*, 146
Wiest's sphinx moth, 13, 15
Wildlife corridors, conservation headlands, 313
Windsor Forest and Great Park, saproxylic habitat loss, 166
Wood pasture
establishment, 160–161
loss and degeneration, 164–165
pollarding, 194–195
Wood-feeding insects, *see also* Saproxylic invertebrates

Woodland Grant Scheme, 198
Woodlands
ancient sites for butterflies and others, *poster presentation*, 425
changes in extent and composition, *164*
changes in status of woodland insects, 167–176
insects of old forest woodland habitats, 168–170
insects of open habitats within woodland, 171–176
collection of standing and fallen dead wood, 165, 166
ecological specialization among insects of mature woodland and open spaces, 176–190
mobility, 187–190
future prospects, 201–202
historical background
age of man, 159–162
last 200 years, 162–167
post glacial wildwood, 156–158
management for insect conservation, 190–199
determination of priorities, 190–193
habitat continuity and maintenance of traditional regimes, 193–195
re-creatibility of woodland habitat — opportunities and limitations, 198–199
simulation of traditional practices within modern wood management, 195–198
microhabitats
humidity, 178–180
position, 182
temperature, 180–181
need for research and education, 200–201
net changes in British types, *163*
nutrient status of dead wood, 181–182
oak forests, threats, 140
pollard wood pastures, 194
re-creatability of habitats, 198–199
standing dead trees, 197
World Heritage Convention and Sites, Unesco, 57
World Wide fund for Nature, XV, and International Union for Conservation of Nature, 7

X

Xanthopan morgani subsp. *praedicta*, 327
Xenandra, possible extinction threat, *388*
Xerces blue, 20
Xerces Society, 12
Xylophagidae, 180
Xylota caeruleiventris, 170
Xyloterus, 183

Y

Yellow jacket, 83
Yucca moths, 327

Z

Zoos, role in conservation biology, and risks, 98
Zygaena lonicerae, *230*, 230
Zygaenids, Hawaii, 84